BOOKS BY DAVE HAGE

*No Retreat, No Surrender: Labor's War at Hormel*
(with Paul Klauda)

*Reforming Welfare by Rewarding Work:
One State's Successful Experiment*

WITH JOSEPHINE MARCOTTY

*Sea of Grass: The Conquest, Ruin, and Redemption
of Nature on the American Prairie*

*SEA OF GRASS*

# SEA OF GRASS

## THE CONQUEST, RUIN, AND REDEMPTION OF NATURE ON THE AMERICAN PRAIRIE

### DAVE HAGE AND JOSEPHINE MARCOTTY

| Random House | New York

Random House
An imprint and division of Penguin Random House LLC
1745 Broadway, New York, NY 10019
randomhousebooks.com
penguinrandomhouse.com

Copyright © 2025 by Dave Hage and Josephine Marcotty

All rights reserved.

Published in the United States by Random House,
an imprint and division of Penguin Random House LLC, New York.

RANDOM HOUSE and the HOUSE colophon are registered
trademarks of Penguin Random House LLC.

LIBRARY OF CONGRESS CATALOGING-IN-PUBLICATION DATA
NAMES: Hage, Dave, author. | Marcotty, Josephine, author.
TITLE: Sea of grass / Dave Hage and Josephine Marcotty.
DESCRIPTION: New York, NY : Random House, 2025. |
  Includes bibliographical references and index.
IDENTIFIERS: LCCN 2024047803 (print) | LCCN 2024047804 (ebook) |
  ISBN 9780593447406 (hardcover) | ISBN 9780593447413 (ebook)
SUBJECTS: LCSH: Prairie ecology—Great Plains. | Prairies—Middle
  West—History. | Human ecology—Great Plains—History. |
  Agriculture—Environmental aspects—Great Plains—History. |
  Great Plains—Environmental conditions.
CLASSIFICATION: LCC QH104.5.G73 H35 2025 (print) |
  LCC QH104.5.G73 (ebook) | DDC 577.4/4097648—dc23/eng/20250122
  LC record available at https://lccn.loc.gov/2024047803
  LC ebook record available at https://lccn.loc.gov/2024047804

Printed in the United States of America on acid-free paper

randomhousebooks.com

1st Printing

First Edition

Title-page and part-title art credit: akkraraj © Adobe Stock Photos

*Book design by Sara Bereta*

The authorized representative in the EU for product safety and
compliance is Penguin Random House Ireland, Morrison Chambers,
32 Nassau Street, Dublin D02 YH68, Ireland.
https://eu-contact.penguin.ie

For our grandkids:
Cora, Simon, George, Angel, Esther, James,
and babies yet to arrive.
They will inherit the Earth we leave behind.

CONTENTS

|                              |      |
|-----------------------------:|-----:|
| *Map*                        | *x*  |
| *Introduction*               | *xiii* |

**PART ONE**

|                                          |     |
|-----------------------------------------:|----:|
| CHAPTER ONE: *Prairie*                   | 5   |
| CHAPTER TWO: *Plow*                      | 26  |
| CHAPTER THREE: *Swamp*                   | 40  |
| CHAPTER FOUR: $NH_3$ *(Ammonia)*         | 60  |

**PART TWO**

|                                  |     |
|--------------------------------:|----:|
| CHAPTER FIVE: *River*           | 89  |
| CHAPTER SIX: *Dirt*             | 114 |
| CHAPTER SEVEN: *Bugs*           | 140 |
| CHAPTER EIGHT: *Water*          | 170 |
| CHAPTER NINE: *Plow II*         | 194 |

**PART THREE**

|                                      |     |
|------------------------------------:|----:|
| CHAPTER TEN: *Prairie II*           | 221 |
| CHAPTER ELEVEN: *Farmers*           | 246 |
| CHAPTER TWELVE: *Ranchers*          | 269 |
| CHAPTER THIRTEEN: *Tatanka*         | 291 |

|                       |     |
|----------------------:|----:|
| *Epilogue*            | 321 |
| *Acknowledgments*     | 327 |
| *Bibliography*        | 333 |
| *Index*               | 351 |

SHORTGRASS PRAIRIE

MIXED PRAIRIE

Before Europeans colonized North America, grasslands covered millions of acres across the center of the continent—lush tallgrass prairie in states with plentiful rainfall, then mixed-grass and shortgrass prairie in the drier states west of the 98th meridian. Today just 1 percent of the original tallgrass prairie remains. The Dead Zone is where the Mississippi River dumps its load of agricultural fertilizers into the Gulf of Mexico. *Graphic by Alexander Hage*

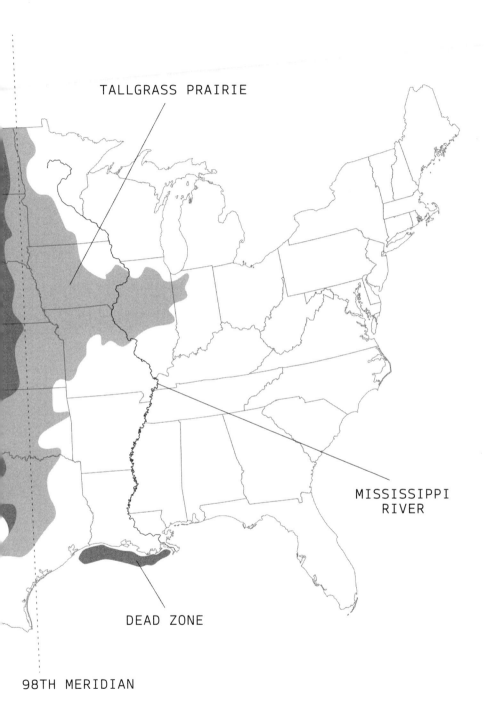

INTRODUCTION

HEADING INTO NORTHEASTERN MONTANA ON U.S. ROUTE 191, YOU could easily think you've entered a barren land. You can drive for miles without seeing a house or a barn, a cow or a dog. Just grass and sage, grass and sage, rolling away in gentle swells under the giant sky in a landscape so empty that it's almost spooky. It's no wonder that eighteenth-century mapmakers called this part of the continent the "American desert."

But pull off the road, step into the grass, and you find yourself in a landscape teeming with life. Delicate white butterflies flutter along the grass tops. Crickets crouch beneath the canopy of stems. A tiny vole, shaggy and brown, scurries across the ground and darts into its burrow. The tan carpet is dotted with flecks of yellow and white where yarrow and prairie aster reach toward the sun. Here and there a prickly pear cactus pokes through, its gold blossoms announcing themselves between the blades of wheatgrass. The silence is stunning until you really listen: The wind sighs over the grass and a bobolink hurls its song into the air.

Take another step. A startled pheasant bursts from a thicket and flaps away noisily. A bit of coyote scat, still showing tufts of fur, explains how one unlucky jackrabbit met its end. Two ducks gliding on a nearby pond take flight and climb away in a long arc. A white-tailed deer appears on a ridge fifty yards distant, stares for a moment, and then bounds off into deeper cover. In every direction the grass stretches off to the horizon in a way that calls to mind the word "infinite."

This is the paradox of the prairie. Feared by pioneers, shunned by tourists, dismissed today as a wasteland best viewed from thirty thousand feet, the North American prairie is nonetheless one of the richest ecosystems on Earth. The 180 million acres that lie between Sioux Falls, South Dakota, and Missoula, Montana—the western shortgrass prairie—contain 220 varieties of butterfly, 1,600 species of grass and flowers, some 1 billion grassland birds, and a roster of creatures synonymous with America's untamed West: bison, wolves, grizzlies, eagles, and coyotes. Farther east, in the mixed-grass prairie of the Dakotas, eastern Nebraska, and western Minnesota, tiny lakes known as prairie potholes lie scattered like sequins across the land, providing nesting grounds for fully half of North America's migrating birds. A recent soil analysis found that one cubic yard of tallgrass prairie sod contains so many grasses, sedges, flowers, burrowing mammals, worms, mites, nematodes, and soil microbes that on a small scale it rivals the tropical rainforest for biological diversity. And, like the rainforest, the prairie showcases nature's prodigious talent for symbiosis. Deer, elk, bison, pronghorn, and other grazers feed on this carpet of grass and return the favor by spreading nitrogen and seeds in their manure as they roam. Bison and prairie dogs crop the grass short, stimulating the growth of new shoots while allowing air and water to reach young seedlings. The grazers in turn feed prairie predators—foxes, coyotes, wolves, and owls—who prowl the land by night, hoping to catch a rabbit or mouse too late in finding its burrow. When done feeding, the big carnivores leave carcasses for crows, vultures, and a million bugs, and when they, too, die they return their own store of organic matter to the living soil. These food chains have operated for millennia but are only now yielding their secrets to science.

The American prairie is one of Earth's four great temperate grasslands—a sibling to the steppes of Central Asia, the pampas of South America, and the veld of southern Africa. These landscapes have played a central role in the course of human history because they support huge populations of grazing beasts—goats, sheep, cattle, horses—who, in partnership with grass, perform one of nature's miracles: turning sunshine into protein.

Yet today the North American prairie is also one of the most threatened ecosystems on Earth. The eastern tallgrass prairie, which once covered great swaths of Illinois, Iowa, Minnesota, and eastern Kansas, is 99 percent gone—lost to the plow in the nineteenth century. The western shortgrass prairie, in states such as South Dakota, Montana, and western Kansas, is disappearing at the rate of one million acres a year as farmers plow up grass to plant corn, wheat, and soybeans. That's an area the size of Connecticut disappearing every three and a half years. With little notice, these grasslands are vanishing faster than the Amazon rainforest.

In addition to devastating the land and its creatures, this is a disaster for climate change. Grasses inhale vast quantities of carbon dioxide through photosynthesis and store the carbon deep underground in their roots and the tiny soil particles that form around them. Because grasslands and their wetlands have been performing this alchemy for millennia, largely undisturbed by humans, their soils now hold more carbon than the planet's rainforests or the atmosphere itself. Every time a plow breaks the sod, it lays that soil open and releases its carbon stores to the atmosphere. Worse, every acre lost to the plow is also an acre that won't capture future greenhouse gas emissions as efficiently as native prairie. The environmental scientist Tyler Lark at the University of Wisconsin has estimated that recent land conversion on the North American grasslands is the climate change equivalent of adding 11.2 million cars to the road every year.

Our own history warns that this is a calamity in the making. Where farmers plowed the tallgrass prairie in the nineteenth century, the ecosystem is in collapse. Rows of corn and soybeans march to the horizon, leaving no habitat for monarchs or meadowlarks. Pollinators have gone into a death spiral, particularly the wild and domesticated bees that play a critical role in cultivating the nation's fruits and vegetables. Chemical fertilizers have so fouled the continent's great rivers—the Mississippi, the Missouri, the Platte, and the Ohio—that fish suffocate and long stretches are unsafe for swimming. The tap water in places like Des Moines and East St. Louis is unsafe to drink without costly treatment, and thousands of private wells are contaminated by farm chemicals. The U.S. Environmental Protection Agency declared not long ago that agriculture is now the chief polluter of the nation's rivers and a leading cause of pollution in its lakes.

In just over a century, Euro-American settlers took a landscape of dazzling diversity and made it a factory for food.

For much of our history Americans told this story in the language of triumph, and they had their reasons. Settling the prairie allowed millions of impoverished European immigrants to own a piece of land for the first time and build a better life for their children. Frontier farmers made the Midwest a breadbasket for the world, a place that fed Allied troops through two world wars and reduced global hunger in the twentieth century. The prairie built great cities—Chicago, Kansas City, and St. Louis—and gave birth to signature American brands such as Quaker Oats, Cheerios, Sears Roebuck, and John Deere.

The prairie also shaped the national character, producing beloved American heroes and an enduring American mythology. It gave us cowboys and wagon trains, Laura Ingalls Wilder and Willa Cather, Buffalo Bill and Wyatt Earp, Sitting Bull and Little Crow,

William Jennings Bryan and Eugene V. Debs, *The Wonderful Wizard of Oz* and *The Grapes of Wrath*.

Of course much of that mythology reflected the fantasies of a young nation, and more recent historians have acknowledged a darker side to the story. We know that Laura Ingalls Wilder had a childhood of tragedy and deprivation, that L. Frank Baum never even lived in Kansas. And the darkest, hardest truth: Settling this land was accomplished through genocide against its Native people—nations whose civilizations flourished on this continent far longer than Europeans have—and left a lasting stain on our character as a people. Converting the grasslands to an industrial model of agriculture left us with environmental problems as big as the landscape itself.

Today these uncomfortable truths have sown doubt about the wisdom of settling the prairie in the first place. In a 1987 essay, the geographers Deborah and Frank Popper described settling the plains as "the largest, longest-running agricultural and environmental miscalculation in American history."

NOW THE PLOWS are moving west again. Until the late twentieth century, American farmers paused halfway across the continent at the 98th meridian, roughly the eastern edge of the Dakotas, believing the western plains mostly too dry for crops and suitable only for grass and cattle. But in the last two decades farming has become profitable on this drier, harsher landscape thanks to the introduction of novel seed hybrids, advanced pesticides, and generous federal farm subsidies that minimize risk. Now the same calamities that befell the eastern prairie in the twentieth century—erosion, flooding, endangered wildlife, tainted drinking water—are turning up in places like South Dakota, Nebraska, and Kansas. A great buffer against climate change is shrinking fast.

The historian Geoff Cunfer has observed that agriculture is the single greatest interaction humans have with the natural world.

And yet much environmental writing—and virtually all federal environmental law—has been silent on farmers and farming, focusing instead on industrial polluters such as factories and power plants. More and more, conservation groups are recognizing that progress toward a healthier planet—be it cleaner water, more abundant wildlife, or fewer greenhouse gas emissions—cannot move far unless we change our relationship with food and the land.

It would be easy to cast farmers as the villains of this story. But in our view that would be unfair and mostly incorrect. The farmers we met while researching this book are generous, curious, and hardworking people who survive in a system not of their own making. The seeds they sow and the crops they raise are largely dictated by a handful of multinational food manufacturers and agrochemical corporations; no one farmer has the power to change consumer tastes or demand a safer herbicide. The prices they receive are set on global markets by huge grain merchants who don't care if hail wrecked the crop or the farmer tried organic fertilizer. In this industrial model of agriculture practically every bushel of corn and soybeans is interchangeable whether it's grown in Brazil or Iowa, and the market rewards the highest possible production at the lowest possible price. To break from the industrial model—to plant a novel crop or raise cattle without chemicals—is a risky and lonely venture.

Because the western prairie is remote and mostly unloved, its transformation is taking place unnoticed by most Americans. A visitor to Hyde County in central South Dakota could easily assume that plows and combines have always worked these fields, never knowing that just a decade ago this place was mostly grass. Who travels to places like Alvo, Nebraska, anyway? And if a place isn't worth a vacation, is it worth protecting?

Starting in the 1990s, however, the world's major conservation groups began to give grasslands a fresh look. A seminal study by the conservation ecologist Jonathan Hoekstra and colleagues encouraged NGOs to focus on those ecosystems that are disappearing most quickly *and* are least protected as national parks or wilderness reserves. To the surprise of many, it wasn't rainforests or oceans that landed at the top of the list. It was prairies. Now conservationists recognize that their value as habitat for endangered species, their capacity to prevent flooding and provide clean water, and their ability to buffer climate change merits worldwide attention. Even the soil beneath the grass is only now acknowledged as a miracle of biology, with a universe of life all its own.

Since then conservationists have launched a growing number of projects to restore or protect the continent's remaining grasslands. In Kansas and Iowa, grassroots farm coalitions are promoting cultivation techniques that protect soil and water. In Montana, the World Wildlife Fund has partnered with cattle ranchers to protect water and habitat. A pioneering coalition of conservationists and leaders of the Sioux, Assiniboine, Aaniiih, and Blackfeet people are establishing new herds of wild bison using tribal land and the iconic buffalo of Yellowstone National Park. In the few places where islands of virgin tallgrass prairie remain, in states such as Minnesota and Kansas, ingenious projects have begun to restore and protect them.

Are these projects enough to save this priceless landscape? We don't know yet. But they offer proven strategies to protect grasslands where they have survived, and reduce the damage of agriculture where they have not. The nascent movement also suggests that more Americans may be embracing what Aldo Leopold called a "land ethic," the idea that the land is more than an economic asset to be exploited; it is an irreplaceable gift to be honored in its own right—a concept that would have been familiar, even obvious, to the people who lived here before Europeans arrived.

In telling this story we have tried to avoid overreach. This book touches down in just a few places in the vast center of America—places that reflect the history of the grasslands and their transformation, and we have dealt only sparingly with the equally important history of the Native peoples swept aside by colonialism. We have also tried to avoid broad generalizations and predictions, be they gloomy or hopeful. Nature regularly mocks our expectations, and it's possible that in a generation or two the rivers, creatures, and soils of the grasslands will heal themselves from the wounds described in this book.

It's worth remembering, however, that Americans aren't the first people to squander the natural riches that helped their civilizations rise. Rome, Egypt, and Mesopotamia all collapsed in part because of aggressive farming practices that overtaxed their land. Like them, the Europeans who swarmed across North America centuries later were swept up in the arrogance of progress—the abiding confidence that technology will solve our problems, that nature is limitless, and that the land is ours for the taking. The difference is that they didn't understand the destruction they had wrought until it was too late. We could be the first civilization with the knowledge to change course in time. If we also have the heart.

# PART ONE

But the great fact was the land itself, which seemed to overwhelm the little beginnings of human society that struggled in its sombre wastes. [The boy] felt that men were too weak to make any mark here, that the land wanted to be let alone, to preserve its own fierce strength, its peculiar, savage kind of beauty.

—WILLA CATHER, *O Pioneers!*

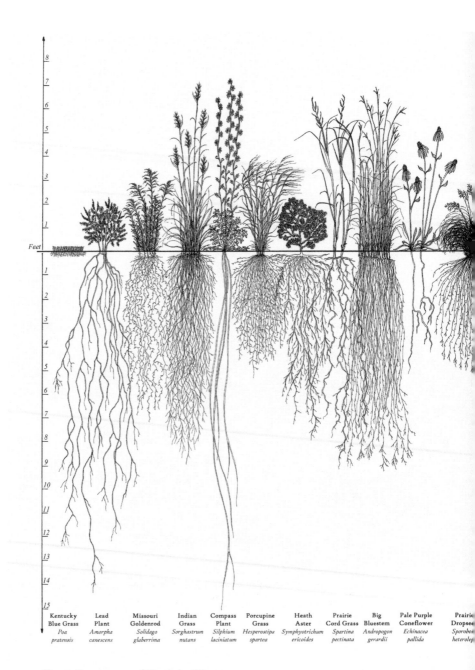

## Root Systems of Prairie Plants

*Heidi Natura and Living Habitats, copyright © 1995*

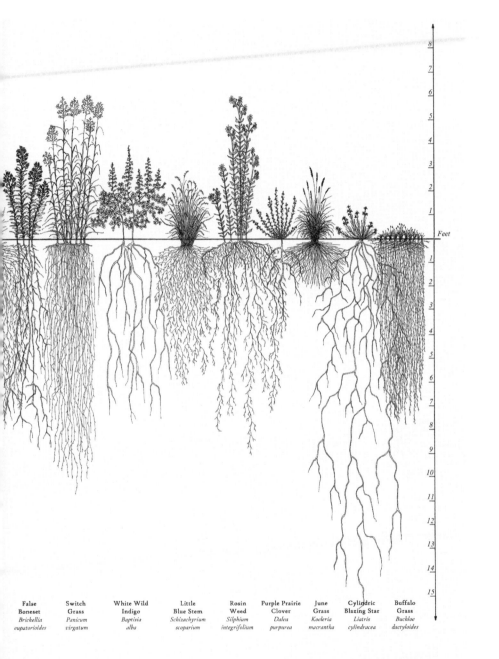

Living Habitats

Heidi Natura 1995

CHAPTER ONE

## *Prairie*

IN THE BARS AND CAFÉS AROUND COTTONWOOD FALLS, KANSAS, REGUlars tell the story of a settler family that arrived in Chase County about 1860, their wagon loaded with the household goods and farm implements required to start a new life on the prairie. They encountered a member of the Kansa tribe, according to the story, and asked where they might find a good place to farm. He gave them a warning: Their steel plow would be useless here. Puzzled but undaunted, they claimed a homestead and hitched horse to plow, only to discover that he had been right. In this part of east-central Kansas, the hard limestone bedrock lies so close to the surface that a plow can't carve a furrow without striking rock. Subsequent settlers got the message and chose to settle farther west. So by a quirk of geology, the Flint Hills region survived as a rare island of grass in a vast sea of corn and wheat. Today it covers some four million acres in eastern Kansas and northern Oklahoma and includes the Tallgrass Prairie National Preserve, the nation's largest remaining patch of native tallgrass prairie. It may be the best place in the country to see the ancient landscape that shaped life here for centuries before Europeans arrived.

Standing on a rise at the center of the preserve you can see for miles and imagine the sea of grass early European explorers described centuries ago. The gently rounded hills roll like ocean swells—nothing but amber grass and prairie flowers to the edge of the sky. To the west, a lone bison grazes in a creek draw. To the northeast, a red-tailed hawk soars on the breeze, gracefully circling in its hunt for prey. The silence is immense until a breeze comes over a nearby rise, playing the grass like an instrument. This is what the prairie must have looked like two hundred years ago, when Lewis and Clark explored the West for President Thomas Jefferson. And five hundred years ago, when the Pawnee, Osage, Wichita, and Kansa people hunted buffalo and learned to turn the stone into weapons. Or even twelve thousand years ago, when this landscape emerged fresh to the sky after centuries under the combined forces of ice and geology.

Between sixty and eighty million years ago, the prairie had not yet made its appearance. A wide sea lay across the middle of what is now North America. As the planet warmed, the waters receded, leaving a moist plain covered by hardwood trees. Then, about 65 million years ago, a massive shift in the Earth's tectonic plates caused a geological event known as the Laramide orogeny, which shoved the western half of the continent toward the eastern half, lifting a giant crust of rock that would become the Rocky Mountains and the Sierra Nevadas. By erecting a high barrier between the Pacific Ocean and the center of the continent, the mountains cast a "rain shadow" over the land to their east and gave the region its defining dry climate. When clouds roll into the West Coast from the Pacific they collide with the wall of mountains and dump most of their moisture on the western front in the form of rain or snow. As a result, the drier, eastern side of the Rockies receives just ten to fifteen inches of precipitation annually, one-half or one-third of what the rest of the continent typically receives. The central hardwood forests dwindled, eventually leaving a dry open plain.

This open expanse also left the wind free to blow, and blow it does—hard and incessantly. The winds dry up much of whatever moisture does fall from the sky and punish any living thing that dares raise itself more than a few feet above the ground. The early climatologist Robert DeCourcy Ward described the dramatic effects of a warm westerly wind known as a chinook: "Evaporation and melting are so rapid that a foot of snow may disappear within a few hours, being sucked up from the ground without [leaving] even a trickle of water."

And because this wide plain lies hundreds of miles from either coast, it lacks the moderating influence of the Pacific and Atlantic oceans. The result is wild temperature extremes, from 30 or 40 degrees below zero in the winter to 110 degrees or more in July and August.

Still, millions more years passed before the prairie grassland made its appearance. Some twenty thousand years ago, much of North America was covered by the Laurentide ice sheet, a towering glacier that stretched from the Arctic Ocean as far south as Iowa and as far east as New England. As it crawled south across the continent, it scraped the land flat, scooping up rocks and boulders and grinding them to fine gravel. When it receded, starting about eighteen thousand years ago, it deposited this gravel as a layer of fine glacial till, leaving behind a vast, mineral-rich plain stretching from the Rocky Mountains to the Mississippi River and parts of Illinois and Indiana. Over the ensuing centuries, rivers flowing down from the Rockies carried an additional fine sediment of eroded rock and spread it in alluvial layers across the plain. Westerly winds dropped an even finer coating of rock dust, adding another endowment of minerals. Then millions of mammals and plants lived and died there, bequeathing their own rich organic material to the ground. The resulting layer of earth was deep, porous, full of tiny microbes, and packed with carbon—the perfect medium to produce one of the greatest ecosystems on the planet.

Even so, the unforgiving climate made this place hostile to virtually every form of plant life but one: the hugely adaptable family *Poaceae*—grass. *Poaceae*'s twelve hundred species are found all over the globe, from Patagonia to Uzbekistan, and account for all the world's leading food grains—corn, rice, wheat, barley, and oats. On North America's great plain more than one hundred species took root, perfectly adapted to a place that was dry and windy. The roots of prairie plants such as switchgrass, big bluestem, and blazing star reach as much as twelve feet underground, where they tap deep stores of moisture to survive dry spells. The same roots create long vertical channels where rainfall can penetrate far into the soil, hide from the sun, and replenish the grass. When winter's frigid temperatures approach, these grasses pull energy from their leaves and stems into their roots, then store it there to produce new growth when spring arrives. These root webs also create a sort of underground forest populated by mites, tiny worms, beetles, ants, bacteria, and fungi—all of them feeding off soil nutrients and nourishing them in return. These immense root networks give prairie sod a density unrivaled in other places. The conservationist and writer John Madson has noted: "One may dig a trowel into a lowland forest floor and come up with soil that is not wholly occupied with plant parts, but in the tallgrass prairie every cubic inch of soil surface is a mass of rootlets." A cubic yard of big bluestem sod, he wrote, can contain nearly thirteen miles of tiny roots and root hairs. And when the roots die, they deposit additional organic matter to enrich the soil.

Aboveground, too, grasses are exquisitely adapted to a dry climate. Their profusion of narrow blades can capture huge amounts of sunlight—one acre of grass can contain ten acres of leaf surface—while leaving little area for moisture to evaporate. In addition, these blades typically are coated with tiny hairs or ridges that protect moisture from the wind. Like other plants, they have stomata—miniature valves that inhale carbon dioxide and exhale water vapor—but unlike most other plants, many species of grass can shut their stomata in the heat of day to retain water, then open them in the cool of the night.

Grasses also flourished because they can turn the region's scourges to their advantage. Wildfires that are common on the prairie—some caused by lightning, some set by people—kill off bigger, woodier competitors such as sumac and red cedar and burn away the layer of dead grass known as thatch that can smother young grass shoots. Grasses survive fire because 80 percent of their biomass lies belowground in the form of roots and horizontal underground stems known as rhizomes, which protect their stores of water and carbohydrates until the flames pass. In much the same way, prairie grasses adapted to survive the grazing animals that have populated the region for centuries. Most plants have nodes where new growth sprouts; in most grasses that node lies close to the ground, even belowground, safe from the teeth of elk, antelope, and bison.

By the time Europeans arrived in the sixteenth century, grasslands covered one-fifth of North America, an expanse exceeded only by the boreal forest.

As they evolved together over centuries, prairie grasses and prairie soils developed a remarkable partnership. Some grasses use only about two-thirds of the carbohydrates they create through photosynthesis and send the rest down their roots into the surrounding soil. This banquet of sugars attracts a rich underground community of bugs, fungi, bacteria, and other microbes that perform tasks the plant can't perform for itself. Tiny creatures such as springtails and nematodes break down plant residues, releasing their nutrients into the soil. These in turn feed worms, beetles, and small rodents who burrow through the earth, creating channels for air and water and churning the soil so that nutrients rise toward the surface. A star in this ecosystem is a class of bacteria known as symbiotic rhizobium, which performs one of the planet's rarest and most invaluable operations: converting atmospheric nitrogen into nitrate and nitrite—nutrients that plants can use—

and then, when the plants are finished, converting the compounds back into a form that can be released into the air. Jo Handelsman, a University of Wisconsin biologist who advised the Obama administration on science matters, describes this invisible underground community as "the most complex habitat on Earth." She notes that the Earth's soils contain fully 25 percent of the planet's known species, from bacteria and fungi to nematodes, moles, and badgers. And because they have been inhaling carbon dioxide for millennia, the Earth's soils now contain one-third of the planet's terrestrial carbon—more than the total released by human activity since the start of the Industrial Revolution.

Over time, the prairie evolved into three more or less distinct zones, each with its own personality, depending on the region's annual rainfall. The eastern edge of the plains, in states such as Iowa, Minnesota, and eastern Kansas, received enough rain to produce the tallgrass prairie, a lush region dominated by immense species such as big bluestem, switchgrass, and prairie cordgrass. In restored tallgrass prairies today, such as the Neal Smith National Wildlife Refuge in Iowa, a visitor can sense what the tallgrass prairie must have looked like to the first settlers who emerged from the woods of Ohio and eastern Illinois. Towering stands of big bluestem brush your shoulders, their magenta sheaves waving alongside the lacy fronds of feather reed grass. Hummingbirds and monarchs flit from flower to flower and wild raspberry bushes snag the trousers of anyone walking by. The refuge is crisscrossed with mowed footpaths because the thickets of grass and flowers would otherwise be all but impenetrable.

On the plains' western edge, the drier climate produced the shortgrass prairie of Montana, Wyoming, Colorado, Oklahoma, and west Texas. Here the dominant plants are western wheatgrass, sage, and blue grama, species that can survive on as little as ten inches of annual rainfall. It's a far more austere landscape, the gravelly brown soil visible between clumps of grass that grow barely shin-high. By midsummer this landscape can be washed out to a pale amber, a thin carpet decorated with small cacti, spiky yucca plants, and tiny red asters. Meandering rivers cut through this terrain, creating shallow valleys that serve the land like arteries, nourishing cottonwood groves and thickets of taller grass and providing shade, shelter, and transportation routes for wildlife and early humans.

Between the two zones a mixed-grass prairie emerged in the region that includes western Minnesota and much of the Dakotas. Here, thickets of little bluestem and Indian grass grow thigh-high—green, aromatic, and dense in midsummer—and broken up by clumps of red twig dogwood and sumac. Part the grass stems and you can find badger holes and the eggs of ground-nesting birds such as quail and killdeer.

The eastern tallgrass prairie, with its generous rainfall and rich soils, fell to the plow quickly in the nineteenth century, and today very little remains. Out on the dry western plains, where row crop farming never took off, as much as 40 percent of the shortgrass prairie survives, mostly on cattle ranches and federal land.

Centuries of evolution didn't merely impart survival traits to prairie plants, they gave the landscape a riotous diversity of life. When the Nebraska ecologist Chris Helzer set out to photograph a year in the life of the prairie, he cordoned off a patch of grass one meter square—and counted 113 species of plants, bugs, and animals, including 18 varieties of beetle, 14 bee species, and more than a dozen flowering plants. Diversity like that is not accidental—it's nature's insurance policy against adversity. If the climate delivers a chilly summer, cool-season grasses such as western wheatgrass and wild rye will thrive; if the year is hot, warm-season varieties such as switchgrass, little bluestem, and side oats grama will flourish. The mix of grasses produces roots of every length—anywhere from two feet to twelve feet—so that the plant community taps soil moisture and nutrients at every level of the root zone below. Diversity also helps protect the community from bugs and pathogens; parasites tend to specialize in certain hosts and can't establish a foothold when their favorite food is just one of dozens of species in an ever changing landscape. In much the same way, diversity helps prairie plants survive grazing animals. Pronghorns like flowers, and where they graze they create space for grass. Bison generally prefer grass, opening space for flowers. The variety of grazing animals, in turn, creates a mosaic of grasses and wildflowers that offer varied habitat for a variety of birds. The close-cropped grass near prairie dog towns attracts longspurs and mountain plovers because they can spot predators at a distance. Bobolinks and eastern meadowlarks prefer taller grasses, where they can build nests and hide from hawks and coyotes. The grazing mammals that surround them make their own contributions: The saliva of some ungulates contains a chemical that stimulates a plant's root growth, and roaming bison spread seeds and healthy bacteria over great distances in their dung. The variety of grazers also feeds a variety of carnivores: foxes and ferrets prey on prairie dogs, wolves hunt bison and deer. The carnivores return the favor: By preying on slow and weak bison and antelope they ensure that only the strongest and fastest contribute their genes to the next generation. The mix of plants and animals shifts every year with the amount of rain, fires, pests, and other stressors. But something is always healthy, always ready to thrive.

In the early nineteenth century, Americans who came upon the prairie were fond of describing this landscape as a timeless and unchanging Eden. Today ecologists believe that it changed constantly and is still changing. The forest line that marks the eastern boundary of the tallgrass prairie, for example, apparently moved back and forth like an ocean tide—advancing west in years when rainfall was sufficient for trees to spread, then receding east in dry years, when tiny seedlings couldn't compete with the hardy grasses. There is some evidence that the tree line was moving westward even before Europeans arrived, apparently because the climate was growing more humid, and the writer John Madson has speculated that if European explorers had arrived in 3,000 C.E.

rather than in the 1500s, they might have found hardwood forests instead of grassland in the middle of the country.

From the beginning, this fertile place supported a remarkable abundance of wildlife. When the first humans arrived—perhaps around 11,000 B.C.E., although the scholarly consensus keeps changing—they found a grassland so rich in animals that the historian Dan Flores has called it "the American Serengeti." Mastodons and mammoths roamed this land, as well as saber-toothed cats, hyenalike dogs, and enormous canines known as dire wolves—all finding prey in the huge population of horses, camels, giant bison, and other grazing animals that lived off the grass. Flores writes: "It was one of the natural spectacles of the world, equaled only by places like the Serengeti plain of East Africa, the Masai Mara, or the veld of South Africa. . . . It was not just the settings, the vast opalescent distances, the blue pyramidal forms of mountains on distant horizons, the intersecting planes of horizontal, yellowed grassland. It was the presence of so much sheer life." This rich land became home to what are believed to be North America's earliest human occupants, the Clovis people and Folsom people of the southern Great Plains. The size and number of the beasts they hunted triggered technological innovation, and they learned to use flint, chert, obsidian, and other stone to make spears sharp and powerful enough to pierce the hides of their prey. Their spearheads can still be found lodged in the skulls of petrified mammoths.

Three millennia later, the megafauna had mostly disappeared, likely because of a warming climate and over-hunting by people who did not know their prey was vulnerable. By 8,000 B.C.E. some three dozen species of giant animals had disappeared. But the grasslands still supported vast populations of smaller grazing mammals and the human communities that thrived among them. While smaller than mastodons and saber-toothed cats, these creatures existed in mind-boggling numbers: an estimated one million elk, ten million antelope, eighty thousand wolves, five billion prairie dogs, and bison in uncounted numbers. Bison are believed to have arrived a few centuries earlier, crossing the Bering Strait on a land bridge from Asia, but were at first held north by ice in what is now western Canada. But when the climate warmed and the ice receded, they had a path into the rich grassland emerging in the south. After the giant mammals that competed with them for grass and space became extinct, their numbers exploded so fast that they transformed the land. Their grazing overwhelmed the region's early grass species—bunch grasses such as needlegrass and feather grass—which gave way to varieties such as grama and buffalo grass, which can grow from rhizomes and survive the intensive grazing. The historian Geoff Cunfer argues that bison may have altered the central grasslands as much as Europeans did centuries later with their plows and horses.

With abundant grass and few predators, bison expanded their range from the Rocky Mountain foothills to the East Coast, and from the subarctic to the Gulf of Mexico, and in some places lived in herds so vast they darkened the land. The Montana writer Richard Manning has estimated that these native grasslands were so productive before Europeans arrived that they produced more protein, in the form of bison, than the American cattle industry does today on the same land. Like the ecosystem around them, bison evolved with changing conditions and avoided an early extinction by becoming nomadic and specializing in shortgrass grazing. They literally shrank, generation by generation, adapting to changes in the climate and predator populations, becoming lighter, faster, and more mobile. Inevitably, the people of the plains began hunting bison as their primary source of food, and they too adapted: Whole Indigenous nations migrated to the plains, either permanently or on seasonal hunting trips, to take advantage of the immense bison herds. The Crow, Omaha, and Kansa abandoned their fields in the Missouri Valley; the Kiowa moved from the Yellowstone Valley. Chief among them were the Sioux Nations—including the Lakota and Dakota—the Cheyenne, and the Assiniboine. New alliances formed, and though many of the new tribes did not give up farming, they also hunted bison, trapping them, driving entire herds over cliffs, or chasing them into box canyons and then killing them with spears and arrows.

This was not the untamed wilderness imagined by some Europeans. In the centuries before Columbus arrived, humans occupied every corner of the American continent, numbering at least five million. Indigenous people established complex societies, built large fortified cities, mastered agronomy, and raised crops such as corn, beans, and squash. In the twelfth century, a great city-state known as Cahokia dominated the central Mississippi Valley. Centered around a massive earthen pyramid across the river from what is now St. Louis, Missouri, it supported a population about half the size of London's at the time. Cahokia eventually collapsed in the thirteenth century, for reasons that may have included overpopulation and contaminated drinking water, and its population dispersed for reasons that remain largely unclear. Farther west, nomadic hunting tribes made strategic use of controlled burns to round up bison, refresh the grass, and convert wooded areas to meadows that would attract grazing prey. Still, historians believe the Indigenous people of the plains maintained a sort of overall balance with the natural world. Although they fought fiercely for dominance over hunting and foraging territory, most Indigenous societies never presumed that the land was theirs to own. Survival itself required understanding and respecting the behavior of animals and the cycles of nature. Their wildfires altered the land, and in places became part of the balance preserving a healthy grassland, but never on a scale sufficient to change the ecosystem for good. And though they killed bison by the thousands, their hunts never endangered the species' survival. Embarking on a hunt, many tribes would honor their prey with ceremonies, acknowledging the animal's sacrifice and treating it as a gift. The resulting belief systems typically spelled out spiritual kinship bonds between animals and humans, a set of traditions that wound up protecting the grassland ecology for millennia. In a stark contrast to the colonialists who followed them, the Indigenous people saw themselves as part of the natural world, not its masters, a concept that persisted among them long after Europeans arrived. "The earth is our mother—we are now on it with the Great Spirit above us," the famous Sauk war chief Black Hawk said in a speech to white Iowa settlers who, in the early 1800s, had forced his people from their home along the Rock River in western Illinois.

When Europeans arrived on the plains in the sixteenth century, they found a landscape unlike any they had ever seen before—vast, terrifying, swarming with insects, and baffling in its monotony. The Spanish conquistador Francisco Vázquez de Coronado, arriving on the southern plains in 1541, reported that his troops got lost constantly because there were no trees or hills to serve as landmarks. Even finding the trail of his army presented a challenge to its rear guard because the dense grass so quickly closed over its tracks.

> Who could believe that 1,000 horses and 500 of our cows and more than 5,000 rams and ewes, in traveling over those plains, would leave no more trace . . . than if nothing had been there. The grass never failed to become erect after it had been trodden down and [became] as fresh and straight as before. . . . [S]o that it was necessary to make piles of bones and cow dung now and then so that the rear guard could follow the army.

Coronado's scouting parties soon learned a trick from their Native American guides. They would shoot an arrow in their desired direction of travel and when they reached it, mark the spot and shoot another. In this way they made their way hopscotch-fashion across the land, leaving cairns so they could retrace their steps.

Even so, scouting parties regularly lost their bearings and became disoriented. More than one set out for a brief reconnaissance mission and was never seen again. Although he found the land beautiful and full of wildlife, Coronado soon wrote to his royal patrons that the grasslands were no place for the Spanish.

> It would be impossible to establish a settlement here, for besides [being remote from both oceans] the country is so cold . . . that apparently the winter could not possibly be spent here because there is no wood, nor clothes with which to protect the men except the skins which the natives wear.

For nearly three hundred years the grasslands and its people thwarted colonization by European powers. Native people possessed a deep knowledge of the land and how to thrive on it, even as Europeans found it baffling and difficult. Their decentralized political systems—defined by kinship and often led by women—perplexed European conquerors who were accustomed to vanquishing a society simply by dethroning its emperor. The Spanish never penetrated deep into the southwest plains, often because they were fooled by Native guides who exploited their greed and arrogance. Near the Great Lakes, the Iroquois League held Europeans at bay for decades by playing rival trading powers against one another and maintaining something like monopoly control of the best fur-trapping territory. In 1675, nearly seventy years after the founding of the seminal Jamestown colony in Virginia, the English still numbered a mere forty thousand huddled along the eastern shore of the continent. In fact, the Europeans' most potent weapon against Native people was accidental (at first)—infectious diseases. Historians believe that far more Native people died from smallpox, measles, and influenza than from bullets and steel. In the 1600s and 1700s, epidemics ravaged the sedentary village tribes east of the Mississippi.

By 1800, after a period of explosive growth, the non-Native population of the United States had reached some five million, but the great middle of the continent remained a remote and forbidding place to most white Americans. Fully 90 percent of the nation's population lived east of the Appalachian Mountains—in no small part because Native people had fought off their encroachment—and most of the nation's states bordered the Atlantic Ocean. The grasslands out west were so alien that the response of white Easterners seems comical today. The editor Horace Greeley (he of "Go West, young man") took his own sojourn west and claimed to have seen wind gusts so strong that they blew the wheels off a wagon and straightened their iron rims into flat straps. That ignorance persisted well into the nineteenth century. When Congress passed the Timber Culture Act a few years later, it ordered settlers to plant forests on the prairie—as if lawmakers could order the rain to fall and require prairie dogs to climb trees.

European writers visiting the plains fell back repeatedly on ocean metaphors for the simple reason that nothing else in the European experience matched the scale of the prairie. And so it became a sea of grass.

Despite its mysterious nature, or perhaps because of it, the continent's interior gradually came to fascinate the eastern public. In 1810, when the explorer Zebulon Pike published an account of his journey along the upper Mississippi River, it became a hit. The diary was a gripping tale of bear hunting, river rafting, negotiating with Native American warriors, and huddling around winter campfires to avoid death by freezing. But Pike also managed to capture something of the majesty of the land.

> We crossed first a dry flat prairie; when we arrived at the hills we ascended them, from which we had a most sublime and beautiful prospect. On the right, we saw the mountains

which we passed in the morning and the prairie in their rear; like distant clouds, the mountains at the Prairie Le Cross.

Pike's journal caused a sensation back East and was translated into several languages for European audiences. Three decades later, the painter George Catlin found similar enthusiasm on a lecture tour with his iconic portraits of Native American leaders. The circuit took him across New England and to France, where the great writer Charles Baudelaire remarked that Catlin's paintings captured the "proud, free character and noble expression" of Indigenous people.

The naturalist John James Audubon, having completed his famous study of North American birds in the 1830s, set out to inventory other prairie wildlife and found himself overwhelmed by the populations of deer, elk, prairie dogs, and beavers: "It is impossible to describe or even conceive the vast multitudes of these animals that exist even now and feed on these ocean-like prairies," he wrote in a journal. "My head is swimming with excitement, and I cannot write any more."

Still, large-scale settlement of the prairie didn't begin until well into the nineteenth century. When Thomas Jefferson signed the Louisiana Purchase in 1803, instantly doubling the size of the United States, the nation's leaders began promoting westward migration. America was growing at an astonishing clip—its population doubling every three decades—and prairie farms promised a way to feed a growing nation while giving newcomers land of their own and staking a national claim to the West. In the minds of leaders such as Jefferson, frontier agriculture was more than a material goal: It fulfilled their vision of a society built on the fortitude and patriotism of independent yeoman farmers. Jefferson and his peers believed that the land conferred a sort of moral virtue on its inhabitants. Those who worked the land were "the most vigorous, the most independent and the most virtuous citizens," he wrote. Having lived through a war for independence from the English crown, Jefferson also saw small landowners as a bulwark against tyranny—be it by government or industry—whose stewardship of the land would be regulated by "the natural affection of the human mind."

To promote the yeoman ideal and westward settlement, between 1796 and 1820 Congress passed a series of measures known as Land Acts. The laws embraced an American philosophy that land is most productive under private ownership and rapidly put the premise to the test. They established a national network of land offices to facilitate property sales, deployed an army of federal surveyors to inventory the land and map it into township grids, and made government land available to farmers and investors on attractive terms. Between 1820 and 1840, federal land sales to private owners—generally farmers and speculators—rose fivefold. By 1840 some 40 percent of Americans lived west of the Appalachians.

The wave of westward expansion triggered a new chapter in the government's conflict with Native Americans. President Andrew Jackson took office in 1829, a military hero and Indian fighter who intended to force tens of thousands of eastern tribal people to move west, a strategy formalized when Congress passed the Indian Removal Act of 1830. Jackson planned to displace 80,000 to 100,000 people from some of the richest farmland in the south and move them west of the 95th meridian, which would become the "permanent Indian Frontier." Thousands of Cherokee and Choctaw people died on "the Trail of Tears," and by the late 1840s, roughly 100,000 Indigenous people had been moved into so-called Indian Territory west of the Missouri River. In the next two decades the removals accelerated as new settlers poured into the region that is now Ohio, Michigan, Indiana, and Illinois; the government actually forced Native people onto railroad cars and moved them west with armed soldiers standing guard.

Pekka Hämäläinen, a Finnish historian specializing in Indigenous North America, describes this period as one of jarring contrasts. On the one hand, eastern and southern tribes were forced into the West under threat of government-sanctioned violence. At the same time, many Nations resisted fiercely and successfully. The Comanche were ascendent in the southwest; the Sioux peoples expanded their homeland on the northern plains using their superb skills with horses, hunting, and diplomacy. But in 1862 Congress passed the Homestead Act, granting 160 acres of federal land to any citizen who pledged to farm it, and settler migration into the Midwest accelerated. In Minnesota tens of thousands of new farmers poured onto the rich tallgrass prairies to carve out farms, occupying Dakota tribal land and disrupting the hunting and foraging of Native people. Food and other supplies promised to the Dakota by treaties frequently went missing, often due to corrupt government officials. The Dakota, starving, went to war, burning farms and butchering livestock, and in the end they killed around five hundred settlers. The federal government in effect declared war on the Sioux, imprisoning thousands and indicting four hundred for atrocities against the settlers. In the end, thirty-eight Dakota men were hanged en masse in Mankato, Minnesota, the largest public execution in American history.

In the following years, conflicts flared across the plains, and the powerful, mobile horse-riding tribes often prevailed. Weary of conflict, Congress initiated a peacemaking phase with the western tribes. In 1867 the Medicine Lodge Treaty gave southern Plains tribes including the Comanche and Kiowa exclusive rights to western hunting grounds south of the Arkansas River "so long as the buffalo may range thereon in such numbers as to justify the chase." In 1868, the federal government signed the Treaty of Fort Laramie, creating the Great Sioux Reservation and effectively promising the Dakota and Lakota people all of what is now the western Dakotas and forbidding white people to settle there. But then, a mere six years later, Lieutenant Colonel George Armstrong

Custer led a scientific expedition into the heart of the Black Hills and reported finding abundant gold. Speculators and miners rushed into the region, often with a tailwind from public opinion in the East. An editorial in *The New York Times* said, "The red man will be driven out, and white man will take possession. This is not justice, but it is destiny."

In the spring of 1876, defying a federal order to move onto reservations, more than ten thousand people from a multitude of Plains tribes followed the Lakota chiefs Sitting Bull and Crazy Horse to the Little Bighorn River in south central Montana, which Indians called the Greasy Grass. In a battle that marked a turning point in American history, they crushed a cavalry regiment led by Custer, killing the entire contingent in less than an hour. Shocked and humiliated by the unexpected rout, Congress and the army began a comprehensive campaign to crush the Lakota and take their land. The military seized their horses and their guns and began forcing the Lakota and other northern Plains tribes onto reservation land. On the morning of December 29, 1890, federal forces surrounded an encampment of roughly three hundred Lakota near the creek known as Wounded Knee. During what was apparently a clumsy attempt to disarm the Lakota men, shooting broke out, the soldiers opened fire with heavy guns, and in a matter of minutes killed 270 Lakota, mostly women and children. Twenty of the soldiers who participated were awarded the Congressional Medal of Honor for an act that Hämäläinen described as an atrocity and human catastrophe. In the following months, small bands of Dakota, Lakota, and Assiniboine tried to continue the resistance but they now faced devastating forces. While the Plains tribes once had the advantage of superior mobility and deep familiarity with the geography of the plains, the U.S. Army now had overwhelming firepower and the logistical advantage of railroads and telegraph lines. Hundreds of Sioux and Assiniboine families were pushed even farther west to reservations in Montana. By the end of the nineteenth century the American campaign had reduced the Indigenous population by 70 percent—leaving a mere 250,000 people.

Although nineteenth-century Americans liked to think of themselves as pioneers rather than usurpers, conquest of the prairie placed them in the long history of empires that expanded by what many historians now call "settler colonialism," a concept often credited to Australian scholar Patrick Wolfe. Rather than enslaving Indigenous people or extracting the land's riches and taking them home, as many previous colonial powers had done around the world, Europeans settling North America eliminated the Indigenous societies through removal and genocide and then occupied the land they coveted. Though it is just one lens through which to see the past—and the present—many historians, Native and non-Native, now consider North America a premier example of settler colonialism and resulting social structures that continue to affect Native people.

In addition to the human and moral tragedy, European settlement of the prairies also represented an ecological event of global significance. Around the world, from the pampas of South America to the steppes of Central Asia, colonial empires were discovering the agricultural potential of grasslands and transforming them to feed their growing populations. What followed was a staggering expansion of human agriculture. Bringing those grasslands under the plow more than doubled the amount of cropland available to humans—to something over two billion acres—in just a few decades. This hugely increased the world's food supply, but it also made an indelible mark on the planet. In less than a century the Earth's fundamental natural rhythms—its carbon cycle, its nitrogen cycle, its hydrology, and its wildlife migration patterns, features that had been in rough equilibrium for thousands of years—changed permanently on hundreds of millions of acres. The environmental consequences were not immediately apparent, but would play out for decades and are still compounding today in ways that are ruinous to the land, the water, the creatures, and the planet itself.

ON A GRASSY stream bank near the outskirts of Wolf Point, Montana, a small patch of ground is marked with a handful of simple white wooden crosses. Though humble and almost hidden, this graveyard marks a dark chapter in the history of the Fort Peck Indian Reservation. In 1878 this corner of northeastern Montana became home to hundreds of Assiniboine and Sioux people who had been pushed west by the U.S. Army. In exchange for giving up their land and nomadic life, they were promised food, money, and medical supplies from the federal government. But in a pattern familiar to any student of Native history, a series of white supervisors—"Indian agents"—proved corrupt or incompetent, and the supply chain broke down time and again.

The autumn of 1883 was extremely dry at Fort Peck, producing poor harvests for Sioux and Assiniboine families, who had been reduced mostly to farming rather than hunting. Despite the privation, officials in Washington, D.C., had cut appropriations for the tribes by 25 percent. Samuel E. Snider, the Indian agent who arrived at Fort Peck that summer, wrote that "with no crop, no game and as yet no supplies, the wolf of hunger is in every lodge." Autumn was followed by a brutally cold winter that made hunting even more difficult. Families were forced to subsist on one-and-a-half pounds of flour and a piece of meat provided every three weeks, or they ate their dogs and horses. Many survived on the quart of soup they could get at the Indian agency headquarters. By January, famine and disease were spreading across the reservation so fast that residents had no time to dig individual graves in the frozen ground. The result was mass burial sites like the one marked by the white crosses at Wolf Point, a tragedy that is still vivid among the residents of Fort Peck today. "They froze. They died. They stacked them like firewood," says Jonny BearCub Stiffarm, an attorney and elder of the reservation.

The graves at Wolf Point mark a shameful and heartbreaking episode in American history, but they also signify an environmental tragedy of monumental proportions. By replacing bison with cattle and grass with row crops, Euro-American settlers not only decimated Native societies, they began the destruction of a remarkable landscape. They brought to bear the might of an industrializing nation—steel plows, mechanized machinery, and other tools of a new, more efficient era in farming—and, with breathtaking speed, remade an ecosystem that had evolved over millennia. In less than a century after 1820, frontier farmers plowed up one-fifth of the continent and transformed a landscape of immeasurable richness and complexity into a gridwork of tidy farms. The sea of grass was about to become a sea of crops.

CHAPTER TWO

# *Plow*

ON A SPRING DAY IN 1837, A DOZEN CURIOUS FARMERS GATHERED IN A field outside Grand Detour, Illinois, to examine a miraculous new implement designed by a local blacksmith. The settlers and their neighbors had been pouring west onto the Illinois prairie for more than a decade, but they soon discovered that the farm implements they had used back East were no match for the heavy soils of their new home. In particular, their cast-iron plows, which worked perfectly well in the sandy, porous soils of New England and Pennsylvania, could not handle the dense, wet loam of the prairie. This "gumbo soil" clung to the plow's rough surface, forcing a farmer to stop repeatedly to scrape the blade clean as he made his way across a field and turning a routine task into an ordeal that could take days or weeks.

That young Illinois blacksmith, John Deere, had an important breakthrough: a plow with a cutting edge made of polished steel. Its sharp blade could sever the tough roots of prairie grasses and its smooth, shiny surface effortlessly shed the heavy soil as it moved down a furrow. Deere's blade glided across a field so quickly that it actually gave off a quiet hum. Astonished farmers soon dubbed it "the singing plow."

Within a decade Deere was selling a thousand plows a year. Soon it was ten thousand.

Deere was just one of countless tinkerers on the prairie frontier, and almost certainly not the first to fashion a steel plowshare. We remember him today because his modest venture grew into

John Deere's innovative steel plow became such a hit with prairie farmers in the mid-1800s that it soon went into mass production. *Photograph courtesy of the John Deere Company*

the world's largest farm implement manufacturer, its signature green tractors rolling across farms and suburban lawns. But the steel plow that he popularized was the first in a series of technological breakthroughs that enabled Europeans to settle the prairie, and it typified a wave of mechanical devices—reapers, threshers, ditchdiggers, planters, tractors—that transformed agriculture in the second half of the nineteenth century. In just three decades, prairie settlers using plows like Deere's broke as much virgin sod as eastern farmers had plowed in the previous two centuries.

John Deere arrived in Illinois on the leading edge of an important wave of migration from the Eastern Seaboard. By the first decades of the nineteenth century, Vermont, Massachusetts, Pennsylvania, and other eastern states were growing crowded and farmland was getting expensive. Many farmers took on debt to buy livestock and tools in the early part of the century, and many then lost their farms in the financial panic of 1837. Then too, a century of careless agriculture had begun depleting the soils of eastern states. Because land was so plentiful in the New World, many farmers took it for granted and failed to use time-honored techniques that preserve the soil, such as resting fields periodically or fertilizing with manure. By one estimate, crop yields in parts of Pennsylvania had dropped by 75 percent in the century since colonial settlement as farmers exhausted the land.

The virgin black soil of the prairie, by contrast, still seemed infinitely fertile. "God has here, with prodigal hand, scattered the seeds of thousands of beautiful plants, each suited to its season," wrote Charles La Trobe, a British adventurer. A German traveler, Ferdinand Ernst, wrote of central Illinois farmers in 1819, "They have only broken up the sod with the plow and planted their corn, and now one sees these splendid fields covered almost without exception with corn from ten to fifteen feet high." Farmers themselves were equally rhapsodic in letters to relatives back East. "I think that I can plow and harrow out hear without being nocked and jerked about with the stones as I always have in Jersey," wrote Ephraim Fairchild in 1857 from Iowa. "If Father and Mother and the rest of the family was out here, they would make a living easier than they can in Jersey."

More important, land was cheap on the frontier. A farmer who might have paid $35 per acre in Ohio or Pennsylvania could buy prairie land from the federal government for as little as $1.25 an acre—often less if the parcel in question had failed to sell at public auction. If they lacked the requisite $100 or $200, a prodigious sum at the time, settlers could simply squat on unclaimed parcels with the expectation that "improving" the land—plowing a few acres, fencing a pasture, and building a cabin—would entitle them to take possession from the government. Under the Preemption Act of 1841, a pioneer who broke the sod, erected fences, or made other improvements to an unclaimed parcel could claim 160 acres of surveyed public land, then buy it before it went on sale at public auction.

Of course this was government land only because the United States had forcibly removed most of the Indigenous people who already lived there. At the Battle of Fallen Timbers near Lake Erie in 1794, U.S. Army Major General Anthony Wayne crushed a confederation of Shawnee, Miami, and Lenape warriors, leading to a series of treaties that extracted huge and bitterly contested land cessions in what are present-day Ohio and Indiana. Farther west, a confederacy of Shawnee, Kickapoo, and Potawatomi tribes held off white settlers for several years under the leadership of the great chief Tecumseh, who held that his people should own land in common and that it was evil to sell it to white settlers. But in 1811, at the Battle of Tippecanoe in central Indiana, they suffered a mortal defeat by U.S. troops led by William Henry Harrison, who rode the victory to national prominence and became the nation's ninth president. At a conference in St. Louis in 1805, the Sauk leader Black Hawk told representatives of the U.S. government, "Never did the French, Spaniards, or British suffer such an invasion of our rights. We did not come to this council fire to beg anything of you. We want nothing but our own."

Military force wasn't the only assistance rendered to white settlers by state and federal governments that were eager to promote territorial expansion and the transfer of public land into private hands. The opening of the Erie Canal in 1825 triggered a flood of westward migration by offering cheap, fast passage from the East to central Ohio; it also gave landlocked midwestern farmers an inexpensive shipping channel to move grain from the Midwest to the port of New York—and on to world markets. The land rush between 1820 and 1860 produced such a migration that five new states of the Northwest territory—Ohio, Michigan, Wisconsin, Indiana, and Illinois—saw their population soar from 800,000 to 7 million. The historian Daniel Boorstin has argued that these measures also established a new relationship between Congress and its constituents: The federal government became a service agency that dug canals, built roads, and surveyed land. And because many of these projects were designed to help pioneers settle the new land and secure their ownership, they fostered an enduring American attitude toward land. This was the radical new idea—a sharp contrast to feudal Europe and the Indigenous societies of the plains—that those who worked the land were entitled to own it. Land quickly became a commodity to be bought and sold, creating an ethic that revered individual property rights—a philosophy that would greatly complicate efforts to protect soil, water, wildlife, and other natural resources for the next two centuries.

Despite cheap land and rich soil, the new landscape puzzled European settlers who were accustomed to the wooded river valleys of New England and rolling meadows of Pennsylvania. Many were dumbfounded by the vast expanses of open grass, and some assumed that the land must be infertile simply because no trees grew there. Early settlers clung to the tree-lined stream banks—partly because they needed timber and water, and partly because the open grasslands terrified them. The historian Allan Bogue cites a Pennsylvania settler who broke sod well beyond the timber line in Livingston County, Illinois, and was "generally pronounced a lunatic."

Yet month by month and year by year, farmers discovered the merits of prairie soil. By trial and error or by copying neighbors, they found that they could sow wheat or corn by hand and produce a healthy crop almost instantly. And while the lack of timber left settlers without wood to build houses and fences, it had its own advantages. The colonial farmers of New England found it necessary to clear their forested land by felling trees and removing stumps—usually by hand, with an axe—a backbreaking chore that might clear as little as one to three acres a year. On the open prairie, a farmer had simply to break the sod and sow some seeds.

The only obstacle was that stubborn prairie sod. Those ancient native grasses with their deep underground roots created dense mats that proved impenetrable to an ordinary plow-and-horse team. In the early decades, plowing virgin land required massive implements known as breaker plows that might weigh as much as 125 pounds and require a team of six or eight oxen. Few farmers could afford to own such a rig; many paid custom teams up to $2 per acre to break the ground—often more than they had paid for the land itself. And even after they had broken the sod, the sticky black soil made plowing miserable. This was the challenge that would make John Deere famous.

Like many others on the frontier, John Deere went west to make a fresh start after a series of failures. Born in Rutland, Vermont, in 1804, Deere apprenticed himself as a youth to a local blacksmith. After a year or two, he set out to establish his own practice mending wagons, carriages, and farm tools in the villages around Middlebury. In 1827 he married Demarius Lamb, the daughter of a prosperous neighbor, and two years later opened his own smithy in nearby Leicester. But the couple were thwarted time and again. Deere struggled to compete with more established blacksmiths; fire destroyed two of his smithies, forcing the family into an itinerant life around central Vermont for several years. In 1835, Deere borrowed $78 in a third attempt to establish his own blacksmith shop. But this venture also failed, probably because Vermont already had plenty of accomplished blacksmiths. Unable to repay his loan, Deere was served a sheriff's summons and arrested. He managed to post bail but found that a judge had placed a lien on his property. At age thirty-four, Deere faced a bitter choice: debtor's prison or escape to the frontier. He chose the latter, boarding a canal boat west and leaving Demarius and four children behind.

In Grand Detour, Deere's luck seemed to change. The village, named for a deep bend in the Rock River, was a thriving settlement that provided plenty of work for a skilled blacksmith. Deere had friends from New England here, including a Vermont acquaintance named Leonard Andrus, who had founded a sawmill and other businesses along the river. Andrus gave him work immediately repairing mill machinery and, with less competition than he had faced in Vermont, Deere established a business repairing plows, wagons, and tools. In three years he had earned enough money to repay his Vermont debt and send for Demarius and the children.

Starting over with new neighbors, Deere could also redeem the social standing he'd lost after his failures in Vermont. With his square jaw and furrowed brow, he earned a reputation for hard work and New England rectitude. He was known to fire up his forge before dawn and toil until midnight, and the steady volume of business noted by historians suggests that farmers trusted his work and work ethic. He was also an aggressive competitor; court records show that he sued Illinois rivals repeatedly for copying his inventions and infringing on his patents. In time he became active in public life, joining Illinois's abolitionist movement as the Civil War approached and becoming an officer in the local Whig party.

The exact circumstances of Deere's brainstorm are uncertain—the historical accounts are assembled chiefly from family yarns and corporate lore. The accepted version is that one day, while visiting one of Andrus's sawmills, he noticed a broken steel saw blade lying in the corner. Deere by this time understood the vexation that cast-iron plows were causing prairie farmers, and he knew that blacksmiths in England and the East were experimenting with plowshares of polished steel. The story has it that he took the discarded blade back to his shop and, using a hammer and chisel, cut off the saw teeth until he had a polished slab about the size of a large serving platter. This he fashioned into the familiar shape of a plow—a curving, concave blade, its sharp leading edge raked back at an angle—and affixed it to the wooden frame of a conventional horse-drawn plow. In one version of the story, he displayed the new plow on a wooden produce box outside his workshop until a passing farmer took notice; by another account, he offered to demonstrate it to curious farmers at the nearby farm of a friend, Lewis Crandall.

Deere almost certainly was not the first blacksmith to forge a steel plowshare; many historians assign that distinction to John Lane, an Illinois blacksmith who is believed to have attached a steel plowshare to a wooden plow frame in 1833 and whose work was probably known to Deere. What seems to have set Deere apart was a combination of perseverance, ingenuity, and marketing savvy. In particular, he seemed willing to tinker endlessly to produce an implement that was durable and easy to use. Deere biographers Neil and Jeremy Dahlstrom write that Deere himself never claimed to have invented the steel plow; rather, his contribution consisted of "constant improvement, quality craftsmanship and superior sales efforts."

Whatever his creative spark, Deere soon realized that producing farm implements turned a better profit than repairing them; he gradually transformed his blacksmith shop into a manufacturing operation. With a growing staff of assistants, he produced forty plows in 1840 and seventy-five the next year. His output reached one thousand plows by 1846 and ten thousand a decade later. In 1848 he moved operations seventy miles west to Moline, a Mississippi River town with abundant water power and steamboat traffic that would carry his merchandise as far north as St. Paul and as far south as St. Louis.

Deere introduced his plow during a period of technological change that, in retrospect, is astonishing. In just three decades, between 1859 and 1889, Alexander Graham Bell invented the telephone; Thomas Edison introduced the light bulb and began generating electricity; Karl Benz built the world's first successful automobile; the transcontinental railroad, completed at Promontory Summit, Utah, joined the East Coast to the West; and oil was discovered in western Pennsylvania, with a byproduct—gasoline—that would revolutionize life in the Western world. The same period, roughly, introduced the sewing machine, the phonograph, the typewriter, a modern camera, and, thanks to Montgomery Ward, the first mail-order catalog. But, at a time when agriculture employed 60 percent of the labor force, it was farm machinery that held a special fascination for the American public. Farm implements were so important to the people of Illinois that the state dispatched a correspondent to the great Paris Exposition of 1867 because one of Deere's plows was competing for a gold medal with the latest European inventions. The plow, an A Number 1 Clipper, was sold to the Prussian minister of agriculture before it could even be entered in the competition.

Novel though it was, Deere's plow was just one of many innovations that, in combination, transformed nineteenth-century farming. For example, it made no sense for a farmer to plow forty acres with a new steel plow if, sowing seed by hand, he could plant only a few acres or, cutting wheat with a scythe, could harvest only a few. But the mid-nineteenth century produced a string of other breakthrough implements that enabled settlers to farm on a scale that matched the landscape. In 1840, Cyrus McCormick introduced a mechanized horse-drawn reaper, an invention that greatly reduced the amount of labor needed to harvest grain and would create the industrial giant International Harvester. Mechanical planters known as grain drills allowed farmers to sow seed faster and more systematically, while horse-drawn threshing machines soon replaced the handheld flails used to separate kernel from chaff. Farming had entered a new age.

The diary of an Illinois settler named Philemon Stout illustrates the way that technology changed frontier farming during the nineteenth century—and how frontier farmers transformed the prairie landscape. Philemon and Penelope Stout arrived in central Illinois from Kentucky in 1836, a year before Deere introduced his steel plow. The couple and their children settled along Sugar Creek, just south of Springfield in a lobe of the tallgrass prairie about two hundred miles south of Grand Detour. They acquired 350 acres, built a log cabin, and began breaking the prairie sod with a plow made of wood and iron. Philemon Jr., the fourth of their seven children, was twelve at the time and began helping his father in the fields after his two older brothers established farmsteads of their own. Father and son planted their corn by hand in the spring, husked it manually in the fall, and weeded it with hoes during the months in between. When his father died in 1846, Philemon Jr. understood that 350 acres was too much for him to farm alone, so he rented a portion to a tenant family and took on three hired hands.

By 1860, Philemon Jr. and his wife, Louisa, were raising four children and farming 440 acres while renting another 160 acres to tenants. They grew corn, oats, and wheat; raised hogs; and pastured their horses on grassy fields they hadn't plowed. The farm was still highly self-sufficient: They ate corn meal milled from their own corn; produced their own eggs, milk, and pork; and fed their hogs with husks and stalks left over from the corn harvest. For a bit of cash income—to pay taxes and buy manufactured goods such as nails and window latches—they drove hogs to market in St. Louis once a year. They sowed their oats by hand, harvested them manually, and threshed them by stomping the kernels on the wooden floor of their barn.

In the fall of 1860 the Stouts acquired a Manny reaper, a horse-drawn machine with a mechanized blade to harvest their wheat and oats. With the reaper they could cut ten to fifteen acres of wheat in the time it would take a hired hand to harvest one with a hand scythe. Later that fall, for the first time, Stout borrowed a neighbor's horse-drawn threshing machine for the essential task of separating wheat from chaff. The machine could clean thirty bushels of wheat in the time it would have taken one worker with a hand flail to separate seven. That fall, the Stouts harvested enough surplus corn to feed forty calves, and they sold the mature animals for a profit of $1,248.

By 1881, Philemon and Louisa Stout owned 2,000 acres and a fleet of horse-drawn implements, farming 250 acres themselves while their children and tenants worked the rest. They had achieved a level of affluence Philemon's parents could never have imagined. They lived in a frame house built of sawmill lumber and outfitted with glass windows, a sewing machine, and a cast-iron stove. On a day off, Stout took the family to a circus and animal show in Springfield, nine miles away, where they bought ladies' magazines and factory-made shoes. A devout Baptist, Stout donated an acre of land to his church to be used for a meeting house and cemetery. He was elected a church deacon, a highway commissioner, and a trustee of the local school district. In a diary he kept from 1860 until 1910, Stout described himself and Louisa as "greatly blessed."

The Stout farm was bigger than most, but it reflected the remarkable transformation of American agriculture during John Deere's lifetime. In 1830 a pioneer family turning the soil with a cast-iron plow and sowing seed by hand might harvest fifty acres in a year, enough to feed themselves and their livestock. By 1860 their children, using a steel plow, a mechanized corn planter, and a horse-drawn reaper, could harvest 100 to 150 acres. Between 1850 and 1880, records from neighboring Iowa show, the median farm grew by one-third, to 160 acres, and the acreage under cultivation on those farms tripled. The historian Douglas Hurt has estimated that in 1820, a typical midwestern farmer could plant and harvest one acre a day; by 1880, a similar farmer could work twenty acres a day. Between 1840 and 1860 the nation's farm output doubled, and then more than tripled again from 1860 to 1890, according to historian David Vaught.

Farms didn't just change in scale, they changed in character.

With one hundred to two hundred acres under cultivation, a prairie farmer could produce much more grain than his family and livestock could consume; he became a producer for the market. Farmers who scarcely had time to catch up on village gossip in 1830 now watched grain futures in St. Louis and hog prices in Chicago. Farming, in turn, accelerated the growth of manufacturing in American cities. Between 1850 and 1890, the typical midwestern farm doubled or tripled its investment in farm machinery and hugely increased its production of cattle and hogs for market—twin developments that explain why Cyrus McCormick and the meatpacking king Philip Danforth Armour became pioneers in large-scale industry on a factory model. The self-sufficient yeoman farmer, if he ever existed outside the imagination of Thomas Jefferson, had given way to a capital-intensive entrepreneur keenly attuned to the marketplace. "Farmers," Vaught notes, "now stood at the center of a vast and complex network of trade and industry."

The mechanization of nineteenth-century agriculture greatly increased farm productivity and probably hastened the plow-up of the prairie, but it also launched two trends that, over time, would come to haunt rural America. Because an individual family could work more and more land—120 acres, then 250, eventually 1,000—rural America needed fewer and fewer farmers. In many rural counties, the population peaked in the 1920s or 1930s, then commenced a long slide from which many places have never recovered. With fewer farmers, small-town commerce suffered from fewer customers; the independent druggist and local haberdasher began to vanish from Main Street, replaced by empty storefronts and, if the community was lucky, a Walmart on the outskirts of town. Remaining farmers understandably felt like a dying breed, left behind in a changing America and threatened by forces beyond their control. Mechanization of farming also fueled the wave of industrialization that was cresting across the American economy at the end of the nineteenth century. Implement factories and slaughterhouses were among the nation's first industrial facilities, preceding Detroit's auto plants by decades. Armour's massive packinghouse at the Chicago stockyards, where hundreds of meat cutters dissected cattle carcasses as they came by on an overhead chain, pioneered the concept of the assembly line—though it was assembly in reverse—and served as a model for Henry Ford's legendary auto assembly plants. Minneapolis became a grain-milling center; in the late nineteenth century it was the flour capital of the world. Chicago built the world's biggest stockyards and was immortalized by Carl Sandburg as hog butcher for the world. But urbanization also meant that the courageous yeoman farmer had to share the national stage with other important characters—the factory worker, the train engineer, the railroad magnate, and the banker. Popular literature of the late nineteenth century began to portray farmers as yokels and hayseeds rather than heroes of the rugged frontier. Bankers and industrialists gained Washington's favor, leaving farmers feeling exploited and frustrated. The presidential candidacy of William Jennings Bryan in 1896 gave voice to this populist protest against railroads, financiers, politicians, and the "cross of gold"—the gold standard, which by limiting the amount of currency in circulation often caused deflationary pressures on the prices farmers received for their crops. By 1920, just 40 percent of Americans lived in rural places, and many began to feel left behind by progress and alienated from their urban cousins—trends that would only intensify later in the twentieth century.

The tools of an ambitious nation—mechanized farm implements, transcontinental railroads, federal land subsidies, and an ongoing war against Native people—pushed American farming west across the prairie with staggering speed. The number of acres under cultivation in the Midwest nearly doubled between 1850 and 1860, and then quadrupled again by 1900, reaching 222 million acres. Six states alone—Nebraska, Kansas, Iowa, Minnesota, and the Dakotas—added more than fifty million acres of farmland between 1880 and 1890. The arc of the nation's frontier—the Census Bureau's line marking the western edge of civilization—moved steadily west, to Iowa and Missouri in 1840, then Minnesota, Nebraska, and Kansas by 1860.

These same decades marked a socio-economic event of global significance. Settler agriculture represented a transfer of land to the people who worked it on a scale that was "unique in the history of the world," according to historian Willard Cochrane. It attracted European immigrants by the millions and offered the promise—often true for only a lucky few—that America was a place where common people could own their own land and determine their own fates. It also turned the Midwest into the breadbasket of the world—a region that fed troops in World War I, succored Europe after World War II, and eventually spawned the "green revolution," whose high-yield grain varieties and synthetic fertilizers would transform agriculture in places as far away as China and India. By the early twentieth century, the United States grew 40 percent of the world's corn and produced 30 percent of its meat supply.

But it also represented the transformation of a natural landscape unprecedented in scale and speed. In just a little over a century, American farmers plowed up 300 million acres of virgin grassland, converting a wilderness of deer and songbirds into a factory for food. Most contemporary writers described this era in the glowing language of conquest and progress, but a few captured the sense of incalculable loss. In 1921 an Iowa lawyer and novelist named Herbert Quick published *Vandemark's Folly,* a rustic narrative whose hero, Cow Vandemark, described plowing virgin ground this way: "Breaking prairie was the most beautiful, the most epochal, and most hopeful—and as I look back at it, in one way the most pathetic thing man ever did, for in it, one of the loveliest things ever created began to come to its predestined end."

By the middle of the twentieth century, just 1 percent of the eastern tallgrass prairie remained.

CHAPTER THREE

# *Swamp*

WHEN THE WORLD'S CLIMATE BEGAN TO WARM SOME TEN THOUSAND TO fifteen thousand years ago, the glaciers that for millennia had stretched from the top of the globe down to what is now northern Iowa began to melt. The land that emerged had been scoured by ice sheets as much as a mile thick and was full of water. In what is now Alberta, Canada, the Dakotas, western Minnesota, and northwest Iowa, the land was distinctively pockmarked with millions of small, round lakes and ponds—the prairie pothole region. These miniature lakes filled with snowmelt and rainwater, overflowing in the spring and shrinking down to their muddy bottoms in the heat of summer. They sparkled across 64 million acres of the midwestern prairie states, each one a tiny bubble of rich and diverse aquatic prairie life—snails, frogs, clams, muskrats, birds, and wetland plants. They were the watering holes for deer, elk, and bison, and every spring the fringes of those ponds formed the nesting grounds for half of the nation's waterfowl. In the fall, the skies would darken as these birds took flight and headed south.

Farther south and east, in Illinois and Indiana, the melting glaciers left a flat plain that held the water like a massive sponge.

Water collected in the low areas, gradually forming new rivers that meandered across the flat land, merging with ponds, sloughs, and wetlands as they found their way to the Mississippi River. Much of Illinois, Indiana, and Iowa was wet prairie, where water stood inches deep much of the year. Grasses soaked up the water and the sunlight and grew eight feet tall. The Midwest region of the United States—the tallgrass prairies—once held a third of the country's 221 million acres of wetlands, an area larger than the size of Texas.

Today it's almost all gone, drained, dried out, and plowed for farmland. Ironically, it's only in recent decades that scientists fully began to understand wetlands' extraordinary ecological value. Wetlands hold the water that refills aquifers. They slowly release it to streams and rivers, preventing erosion and reducing flooding. Because water flows through them slowly, the sediment it carries can settle on the bottoms rather than clouding streams and lakes. Nutrients from fertilizer, manure, and sewage are taken up by plants and microorganisms growing in the soil—filtering functions so powerful that now engineers replicate them to treat the wastewater produced by large cities. Wetlands are among the most biologically rich ecosystems in the world, comparable to tropical rainforests and coral reefs. They are incubators for fish, home to 31 percent of the nation's plant species, and dozens of bird species rely on them for food or nests.

Historians, geographers, archaeologists, and the oral histories handed down among Native Americans can give us only a glimpse of what they once were. More than half of what is now Minnesota and Wisconsin were marshes, shallow lakes ringed with cattails and thick with wild rice, swamps filled with stands of pale green soft-needled tamarack trees, wet scrublands that extended for a hundred miles. Native American settlements flourished along the edges of these wet ecosystems throughout the upper Mississippi River Valley and around the Great Lakes. Hunters could find deer, moose, muskrats, beaver, and fish. They harvested raspberries and other wild fruit, nuts, wild rice, and medicinal plants. Trees sustained by the moisture provided wood and shade. Indigenous people wove baskets from the reeds, and their canoes, designed for the shallow marshes and narrow rivers, made the waterways into highways.

The new European arrivals, on the other hand, came to hate and fear these marshy places. Wetlands in Europe had been drained or channeled centuries earlier, so the massive wooded swamps and coastal wetlands of the New World were alien spaces to them. They saw them as "wicked," their fears inflamed by the Indians who, in the late 1600s used swamps as refuges that Europeans feared to enter. Nathaniel Saltonstall, the judge who oversaw the Salem witch trials and who later published an account of the early Indian wars, wrote to a friend in London, describing the way Indians could hide within arm's reach in a swamp. While an Englishman could not even stand on the soft ground, "these bloody savages will run along over it, holding their guns and shooting too." The clashes between colonists and Native Americans grew, and finally erupted into a conflict known as King Philip's War. In the winter of 1675 it culminated in a vicious battle on a frozen swamp in Rhode Island. There, white soldiers set fire to a Native encampment, killing three hundred warriors and an equal number of women and children in what came to be known as the Great Swamp Fight. Around that same time John Bunyan published his hugely popular book *The Pilgrim's Progress,* which used the fearful imagery of swamps as a metaphor for obstacles in the path of Christians seeking salvation.

When Europeans moved west onto the prairies with their heavy wagons, their distaste for wetlands only grew. The million-acre Black Swamp in northwest Ohio, an impenetrable region of mud and marsh that once functioned as a massive filter for Lake Erie, became legendary among westward travelers. In 1827, it contained "the worst road on the continent," where horses and wagons foundered in the black muck. A towering yellow prairie flower that grew as much as seven feet tall, anchored with a root that ran sixteen feet deep, was named compass plant by settlers because its leaves always aligned north and south. They tied flags on the tips of the leaves to mark the safest way through the wet prairies for those who followed.

In 1821, Samuel Burton described the green-headed flies that would swarm over horses as they slogged through the wet grasslands of Illinois. They were so thick that they had to be "skinned off with a knife," leaving the horse covered in blood. Charles Dickens took a steamboat journey through southern Illinois in 1842 and wrote terrifying descriptions of the wetlands at the confluence of the Ohio and Mississippi rivers, describing them as

> a dismal swamp . . . teeming, then, with rank unwholesome vegetation, in whose baleful shade the wretched wanderers who are tempted hither droop, and die, and lay their bones; the hateful Mississippi circling and eddying before it, and turning off upon its southern course a slimy monster hideous to behold; a hotbed of disease, an ugly sepulcher, a grave uncovered by any gleam of promise: a place without one single quality, in earth or air or water, to commend it.

As John Deere's plow, Cyrus McCormick's reaper, and other mechanized farm implements began to transform the dry prairies in the middle of the nineteenth century, the vast wetlands remained largely untouched. Even as the Stouts in central Illinois were prospering on their 440 acres of rich, dry land, nearly a quarter of their state remained a wet, untraveled wilderness. In 1863 a European explorer traveling the Kankakee River south of Chicago wrote that marshes full of iris and alder bushes stretched as far as the eye could see.

Early settlers often made the mistake of building on what appeared to be dry, flat land in the summer, only to find themselves inundated with high water in the spring. In 1837 a group of Norwegian settlers bought farmland south of Chicago and built their new homes there. Come spring, the land flooded, and with the water came malaria. By fall many of the Scandinavians were dead and the settlement was abandoned.

The true carrier of malaria—the female *Anopheles* mosquito—would not be known until late in the nineteenth century. But European settlers easily made the connection between the disease and what was then called "wet lands." They were viewed as "miasmic" and unhealthy landscapes, the source of ague and swamp fevers. As the population of settlers grew, so did the number of deaths from insect-borne diseases. Malaria was a leading cause of death throughout the Midwest in the 1800s. During the 1820s a powerful strain of malaria killed nearly every resident of Pike County in Illinois between the Mississippi and Illinois rivers. One doctor described that wet boggy area as "a gigantic emporium of malaria."

At first settlers just avoided the wetlands. They snatched up the drier lands that bordered the forests, using the timber to build their homes and their fences. When that land was taken, they followed the wooded banks of the rivers that wound across the Midwest. Only after that land, too, was occupied did they move out onto the prairies, building sod houses and making fences from living bushes instead of cutting them from logs. By the middle of the nineteenth century, European Americans occupied most of the fertile lands in the Midwest and the rapid advances in mechanization allowed them to turn it into productive farmland at ever faster rates. But the plows, reapers, and threshers were useless on the wet prairies.

Inevitably, though, the national ambition and agrarian ideals that drove settlers west took root in the wetlands as well. In 1849 Congress passed the first Swampland Act transferring federally owned wetlands in Louisiana to the state government. For centuries the Mississippi River had collected water and snowmelt from across its basin in the middle of the country, delivering it to the

Gulf of Mexico. Every spring, the big river overflowed its banks, depositing soil, recharging wetlands, and creating a floodplain thirty miles wide. But by the 1800s, a century's worth of development and farming in the upper Mississippi River basin had already removed much of the natural flood control on the landscape—the trees and grasslands that soaked up rain and snowmelt. Landowners up and down the Mississippi River built levees to protect their property from the higher flows, increasing flooding downstream, especially on the sugar plantations in Louisiana and in the low-lying city of New Orleans. The first federal swampland transfer act was driven by Senator Solomon Downs of Louisiana, who argued that draining and plowing wetlands in Louisiana would help advance the nation through the "increase of population, the augmentation of wealth, the cultivation of virtue, and the diffusion of happiness." Supporters of the bill described swamplands as "sterile," barren lands that could not support crops and needed more human ingenuity to reach their true potential as farmland. The bill authorized Washington to transfer ownership of federally held swamplands to states with large amounts of wetlands. The states could sell their swamplands to private owners to generate funds to build public works like drainage channels and levees to hold the rivers within their banks and prevent flooding.

The bill passed and was followed in the next decade by similar legislation granting federally owned wetlands and swamps to fifteen states, including fifteen million acres in the Midwest. The idea was that states would sell public lands to private landowners, and use the proceeds for the common good—to build ditches, dredge rivers, and build levees that would create new land for farming. Increasing private ownership of public land was regarded as the best way to increase agricultural productivity while advancing self-reliance and civic virtue.

It took another twenty years before the plan began to work. States sold the land, but it went for $1 or less per acre, and much of it was purchased by speculators who, despite the intent of the law, had no interest in spending money to drain it. Even with these sales, neither local governments nor small farmers had the enormous amounts of capital needed to dig ditches, and the development of large-scale machines for large-scale drainage was in its infancy. Nor did they have the civic structure they needed to manage, for the collective good, the movement of water across the land.

Only the wealthiest landowners could afford to dig the ditches needed to drain their land, and they had to be creative. In the 1850s in Illinois, one owner of 5,000 acres of wet prairie used a specially built plow pulled by 40 head of cattle to cut 250 miles of ditches. But the ditches could dry only narrow strips of prairie that ran alongside them, and often there was nowhere for the water in the ditch to go—except onto a neighbor's land. Worse, they required constant dredging because they filled in quickly with dirt. As a result, none of the midwestern states honored their federal obligations established by the bill to drain wetlands for agriculture, or to protect the public from swampland diseases. Instead, the states used the proceeds from selling wetlands for other purposes—fighting legal actions, paying off state debts, constructing public works unrelated to drainage, or out-and-out corruption. In 1859, a special Indiana state committee found that only 10 percent of contracted drainage projects had been completed.

It was only in the last quarter of the nineteenth century that drainage began in earnest. The instigator was John Johnston, a Scottish immigrant who came to the United States in 1821. Johnston acquired 112 acres for $1,200 near the Erie Canal in Geneva, New York. The soil was "heavy gravelly clay with a tenacious subsoil." And wet. It was viewed by his neighbors as some of the least productive in the county. The drainage Johnston envisioned was far more complex and effective than digging ditches that drained only small parcels. He wanted to drain the water from beneath his wet fields in their entirety. In 1835, after reading about the success of drainage in England and Scotland, Johnston ordered a pattern for the kind of long, horseshoe-shaped clay gutters used in Europe at the time. These would be buried end to end in trenches, open side down, to capture groundwater that flowed through gaps between the tiles and carry it away before it could drown the roots of crops. They carried the name that agricultural drainage systems have to this day—tile lines. A neighbor who made earthenware crocks and jugs agreed to make the clay tiles for Johnston, and to the amusement and ridicule of his neighbors, Johnston laid them in regularly spaced lines beneath his fields. "John Johnston is gone crazy—he is burying crockery in the ground," one said.

John Johnston, considered the father of tile drainage in the United States, from *History of Tile Drainage in America* by Marion M. Weaver. *Courtesy of the Geneva Historical Society, Geneva, New York*

Although they were a novelty in rural New York during Johnston's lifetime, underground drainage tiles made of clay, stone, or bamboo have been used for thousands of years in the Middle East, Europe, and China. Farmers have long recognized that draining the ground beneath the roots of their crops increased their yields. In spring, the groundwater rises—much the way water rises from the bottom of a bucket that is filled from above. When water saturates the roots of young seedlings, they stop growing. When the water table drops, the roots dry out and remain stunted, crippling the plant's growth throughout the rest of the season. In a field with tile drainage, the groundwater entering the tubes is carried away underground to an outlet in a ditch or stream—a design similar to the way human ribs connect to a backbone. The roots of the plants are forced to grow deeper, stronger, and healthier in order to reach the lowered water table, helping the crops thrive throughout the season. And if wet periods recur later in the summer or fall the drain tiles prevent the soil around the roots from becoming saturated. The results in Johnston's fields were immediate and dramatic. He ditched and tiled a ten-acre field, and his wheat crop exploded from five to fifty bushels per acre. By 1851 Johnston had laid 16 miles of clay tiles under his 320 acres, attracting curious farmers from all over the country. His farm was featured regularly in the agricultural journals that were popular at the time and offered a critical source of information for farmers across the country. Soon the magazines were full of testimonial letters from other farmers describing the improved productivity from drain tiles. A photo of Johnston taken about the time shows a thin man with flyaway gray hair in a heavy black suit. He's staring intently into the distance—as if he could see the hydrological revolution he was about to launch unfolding before him. He became known as the "Father of Tile Drainage in America," and today a small museum in his original farmhouse in Geneva, New York, commemorates him and his legacy.

After his early successes, Johnston and a neighbor imported a tile-making device, the Scragg's Patent Tile Machine, which could turn out twelve thousand tiles daily, enough for more than three thousand feet of drainage lines. Soon, clay tile factories were popping up all over Upstate New York. The factories moved west with the frontier farmers struggling to turn a wet wilderness into orderly cropland. In 1867 Ohio had five hundred steam-powered tile-making machines, enough to make two thousand miles of drain tiles annually. By 1880, Ohio, Illinois, and Indiana had more than eleven hundred drain tile factories, with one hundred more in Michigan, Iowa, and Wisconsin. By 1882, farmers in Indiana had laid more than thirty thousand miles of drainage tiles, and in the next twenty years the practice grew even faster. By the end of the nineteenth century, Illinois had some 126,263 miles of drainage tile laid end to end under the farming countryside. Finally, the kind of technological advances that had driven the expansion of frontier agriculture and urban industry were mastering the wet prairies.

Still, digging ditches and laying tile was backbreaking work. Farmers hired immigrant laborers—many from among the hundreds of thousands of Irish immigrants who were fleeing the potato famine in their homeland. With pickaxes and shovels, they dug the trenches for tiles and the ditches that carried water away into nearby streams and rivers. In about 1850 farmers began using a device called a mole plow to dig the trenches for tile lines. This was a wedge of iron thrust three or four feet into the earth and pulled through the dirt horizontally by a team of horses or oxen at the pace of one half mile per day. Later, they used other horse-drawn machines that plowed up the soil and deposited it in mounds on both sides of a trench. But the devices were hard to find and expensive and came into wide use only in the 1880s. Then, even as other farm machinery was growing in size and power, an inventor introduced the Blickensderfer Tile Ditching Machine, a horse-drawn implement that could dig a tiling trench four feet deep in one pass. As steam power replaced horsepower, the Blickensderfer machine gave way to the Buckeye Trencher from Findlay, Ohio, a machine that in a single day could cut fifteen hundred feet or more of trenches four and a half feet deep. And because water flowing out of tile lines had to be carried away somewhere, farmers soon began using an even bigger class of machines to dig the larger ditches that would collect the water and send it downstream to local rivers. Giant ditching plows, or Swamp Angels, first were drawn by huge teams of horses or oxen. By the end of the nineteenth century, engineers had begun using steam-powered dredgers mounted on barges to deepen and straighten natural waterways in order to handle the ever greater volumes of water that flowed off the growing expanse of drained farm fields.

The drainage revolution changed the public perception of wet-

P. G. Jacobson's ditch-and-tile machine, working near Madison, Minnesota, in the early 1900s, shows how mechanization was accelerating what had been a slow and labor-intensive job.
*Photograph by M. J. Viken, courtesy of the Minnesota Historical Society*

lands dramatically. The experience of a few farmers like Johnston made it clear that once drained, wetlands were the most productive soils in the Midwest. Wetland plants had grown and died over thousands of years, building up rich organic matter beneath the roots of successive generations. Once dry, the soil held huge stocks of nitrogen and other nutrients, and produced more grain per acre than any other arable lands in North America; corn yields on former wetlands could be as much as 50 percent higher compared to dry land. With that kind of soil, farmers had far more flexibility to grow what they chose—wheat, corn, oats, barley, hay—and could respond quickly to shifts in market demand.

Soon draining the land became a sort of public gospel, spread by farming journals, books, newspapers, and railroads seeking to sell marginal land to settlers. Drain tile, they promised, would work miracles—they would remove stagnant water, extend the growing season by drying and warming the soil earlier in the year, and improve the quantity and quality of crops. These publications also detailed the latest, best techniques for laying tile. By the late 1800s technology had advanced to the point that laying tile correctly required skilled engineers and expert diggers. Johnston's horseshoe tiles had evolved into cylinders two to eight inches in diameter, depending on how wet the field was, and a foot long. The distance between each line depended on the type of soil, the gradient of the field, and the depth of the trench.

With each advance in technology and with every new landowner who tried it, drainage became accessible to more farmers. They in turn earned greater profits from better crops and the sale of "improved" land, which they reinvested in more machines to convert more wetlands. The practice became yet another way that nature could be "improved" upon in order to reach its full potential in the service of men. "It is quite evident that our world was not finished on the day of creation," horticulturalist William P. Pierson said in a speech at the 1868 annual convention of the Illinois State Horticultural Society.

> This is a task that has been assigned to man to do. Let nature . . . do her legitimate work, and the results will be witnessed here on these prairies, in the grain field, the meadow, the orchard, the vineyard, that will astonish the world, and that will gladden the hearts of the tillers of the soil, and of the worshipers at the shrine of Horticulture.

By the close of the nineteenth century, tile drainage had transformed wetlands from New York to western Iowa.

Tile drainage on such a scale was just one example in a period of remarkable engineering breakthroughs that revolutionized the economy and captured the public's imagination. Railroads had tied the continent together, steam power was replacing livestock, and electricity was transforming rural and urban life alike. In 1893 more than twenty million visitors flocked to the World's Columbian Exposition in Chicago, where the core theme was science and technology. Visiting celebrities included Thomas Edison and Nikola Tesla. Two American corporations in their infancy, Westinghouse and General Electric, competed ferociously for the contract to provide the fairgrounds with electric lighting, a novelty at the time. Visitors jammed the vast exhibition halls to witness dazzling inventions such as the telephone switchboard, the telegraph keypad, and an electric incubator for chicken eggs.

The city of Chicago itself had just embraced hydrological engineering on a huge scale, though for a very different reason than farmers. Between 1840 and 1890 its population exploded from 4,500 to nearly 1 million as the city evolved into the industrial hub for marketing and shipping the grain and livestock produced on the surrounding prairies. But that extraordinary growth also created a serious sanitation problem. In an ambitious feat of engineering starting in 1856, city engineers used jackscrews to raise the buildings of the central city as much as ten feet to make space for sewer pipes to run beneath them and downhill into Lake Michigan. But because the lake also supplied Chicago's drinking water, the city's engineers were forced to tunnel two miles out into the lake with pipes that could collect clean water for drinking. By 1900 they recognized that turning the lake into an open sewer had been a mistake, and, in a second extraordinary feat of engineering, reversed the flow of the Chicago River so it would carry waste away from the city and Lake Michigan and into the Mississippi River.

Both projects required collective public action for the common good of the city, something that farmers and rural governments soon recognized as critical in the effort to drain wet farmland. It was pointless, after all, to drain just one farmer's field: The water that flowed off had to go somewhere—usually onto a neighbor's land, potentially drowning his crops. Truly effective drainage across a watershed would work only with cooperation among neighboring landowners—or the use of legal force against those who wouldn't cooperate. In short, all that technological ingenuity was useless without a powerful social institution to govern the work and keep the public good in focus. This turned out to be an entirely new unit of government called a drainage district. Michigan and Ohio passed laws governing the management of ditches and drainage as early as 1847. In the next twenty-five years other midwestern states followed suit, and many of the new laws regulating regional hydrology were tested and upheld by the courts. In Illinois and Iowa, laws supporting drainage were written into the state constitutions. Legislatures and courts reasoned that expanding agriculture benefited everyone, even though it was private landholders and businesses that benefited most. The Minnesota Supreme Court, for example, ruled that "the fact that large tracts of otherwise waste land may thus be reclaimed and made suitable for agricultural purposes is deemed and held to constitute a public benefit." Some historians argue that the courts and legislatures seized on the perceived public health benefits of reducing insect-borne diseases as a pretext for passing laws to benefit business and agriculture. Those who objected, whether neighboring landowners or the few who saw ecological value in wetlands, were regarded as obstructionists in the way of economic progress.

The new government bodies were controlled by district boards, with members appointed by county commissioners, judges, or other local officials. A majority of landowners in an area could petition the boards to authorize a new drainage project. The board had the power to finance projects, acquire land—through eminent domain, if necessary—build works, and tax the farmers who benefited. Most laws stipulated that the economic benefit of the project had to exceed its cost, a feat that was easily accomplished because the drained lands, with their far richer soils, immediately rose in value. Between 1880 and 1900 the cost of typical midwestern farmland increased by up to 40 percent. During that same period, the value of drained wetlands often increased by up to 500 percent. In those twenty years, Indiana farmers quadrupled the amount of drainage tile beneath their lands, to 126,000 miles. Illinois and Iowa, in the heart of the Corn Belt, lost over 85 percent of their prairie potholes and marshes. In Ohio, Indiana, and Michigan more than 80 percent of the swamps were converted to farmland, including the million-acre Black Swamp in northwest Ohio and the wetter parts of the Illinois Grand Prairie. Altogether, those feats of hydrological engineering required all three elements of drainage to finally come together: machine technology to do the backbreaking work of digging across miles of prairie, private investment by farmers on their own land in pursuit of commodity agriculture, and the creation of local government entities with a shared mission to regulate land and the movement of water across the continent—"for the public good."

Laying tile in a drainage ditch, early 1900s, Minnesota. *Photograph by Harry Darius Ayer, courtesy of the Minnesota Historical Society*

SOME OF THE last of the tallgrass prairie to be settled by Europeans lay in northern Iowa just west of the Mississippi River. The story of that small region in the last half of the nineteenth century illustrates how the much larger transformation of the country's midsection occurred settler by settler, foot by foot, and, especially on that wet prairie, clay tile by clay tile. Today that region is known as the Des Moines Lobe, named for the last glacier to cover North America. The settlers who moved there were part of a wave that came directly from northern Europe—Germans, Danes, Swedes, Norwegians, and Finns—seeking the cheapest available land. In 1857, a settler and newspaper man named Charles Aldrich joined the migration to northwest Iowa. He took the railroad as far as it went at the time—thirty miles west of Dubuque. From there, he and a group of companions took a lumber wagon carrying the equipment for a printing press, and bumped along a track through the grasslands for another 150 miles. Time and again, they had to stop to unload the wagon and carry the cargo by hand around and through marshes, called sloughs, and shallow lakes. Aldrich was one of the few at the time to write about them as natural wonders: the yellow-headed black birds that glistened like polished gold; the clever marsh wrens that disguised their nests to protect their young from foxes and prairie wolves; the dragonflies—"devil's darning needles"—that darted over the water. He wrote, "No two prairie sloughs were alike. We had ponds or lakelets, where the water was open, in rare instances abounding with fish, and others, where the surface was covered with dense growths of bulrushes and coarse grasses, which looked black when seen from a little distance."

Like the Indigenous people who preceded them, many of the new Iowans initially had to depend on the ponds for survival. Louis Hersom, who farmed in northwest Iowa, recalled in an oral history for the Iowa Historical Society how his grandfather moved to the area with a covered wagon, a team of horses, his wife, and his infant son. He earned the money to build his first house by selling "rat" skins, the nickname settlers gave to muskrats that lived in the ponds. The pelts sold for 10 to 15 cents each, and in the late 1800s, when cash was scarce, settlers used them as money. Demand for muskrat skins was so high that settlers fought fierce battles over claims to ponds and wetlands—all of them dotted with muskrat nests that rose up like tiny haystacks in a field.

In another oral history, a farmer named J. Bruce Haddock recalled how his grandparents arrived in northwest Iowa in 1876 and spent their first winter in a cave. The region was all wild prairie then, the grass so high they had to stand on the back of a horse to find their cows. The stream that crossed their land twisted and turned so much that in springtime it would overflow its banks and flood the adjacent fields, enriching the land with new soil. His grandmother set up a fish trap in the slough near their house and caught enough perch and catfish to feed the family and their neighbors.

But for most of the immigrants, the wetlands were at best a nuisance. Travelers had to halt their wagons at the edge of a prairie slough and plan a way to cross it. Women, children, and livestock would wade across. Men would gather on the far side with teams of horses and oxen, then drag the wagons through the muck with long ropes. Pioneers gave these sloughs names that reflected the dread they inspired—Purgatory, Devil's Island, Little Hell, and Big Hell.

A decade after settlers arrived in that part of Iowa, a Pocahontas County mapmaker and surveyor named Alonzo Thornton published a proposal in the local newspapers urging his neighbors to join the drainage revolution. Thornton knew that agricultural drainage had proven itself farther east, and he promised it would eliminate farmers' greatest risk: a wet season. More important, it would increase the amount of tillable land, attract more settlers, and propel Pocahontas County to the forefront of Iowa agriculture. Two meandering rivers, the Raccoon and the Des Moines, would carry the excess water away to the Mississippi River on the eastern border of the state. "We would realize that the goose that lays the golden egg for Pocahontas County is DRAINAGE," he wrote.

Iowa's first tile factory popped up in the early 1900s in the town of Fonda in Pocahontas County, and farmers like Hersom's father started draining their fields. He hired "ditchers" to do the work, and the family put them up in their home while they worked. His mother cooked for the men in their muddy clothes on a stove that used corn cobs for fuel. "I don't know how my mother done it," Hersom said.

The ditchers, using spades eighteen inches long, cut three hundred to four hundred feet of trenches per day. They worked in pairs: The first would remove the top layer of soil and the second would stand in the shallow trench and dig another eighteen inches deeper. He was the one who required true skill—he could recognize the "fall" of the land—usually a drop of one inch per one hundred linear feet—so water in the tile lines would flow downhill.

Iowa, which lies just east of the 98th meridian, was one of the last tallgrass prairie states to be colonized, and one of the last to embrace tile drainage. But it was acknowledged as one of the most remarkable transformations on the American frontier. In 1909, *The New York Times* took measure of the assembled steam power, the hundreds of laborers, the thousands of tons of earth moved, and pronounced Iowa's drainage project—just Iowa's—an engineering feat on a par with the digging of the Panama Canal. In the next fifty years, farmers would re-plumb one quarter of Iowa—more than ten million acres—converting 95 percent of its wetlands into some of the most fertile soil on earth. Alonzo Thornton's vision of the golden egg of agriculture came true: Today the Des Moines Lobe is a lush green expanse of corn and soybeans worth billions of dollars. But it came with an enormous cost in polluted rivers and contaminated drinking water that, decades later, would make Iowa a national test case in the evolution of environmental law.

It is difficult, perhaps impossible, to know precisely how much of the Midwest has been transformed with artificial drainage; there are no comprehensive records. But according to the 2022 Census of Agriculture, some 96 million acres of cropland in the United States are drained by ditches and underground tile lines—85 percent of them in the Corn Belt states. That includes half the cropland in Iowa and Indiana and a third in Minnesota. And, in part because of the heavier rainfalls driven by climate change, tile drainage continues to expand. Today instead of clay tiles, farmers use miles of perforated corrugated plastic tubes that are buried four feet deep with specialized drainage plows. Between 2012 and 2017 tiled cropland in the United States increased by 14 percent, or seven million acres, though the rate has flattened out since.

After the 1970s, in the wake of the Clean Water Act, the pace of wetland losses in the United States slowed appreciably. The Environmental Protection Agency recognized that wetlands are intrinsically connected to the nation's lakes and rivers—and the health of their water—and enacted regulations to protect them accordingly. In addition, long-standing provisions in the federal Farm Bill granted some protection.

Now, however, even these protections are at risk. In the spring of 2023 the U.S. Supreme Court severed freestanding wetlands, those like many of Iowa's sloughs and the prairie potholes of the north central states, from the Clean Water Act. It threw out half a century of legal and scientific consensus, concluding that unless the surface water of a wetland is directly connected—or "adjoined"—to a lake, stream, or ocean, it is not legally protected by the Clean Water Act. Wetlands now make up about 5 percent of the nation's land surface, and an estimated half of that falls outside the Supreme Court's new definition. As the dissenting justices observed at the time, the decision has had grave consequences, including more frequent and larger floods, greater damage from droughts, worsening drinking water quality, and the loss of even more ecosystems and the thousands of species they support.

Even if the courts reversed course and federal wetland protections were restored, the flow of water across the country has been forever changed. Of the wetlands that dotted the United States when Europeans arrived, more than half are gone, and far more have vanished in what was once tallgrass prairie. Except for a few fragments, wetland complexes that were hundreds of square miles in size, like the Grand Kankakee Marsh in Illinois and Indiana, and the Black Swamp in northwest Ohio, exist only in historic records. In addition, midwestern streams and rivers are much wider now, with powerful currents that strip the banks of soil and cloud the water. Fertilizers and other farm chemicals carried through drainage tiles are now leading causes of water pollution in the Mississippi River basin. They are a major threat to freshwater wildlife—fish, crayfish, frogs, dragonflies, mussels, and the tiny organisms that form the base of the food chain.

The Europeans who colonized North America in the nineteenth century transformed the continent's hydrology as thoroughly as the glaciers—but, remarkably, they did it in less than one hundred years instead of tens of thousands. They completely redesigned the storage and flow of a continent's water, converting the wild, wet tallgrass prairies into the neat, orderly farms that they held in their imaginations. In 1954 John Weaver, often described as the dean of America's grassland ecologists, wrote what reads like a eulogy to America's prairies and the wetlands that were intrinsic to them.

> The disappearance of a major natural unit of vegetation from the face of the earth is an event worthy of causing pause and consideration by any nation. Yet so gradually has the prairie been conquered by the breaking plow, the tractor, the overcrowded herds of man, and so intent has he been upon securing from the soil its last measure of innate fertility, that scant attention has been given to the significance of this endless grassland or the course of its destruction. Civilized man is destroying a masterpiece of nature without recording for posterity that which he has destroyed.

CHAPTER FOUR

# $NH_3$ (Ammonia)

ATMOSPHERIC NITROGEN, A SIMPLE GAS WITH THE CHEMICAL SYMBOL $N_2$, represents one of the great paradoxes in nature. It is hugely abundant on our planet. It comprises nearly 80 percent of our atmosphere, an amount so large that it makes the sky blue from the refraction of sunlight through its molecules. It is the fourth-most common element in the human body, after carbon, oxygen, and hydrogen. It is also one of the building blocks of life—present in every living cell, necessary for the creation of proteins, DNA, and RNA, and essential to all living things from single-celled organisms to African elephants.

But unlike oxygen, carbon, and hydrogen—which are there for the taking in the air and from the Earth—nitrogen is virtually inaccessible to the living things that need it. In its most common form, two tightly bound atoms that make a molecule of nitrogen gas, it is useless to plants and animals. The immensely powerful triple molecular bond that holds the two atoms together makes it nonreactive, or inert, meaning that it doesn't combine easily with other elements such as hydrogen and oxygen to make the nutrients that plants can absorb. For millennia, only two forces on

Earth could break that sturdy triple bond. One is lightning, which can break apart small amounts of nitrogen with its tremendous electrical charge, sending the individual atoms to earth in rain and snow. The other is the remarkable "nitrogen-fixing" soil bacteria that have the rare ability to first break the nitrogen bond with an enzyme, and then combine single atoms of nitrogen with oxygen and hydrogen to create compounds that plants use. But in the natural world, such useful nitrogen is scarce. So scarce that scientists describe it as a "limiting factor" because it puts a cap on the amount of life that can occur naturally on Earth.

No wonder, then, that the long history of agricultural advancement has often featured the sometimes-violent, never-ending quest to discover better ways to get more nitrogen into the soil to grow more crops to feed more animals and people. Humans have struggled for millennia not only to provide enough nitrogen to grow crops, but also to replenish it once it's been removed from the soil by the animals and people who consume those crops as food. Throughout most of history, the planet's human population increased only as long as people could expand onto new arable land—a dynamic that helped drive colonialism and the transformation of forests, savannas, and wetlands into farmland. But by the time Europeans completed their takeover of the American prairies, humans had largely occupied the richest arable land in the world. By the twentieth century, the limited supply of farmland, and its limited supply of nitrogen, became an ironclad limit on the amount of human life on Earth.

Though they didn't understand why, early farming cultures around the world knew that crops would fail if planted in the same spot year after year. After a few harvests, crops exhausted the soil's innate stocks of nitrogen; the tilled land needed a couple of decades to heal as the growth and decay of natural vegetation released stored nitrogen back into the ground. The earliest human societies also knew that adding human and animal waste and decaying plants to soil allowed crops to grow in the same field over and over again. And, remarkably, pre-industrial farming cultures across the globe all learned independently of one another that they could sustain soil fertility by alternating grains with peas, clover, and plants we now know as legumes—a class of plant that has a symbiotic relationship with the soil bacteria that transform $N_2$ into nutrients.

As the world's population grew over the centuries, so did the demand for manure and other sources of nitrogen. Collecting human and animal manure became major pre-industrial endeavors. Early Chinese and Japanese governments mandated the collection of human waste for farming. The Romans had a god for manure. In Victorian England, "night soil men" kept the cities clean of excrement and the crops replenished. The largest application of manure ever recorded came in the nineteenth-century Parisian suburbs that provided the city with vegetables. Small plots of land, many in glass buildings, were supplemented every year with millions of tons of manure left behind by the horses that powered the city's transportation—recycling their waste to feed the farms that fed the people. At their peak, the city's farms used up to three hundred tons of manure per acre.

The highest nitrogen inputs overall, using human and animal waste as well as other sources, were achieved by a traditional agriculture system in South China in the late nineteenth century, where the climate is warm enough to grow crops year-round. In the Pearl River Delta, farmers scrupulously recycled human and animal excrement, planted nitrogen-rich cover and legume crops, and applied the bacteria-rich mud of rice paddies to produce some one hundred pounds of nitrogen per acre. Another farming culture in South China developed elaborate dikes and pond aquacultures to both grow food and make fertilizer. Carp fertilized ponds with their waste, while farmers used the surrounding dikes to grow sugarcane, rice, vegetables, and fruit using hundreds of tons of manure from pigs, ducks, and water buffalo, plus more from decaying aquatic plants and pond mud.

But across the world, crop production was limited by the amount of waste produced in local communities that were essentially closed, recycled nitrogen systems. The food supply, and therefore the population, was limited by the amount of nitrogen the community could produce in waste. Like the farmers in South China, most farmers around the world could collect enough waste and organic matter to equal only about one hundred pounds of nitrogen per acre of cropland. That was enough to produce up to one hundred pounds of plant protein per season, which was enough to feed about half a dozen people with a bland diet of beans, grain, and a little meat. In reality, however, crop production was usually lower. Farmers suffered from uncooperative weather, often had limited water supplies, and needed some land to grow non-food crops. On average, traditional farming communities could support only about four to seven people every 2.5 acres.

As societies converted more land for farming, the world population expanded, and by the late 1700s it had reached nearly one billion. But it was largely unrecognized that the Earth's population would be limited by the amount of land available to grow food. Most leaders and academics in advanced countries believed that continual economic growth would eventually support the world's population without conflict. Then in 1798 the British cleric and political economist Thomas Malthus sparked wide consternation when he pointed out that the world's population always exceeded its food supply—making war, conflict, and poverty inevitable. In his most famous work, he pointed out that if every couple produced four children, the population would double in twenty-five years, and then keep doubling. The food supply could not keep up, he said. He correctly predicted—a hundred years before it happened—that clearing new land for farms and incremental increases in crop yields would never suffice to outpace population growth. Famine and war, he warned, would always be a certainty. He viewed humans as subject to the same forces in nature as other animals, a concept so powerful that Charles Darwin adapted it to his theory of evolution.

At the time, scientists were beginning to unlock the secrets of the elemental forces at work in the natural world and had identified nitrogen as one of them. Saltpeter, or potassium nitrate, had long been recognized as a source of fertility, and also was used to manufacture gunpowder in the fourteenth century. In the 1700s scientists began to identify the different gases in the atmosphere, one of which was odorless, colorless, nonflammable, nonexplosive, nontoxic, and nonreactive. In 1790 a French chemist named this gas *nitrogene,* because it was also present in the compound *nitre* in French, or potassium nitrate. A few decades later one of the founders of organic chemistry, Justus von Liebig, outlined some of the primary chemical processes that drove agriculture. He was one of the first to popularize the concept of the law of the minimum—that plant growth and abundance is limited by the scarcest nutrient, not the total amount of available nutrients. He identified nitrogen, phosphorus, and potassium as essential to plant growth, and argued that plants feed on nitrogen compounds derived from the air. And he recognized the role of decay in releasing nitrogen from dead organisms back to the soil. "And thus death," he wrote, "the complete dissolution of an existing generation—becomes the sources of life for a new one."

Scientists were also slowly coming to a realization that would eventually alter one of the planet's natural systems: The only way to feed the world's growing population was to increase food production beyond the limits of biology and Earth's elemental cycles. And the only way to do that would be to find a new and significant source of nitrogen beyond what was provided by waste and the decay of plants and animals—the death that von Liebig described so poetically.

The first such source appeared in the early 1800s on a cluster of tiny islands off the coast of northern Peru—the Chinchas—once the sacred islands of the Incas. The region is unspeakably dry—it rains only once every few years—and the interior regions are deserts nearly devoid of life. But the surrounding ocean is one of the most productive in the world. Fed by the cold Humboldt Current, which moves north along the western edge of South America from Antarctica, it is rich in sea life—shrimp, fish, and plankton—that feeds seals, whales, and the human fishing communities that survived on the coast.

The current's abundant sea life also feeds the local seabirds—cormorants, pelicans, gannets, gulls, and others—which gathered by the millions on the rocky Chincha Islands and over centuries covered them with layer upon layer upon layer of guano. Untouched by rain, the islands rose from the blue seas, as glittering and white as the peaks of the Andes just to the east. Long before the Incan Empire, local people collected this white earth and used it to fertilize crops along the coast of South America. Incas called it *huano* and considered it to be one of the gods' greatest gifts, on par with gold. Killing a seabird was an offense punishable by death.

Early European explorers were impressed by the fertile fields of those Peruvian farmers, and scientists who studied the soil samples they brought home confirmed that they were high in nitrogen. The famous German naturalist Alexander von Humboldt, for whom the Humboldt Current is named, brought home a lump of guano in 1804. But it wasn't until 1824 that the remarkable substance made it to the United States as fertilizer. The editor of the influential journal *The American Farmer* wrote that a seaman had arrived in Baltimore with two casks of guano from Peru, which possessed "astonishing fertilizing properties." The governor of Maryland tried it on his farm and found it to be miraculously effective on corn. Word spread in Europe and the United States.

Two decades later guano mania exploded across much of the world, and the stuff attained almost mythic stature. One farmer reported yield increases of 30 percent. Others said guano made barren fields productive again and made trees bloom twice in one year. American plantation owners found that guano could restore their exhausted tobacco and cotton fields—and their fortunes. A British farmer wrote that it had become indispensable, "like a necessity of life." "If ever a philosopher's stone, the elixir of life, the infallible catholicism, the universal solvent, or the perpetual motion were discovered, it is the application of guano in agriculture," wrote *The Farmer's Magazine*.

Peru, which had a series of precarious governments after achieving independence from Spain in 1821, quickly recognized the value of its guano endowment. The government nationalized the islands to ensure control of the substance, then contracted with foreign companies to develop the guano trade in Europe and North America. What emerged was an appalling business of digging up the foul-smelling powder and loading it into clipper ships for shipment around the world. Hundreds of ships would congregate around the islands and wait for months for their turn to load their holds with the priceless cargo. On the islands, gangs of mostly slaves and indentured workers labored in the hot sun and choking dust, digging down through several hundred feet of bird dung, piling it into wheelbarrows, and sending it down chutes to the waiting ships below.

Though the industry horrified some observers, massive fortunes were being made. At first the traders sold guano by weight, but as the business became more sophisticated they priced it according to the variable nitrogen concentrations it contained. The government of Peru kept a third of the resulting revenues, which made up three-fourths of the national budget, and used it to leverage foreign loans, which, for a brief period, created enormous wealth for the country.

The supply of guano became so critical to farming that it turned into an issue of national security in America. During the 1850s the United States imported about 760,000 tons at a price of $76 per pound—a quarter of the price of gold. President Millard Fillmore noted in his first State of the Union address that the federal

$NH_3$ (Ammonia) | 67

Chinese laborers working guano in the Chincha Islands, Peru, 1865. *Photograph by Henry de Witt Moulton, from the New York Public Library*

government should "employ all means properly in its power" to import guano. A few years later Congress passed the Guano Islands Act, which allowed any American citizen to claim any uninhabited bird excrement–covered island in the world as a U.S. territory. Ninety-four islands were eventually claimed under the act, but none produced fertilizer of any note. Both the United States and Spain tried, and failed, to claim some of the Chinchas as their own. But they were small islands and their stock of guano, accumulated over centuries, was finite. By the 1870s, thirty years after the craze began, the diggers had reached rock, and it was over. In all, eleven million tons of the best natural fertilizer ever known had been shipped around the world.

As the supply of precious guano faded, governments and businesses began to focus on another nitrogen source that had been identified around the same time, in 1821, by a Peruvian naturalist. It, too, turned up in an unspeakably dry and inhospitable part of Peru—the Atacama and Tarapacá deserts, a plateau west of the Andes and east of the coastal foothills that was later seized by Chile. This seven-hundred-kilometer expanse was encrusted with strange chemicals and salt-based minerals that, without rain to dissolve them or microbes and plants to use them, accumulated over eons in a layer called *caliche*. Among them are huge deposits of sodium nitrate—saltpeter, a chemical compound made up of salt and nitrogen and known by local people as *salitreras*. After digging it up, crushing it, and distilling out the salts, early industrialists used it as an inferior gunpowder. But that changed dramatically in the late 1800s as the guano trade began to fade. A chemist figured out how to make *salitreras* into true saltpeter, the key ingredient for high-quality gunpowder. Simultaneously, people recognized that its high nitrogen content made it a valuable fertilizer. Those two remarkable uses—for food and weapons—made the mineral extraordinarily valuable. Saltpeter had been used for centuries, primarily for fireworks and gunpowder, and now the Peruvian deposits held the world's largest supply, a layer one to three meters thick just beneath the surface across hundreds of miles of desert. The Peruvian mineral had lower concentrations of nitrogen than guano, but it rose in value as guano became scarce. Nitrate-crushing operations blossomed in the desert, drawing thousands of workers and creating wild boom towns at coastal shipping points. Miners blasted the nitrate layers out of the ground with dynamite and hauled the ore to factories for processing. Once again foreign companies controlled the industry—Chilean, British, German, and French companies built the processing plants and hired the workers—while Peru's government took its share in the form of taxes and fees. But in 1879, after the guano trade collapsed, the Peruvian government nationalized the industry, setting the stage for what became known as the Nitrate War between Peru, Bolivia, and Chile. Chile's victory over its neighbors in 1883 gave it complete control over the Atacama Desert and its rich deposits. Historians describe it as a desert worth fighting for, one that ultimately generated up to $10 million a year for Chile.

The depletion of guano and the conflict over sodium nitrate in the Atacama Desert taught the world's most powerful governments in Europe and the United States just how vulnerable they were. Their food supplies and their ability to wage war depended on a finite resource under the control of a foreign power. And the world was changing dramatically, in ways that would only increase the demand for fertilizers. By the end of the nineteenth century, the Industrial Revolution was well underway. In the United States, mechanized farm implements and mechanized land drainage were driving a rapid expansion in food production to feed a rapidly expanding population, placing new demands on soil fertility. In the last half of the nineteenth century the population in northern Europe and the United States grew by two-thirds, from 300 to 500 million. Growing urbanization and rising incomes from the increasingly productive economy changed the common diet based on beans and grains to one that included more meat and sugar. Daily caloric intake rose from about 2,500 to 3,000 per person, with most of the increase coming from meat and sugar.

For a time Western powers met the rising demand for food by simply plowing more acres, resulting in an epic expansion of land under human cultivation around the globe. European colonization of the New World had converted tens of millions of acres of grasslands to farmland. Much the same happened in portions of Russia, South America, and Australia. As a result, between 1800 and 1920, the amount of pasture and farmland more than doubled globally, to more than 8 billion acres, one of the biggest land transformations in world history.

On the American prairies, where European settlers displaced Native Americans and plowed the grasslands, farmers could rely for a time on the soil's extraordinary natural fertility. In contrast to much of the world—farmland in Europe that had been cultivated for centuries, the poorer soils in the eastern U.S. and the nitrogen-depleted plantations in the South—the freshly plowed prairies contained tons of innate nitrogen in every acre. Farmers had no need yet for synthetic fertilizer. Grain yields in Kansas, for example, were twice what they were in Europe.

As European colonists filled up the New World, scientists continued to expand their understanding of nitrogen and how it worked in nature. Nitrogen and hydrogen had been identified in 1785 as the gaseous components of ammonia, the most important nutrient for crops. In 1838, the year that John Deere introduced a new plow and guano sales were in full swing, a French chemist proved that legumes—clover, peas, beans—could provide nitrogen to plants and restore it to the soil. In the following decades, scientists began to understand exactly how that happened when they discovered the first species of rhizobium bacteria, one that had evolved to have a symbiotic relationship with legumes. Within five years, scientists were discovering more species of symbiotic bacteria, as well as other types that could turn nitrogen from the air into nutrients independently of legume plants.

These were the first glimmers of insight into the natural process now known as the nitrogen cycle—the compound's movement through the living and nonliving systems on Earth. Certain bacteria, called diazotrophs, common in all soils, contain an enzyme called nitrogenase that acts as a catalyst at a molecular level to break the triple bond binding atmospheric nitrogen atoms. They then combine nitrogen with hydrogen gas to make ammonia ($NH_3$), or fertilizer. It's called fixing nitrogen, or fixation, a process that simultaneously creates energy for the bacteria's own life cycle. Other bacteria then break the ammonia into nitrogen dioxides, nitrite and nitrate, which plants also use to grow and to make amino acids and nucleic acids. Animals use those acids to create their own proteins and DNA after they eat the plants—and each other—moving it up the food chain. Nitrogen is then returned to the atmosphere in two ways. Nitrate and nitrite in the soil that are not used by plants are converted back to nitrogen by another group of bacteria in a process called denitrification. And in the decomposition of organic matter, bacteria convert nitrate and nitrite back into the triple-bonded nonreactive nitrogen gas that makes up most of the atmosphere.

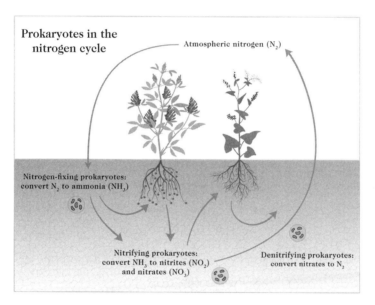

In the natural cycling of nitrogen ($N_2$) through the world's ecosystems, certain prokaryotes in the soil—primarily bacteria—use an enzyme to break the powerful triple bond between two atoms of atmospheric nitrogen, converting them to ammonia or nitrates that plants use to grow in a process called nitrogen fixation. Some plants called legumes, which include clover and soybeans, form nodules on their roots that house bacteria that "fix" nitrogen in a form the plant can use. Other bacteria, also prokaryotes, denitrify nitrogen compounds, converting them to nitrous oxide and back to atmospheric nitrogen. *Design credit: Sara Bereta*

But they still hadn't solved the greatest mystery of all—how to do what bacteria do and distill an unlimited supply of fertilizer from the atmosphere. As early as 1788, chemists had tried to create ammonia by using heat and pressure sufficient to break the triple bond in $N_2$, and these experiments continued unsuccessfully throughout the eighteenth century. Despite rapid progress on other fronts in chemistry and engineering, the solution to breaking that bond was so elusive that those in the scientific world began to think it just wasn't possible. But by the mid-to-late 1800s, the end of imported fertilizer from Peru was in sight, the political precariousness of the supply from Chile was increasingly evident, and European agricultural production had started to decline. All the best farmland in the world was already under cultivation, or very soon would be. And populations continued to grow.

In 1898 the newly knighted British physicist Sir William Crookes decided it was time for the world to pay attention to the coming population crisis. The venue he chose was a meeting of the British Association for the Advancement of Science, where he was to give his inaugural speech as the incoming president. Crookes, a highly regarded scientist, was famous for his discovery of a new element—the metal thallium—and his work on the predecessor to the cathode ray tube. At the time, English newspapers were vigorously debating whether the government should establish stockpiles of wheat as insurance against famine and the political upheavals that could interrupt the supplies of sodium nitrate used for fertilizer and gunpowder. Crookes decided to tackle the "wheat problem" head-on.

"England and all civilized nations stand in deadly peril of not having enough to eat," he declared. Wheat consumption was growing, and England imported 75 percent of what it consumed, much of it from the new farmland in the United States. The British government's proposal—to stockpile 64 million bushels of wheat in national granaries—would add only another 14 weeks to "the life of the population," he observed.

Crookes went on to note that Germany, France, and other European nations faced the same problem. The bread-eating world depended on farmers in the United States, which exported 145 million bushels of wheat a year. But the United States had no more uncultivated land, he said, and soon its own population would grow to the point where it would no longer share its bounty with other countries. Similarly, the rest of the world had no more uncultivated land. The survival of what he called "Caucasian," or white, wheat eaters like the British depended upon the ability to add fixed nitrogen to the soil to increase productivity. Yet the guano fertilizer from Peru was depleted, and though many believed the saltpeter from Chile was inexhaustible, that too would run out within a few decades, he said. In just thirty years a world crisis over food would be upon them, he warned.

But, Crookes continued, the answer was all around them—in the tons of nitrogen that pressed down on them in the air. "The fixation of atmospheric nitrogen therefore is one of the great discoveries awaiting the ingenuity of chemists," he proclaimed. It is "vital to the progress of civilized humanity." Expressing a racist, colonialist perspective common in Great Britain at the time, he said that without such a discovery, "the great Caucasian race will cease to be foremost in the world, and will be squeezed out of existence by races to whom wheat bread is not the staff of life." It was up to the chemists in their laboratories, then, to "develop and guide Nature's latent energies, we must utilize her inmost workshops."

Crookes's remarks, which he turned into a popular book, spread internationally and launched a debate about the limits of the world's resources. Some agreed with him, others questioned his numbers and his predictions about the supply of Chilean fertilizer. But he reignited the scientific search for a solution to a world population explosion that suddenly appeared imminent. And a decade later, the solution was found.

The formula was well known and deceptively simple. One atom

of atmospheric nitrogen ($N_2$), plus three atoms of hydrogen ($H_2$), produces ammonia ($NH_3$), the critical ingredient for growing crops. After a hundred years of studying the chemical processes of plants and the nitrogen cycle, scientists had a basic understanding of the natural processes that created ammonia in nature. They also knew that the tremendous heat and force of lightning could break the powerful chemical bonds of nitrogen, and electricity was something they understood. Throughout the Industrial Revolution, most of the efforts to create ammonia from nitrogen focused on creating electric arcs, like miniature lightning bolts, inside containers that would create enough force to break nitrogen atoms apart. But doing so on an industrial scale was unwieldy, and the electricity needed was hugely expensive. Then in 1908 a German chemist named Fritz Haber wrote a letter to BASF, the giant German chemical company, proposing a different path.

Haber was a German Jew who had been born at the start of the Prussian Reich, and he grew up during the country's unprecedented economic expansion. Prussia had become the most powerful country in Europe, largely because of its scientific and technological prowess, particularly in the field of chemistry. Haber became a professor of physical and electrochemistry at the Technische Hochschule in Karlsruhe. He had traveled throughout the United States to study American industry. He was respected for creativity and attention to detail. And he was driven to succeed.

Haber had experimented with a high-pressure ammonia system, using heat and electricity and iron as a catalyst, in a metal container. In time, he gave up on the approach when it produced too little ammonia to be worth the trouble. But world history pivoted because of an incident in which Haber was publicly humiliated by a fellow scientist. And not just any scientist. Walther Nernst was one of the most respected physical chemists of the day, considered a future Nobel Prize winner for his work in thermodynamics, and a faculty member at one of the most illustrious universities in Germany. He was wealthy—and of considerable importance to Haber given the anti-Semitism of the times, he was a Gentile. At the 1907 annual meeting of the Bunsen Society for Physical Chemistry in Germany, he publicly insulted Haber and his experiments with ammonia. Haber had found that even with a catalyst, the system required temperatures and pressures so high that most of the ammonia was destroyed in the process. Nernst ridiculed Haber, saying that his findings were "highly erroneous" because the amount of ammonia produced had to be even less than Haber had reported. That was unfortunate, he added in another insult, because if Haber had been right, he might have produced a solution to the chemical problem that had thwarted scientists for decades and synthesized ammonia in commercial amounts. "I would like to suggest that Professor Haber now employ a method that is certain to produce truly precise values," he told the assembled scientists.

Haber was crushed, his reputation in tatters. His wife wrote to friends that he was so upset that he suffered from skin ailments and digestive problems. Nonetheless, he was determined to vindicate himself. He and an assistant continued to make improvements on their system, gradually increasing the synthesis pressure and using different catalysts to trigger the reaction, which allowed them to decrease the temperature and increase the amount of ammonia produced. But when he realized that his process still couldn't produce enough ammonia to justify scaling it up to commercial size, he contacted BASF, which was already working, without success, on ammonia synthesis. They made a deal. With BASF's funding Haber was able to buy more elaborate equipment, including containers that allowed him to increase air pressures up to two hundred times normal atmospheric pressure and raise the temperature to 600 degrees Celsius, while still producing ammonia. After a year of tweaking the system in his laboratory, trying different catalysts, different temperatures, and different high-pressure containers, Haber produced a machine small enough to fit on a tabletop, yet capable of converting 6 to 8 percent of the nitrogen going in one end to ammonia coming out the other. BASF executives were horrified at the idea of an industrial process requiring such high pressures. But Carl Bosch, the BASF engineer who oversaw the project—himself a metallurgy expert—insisted that the steel industry could create materials strong enough to withstand those pressures. "I know exactly what the steel industry can do," he said. "It should be risked." In the end, the brilliance and persistence of the two scientists, the chemist and the engineer, made the synthesis of ammonia a reality.

First, however, Haber's two-and-a-half-foot machine had to be scaled up into a factory capable of producing tons of ammonia. That job fell to Bosch, and in a remarkable feat of engineering, he accomplished it in five years. He and his engineers figured out how to extract large volumes of nitrogen from the air and hydrogen out of water. They experimented with multiple catalysts that triggered the chemical transformation until they found one that was cheap and effective. In 1913 BASF built its first ammonia synthesis factory in Oppau, Germany, a plant capable of producing twenty tons of ammonia daily.

Farmers, however, were not about to see the benefits. The next year war broke out across Europe. Like saltpeter before it, ammonia was a valuable ingredient for explosives as well as for fertilizer—a fact that altered the course of history. In a matter of months Germans converted the Oppau plant from producing fertilizers that could feed Europe to manufacturing nitrates for explosives that would influence the course and duration of World War I. In 1916, Bosch, under the direction of BASF, launched the construction of a second, much larger plant, marking the beginning of what would become known as the military-industrial complex.

Haber, who had converted to Christianity as a young man, was deeply loyal to the German government and became one of the scientists integral to its war effort. He eventually took over the military's experiments with gas warfare and refined the development of chlorine gas for use in the trenches. In 1915, thanks to Haber's ingenuity, Germany deployed chlorine gas during the Second Battle of Ypres, in Belgium, which injured and killed thousands of French soldiers. By the end of the war, casualties from gas warfare totaled 1.3 million. In 1919 Haber received the Nobel Prize as the inventor of the Haber-Bosch system of ammonia synthesis, though he was attacked in France, England, and the United States as the "inventor of gas war." Despite his contribution to German munitions research in World War I, Haber's career came to an end in 1933 after Adolf Hitler came to power. He resigned from the Kaiser-Wilhelm Institute, which he had led for two decades, after new race laws would have forced him to fire all his employees of "non-Aryan descent." A year later he died of heart failure while on his way to a job at the newly formed Daniel Sieff Institute in Palestine. He was sixty-five.

Though Haber and Bosch are credited with the discovery and the first production of ammonia synthesis, the nitrogen breakthrough was born, like many great discoveries, from the accumulation of scientific research over time. Haber himself, while accepting the Nobel Prize, acknowledged the work of the many scientists who had preceded him. The invention, however, carries his name, and that little machine that hissed and dripped away in his laboratory in 1909 is now viewed, for better and for worse, as one of the greatest scientific discoveries in human history. The synthesis of nitrogen and hydrogen into ammonia and synthetic fertilizers not only supports a human population far beyond what nature could support on its own, it also fundamentally altered one of the Earth's atmospheric cycles in ways that in a generation would prove to be enormously destructive to its ecosystems.

In North America, it was only after World War II that farmers of the Midwest began to need synthetic nitrogen to keep their yields up. Until then they had essentially been mining nitrogen that had built up in the rich black soil under the grasslands for centuries.

And yet, despite using crop rotations and manure like farmers everywhere, they had slowly depleted the natural stores of nitrogen, and by the end of the war, the decline in yields was apparent. To support its own war effort, the U.S. government had built ten plants to produce ammonia for weapons, all located in the interior of the country near pipelines that carried the natural gas they needed for fuel and the hydrogen they needed for the chemical synthesis. By the end of the war, the plants were producing 730,000 tons of ammonia annually and had a capacity to produce 1.6 million tons. In another stunningly swift transition, this time from war to peace, plants that had manufactured a chemical used to kill millions were converted, finally, to address the challenge Sir William Crookes had issued a generation earlier—growing food to feed billions.

The first research on the use of nitrogen fertilizer emerged from the largest ammonia plant in the U.S., the Tennessee Valley Authority complex at Muscle Shoals, Alabama, and it soon triggered a major social, ecological, and scientific transformation in farming. Agricultural colleges across the Midwest joined in, advising farmers on how to use fertilizer and deepening their understanding of the nitrogen cycle. Farmers quickly boosted their soil fertility with stunning results. No longer forced to find new land when their soil was exhausted, or do the backbreaking work of raising livestock and hauling their manure to their fields, they could increase yields on the same acres with synthetic fertilizer. All they needed was a tractor.

The wide adoption of nitrogen fertilizer transformed one of the fundamental relationships between people and nature. It allowed farmers to altogether bypass the natural cycling of nitrogen between the atmosphere, soil, and plants that had limited the amount of life on the planet. Today nearly four billion people—half the world's population—would not be alive if it weren't for nitrogen fertilizer.

## $NH_3$ (Ammonia)

GARY PRESCHER FARMS in southern Minnesota and remembers when his family still raised livestock and used the manure to fertilize their fields. When he was a kid they also grew a greater variety of crops, including legumes such as alfalfa and clover for animal feed, which added nitrogen in the soil for the corn that would follow in the next season. But starting in the 1950s, when he was about six years old, and just as the gospel of synthetic fertilizer was taking hold across the country, farming changed. Farm families got smaller and the cost of labor rose, making the care of livestock too expensive for small operations. Agriculture became more specialized—cows and pigs were increasingly raised in large, specialized feedlots, and they largely disappeared from farmyards and pastures. With them went the feed crops. With the help of the new synthetic fertilizers other farmers, like Prescher's father, began to specialize in row crops. One of Prescher's first jobs as a kid was to help his dad load the seed planter with five-gallon buckets of dry fertilizer. He remembers the sharp smell of the chemical and the dust that burned his eyes. With the help of that fertilizer and other advances in agriculture, his father gradually transitioned to the exclusive corn-soybean rotation that Prescher still uses today, and that is ubiquitous across much of the Midwest.

During Prescher's lifetime, the world's use of manufactured nitrogen fertilizer increased by 860 percent, and by 50 percent in the United States. By 1975, Minnesota farmers like Prescher's dad were using ninety-one pounds of nitrogen fertilizer per acre of corn. In Iowa it was 102 pounds, and in Illinois it was 117. The U.S. average that year was 105 pounds per acre—or a total of 3.9 million tons just for corn, the most nitrogen-dependent crop. The amount rose steadily until the mid-1980s, when rising environmental concerns about farm runoff and the ever higher cost of fertilizer forced farmers to use nitrogen more sparingly. In the last 10 years nitrogen use in the United States has plateaued at about 13 million tons every year, about 149 pounds per acre on average. More than half of that goes on nearly 100 million acres of corn, mostly in the Midwest. Today, Prescher can choose among a wide variety of fertilizers, depending on the price, the season, and the weather. At the same time, the amount of corn produced per acre has risen steadily, by about two bushels per year, a testament to ever more sophisticated seeds and farm technology. Prescher is also an agronomist and a member of the Minnesota Corn Research & Promotion Council, which advises farmers across the state. "We feed the world" is their frequent refrain. Indeed, much of the developed world has a diet rich in meat and dairy protein, thanks to the nitrogen fertilizer used to grow feed for livestock. But the world is paying a steep price for it.

In just a single generation of human agriculture, the Earth's nitrogen cycle has been hijacked, perhaps more than any of the Earth's other elemental cycles—more even than the carbon cycle from fossil fuels. Today the Haber-Bosch process, combined with the production of fossil fuels and the increase in legume crops like soybeans, creates two to three times more nitrogen than the Earth's natural forces—lightning and those tiny bacteria working away in the soil. Agriculture produces five times more nitrogen than fossil fuels. Altogether, it is far more than the planet's natural systems can use or return to the atmosphere as inert $N_2$, leaving the Earth overrun with nitrogen compounds.

When the Haber-Bosch process breaks the powerfully bound atmospheric $N_2$ into individual, reactive nitrogen atoms and the nitrogen is applied to crops or urban lawns as fertilizer, the individual atoms become remarkably versatile, or "slippery," as scientists describe it. Crops, lawns, and gardens are able to take up only about half of the synthesized fertilizers that are applied each year, meaning that, overall, the amount of nitrogen used to produce food is far higher than the amount ultimately consumed by humans and livestock. The rest of those leftover "slippery" atoms attach easily to other elements, creating other chemical compounds in a series of environmentally destructive transformations called the nitrogen cascade. A single atom of nitrogen combines with oxygen to become air pollution, and then returns to earth where it is converted to nitrate that pollutes drinking water or causes algae blooms in a lake, careening through the Earth's biosphere as it damages biodiversity, contributes to climate change, and damages human health.

One byproduct of excess ammonia in the environment is nitrous oxide ($N_2O$), a powerful greenhouse gas produced as a byproduct when soil bacteria process fertilizer. Though it makes up only about 6 percent of the planet's total greenhouse gases, it is 265 times more effective at trapping heat than carbon dioxide, and the amount in the atmosphere has been increasing for decades. Nitrous oxide also contributes to the creation of ground-level ozone, smog, and small particulate matter, all of which cause significant health impacts, from asthma to heart disease. The Corn Belt states that were once the heart of the tallgrass prairie produce about 10 percent of the world's nitrous oxide pollution. According to one study by University of Minnesota researchers, the air pollution created by growing corn in just five midwestern states causes thousands of premature deaths across the country.

Another significant portion of the nitrogen applied as fertilizer ends up in the water that flows through the vast underground tile drainage systems farmers have installed beneath their crops in the last two centuries. As water flows through the soil and into drainage lines, it carries the excess nitrate created by soil bacteria into ditches, streams, rivers, and, ultimately, the Mississippi River and the Gulf of Mexico. Along the way, the excess nutrients create massive blue-green algae blooms of single-cell organisms called cyanobacteria that produce toxins harmful to people and animals. Perhaps most alarmingly, excess nitrate in ground- and surface water contaminates drinking water across the Midwest.

These issues crop up regularly during meetings of the Minnesota Corn Research & Promotion Council and other farm groups across the country, Prescher says. Those discussions quickly move to the paradox that underlies modern world agriculture:

What would life be like without synthetic fertilizer? "We would be farming like my grandfather," Prescher says. And feeding the world at its current and future population would not be possible.

As a result farm groups have launched projects that teach farmers about the judicious use of nitrogen fertilizer. Many farmers still use more than their crops can use—15 to 20 percent more on average—an excess that gets lost to air and water. Brad Carlson, a soil scientist and an expert on fertilizer use from the University of Minnesota Extension, spends his days crisscrossing the state, going from events to classrooms to farm fields to teach farmers "the four R's." That's the industry abbreviation for the "right source, right rate, right time, right place." On a warm Tuesday in August his classroom was a giant auditorium at a farm auction facility an hour south of Minneapolis. Giant tractors and combines were displayed on a windblown tarmac, and agronomists staffed booths selling seeds and farm products. A handful of farmers showed up for Carlson's class, sitting scattered among the bright orange seats in the auditorium, their arms crossed over their chests as they watched Carlson go through his slideshow.

He described how the Haber-Bosch process takes nitrogen out of the air and turns it into ammonia, and then fertilizer. He explained the nitrogen cycle—how microbes turn the fertilizer into nitrate that the plants can use—but that plants can't use all of it, and the excess leaches away in water and into the air. The process is complicated, subject to the whims of rain, temperature, and the organic matter in the soil. He points out an obvious irony: Knowing how much to use is easy—but only after the growing season is over and farmers know how much rain there was to wash it away or carry it to their tile lines, and how well their corn grew. But whenever farmers lose nitrogen to water, they are also losing their own money; it's not unusual for a midsized midwestern farm to use an amount worth $50,000 to $100,000 annually on nitrogen fertilizer. The most profitable farms, Carlson points out, are invariably the ones that spend the least on fertilizer. Efficiency—

using only what's needed for the crop—is what counts for the bottom line. "Your yield," he said, "does not correlate to how much nitrogen you supply." Yes, he said, it's tempting to use more just for "insurance"—to make up for planting in a bad spot, or because of anxiety about what the season might bring. "But what are you trying to accomplish?" he asked. You can't fertilize your way out of a problem, he said.

About twelve hundred farmers have taken Carlson's class since he started teaching it for the Minnesota Corn Growers Association. His students are the ones who care about water and air pollution, or who are competitive and want to be the best at what they do. No question, if every farmer in the country followed the four R's to perfection the Midwest's air and water would be less polluted.

But using synthetic fertilizer is inherently inefficient, and there would still be an excess of nitrogen cycling through the natural world. In *Enriching the Earth,* his seminal book on the history and science of nitrogen fertilizer, Vaclav Smil concludes that worries about nitrogen pollution don't get much of the public's attention—certainly not compared to climate change. Smil, one of the foremost thinkers on energy, points out that with human will, climate change is a solvable problem, and the world's transition to nonfossil fuels is likely inevitable. In contrast, there is at present no way to grow enough food for humankind without nitrogen fertilizer, and there is no substitute for it on the horizon. In 1920, when the first Haber-Bosch factory began pumping out chemicals for war and food production, the global population was at about two billion. It's now eight billion. In the next generation, another three billion children will be born. Their survival will also depend on the synthesis of ammonia from the most common gas on Earth.

There may be no better way to understand how agriculture has transformed the sea of grass since John Deere's time than to look down from an airplane traveling between Indianapolis and Denver. On a clear summer day, viewed from thirty thousand feet, the land appears as a checkerboard of green squares marching uniformly from east to west and north to south, interrupted from time to time by a metropolis like Chicago or the wavy line of a river like the Ohio or the Missouri. The plows have created the geometric pattern, the tile lines and ditches have removed much of the water, and fertilizer has created the vivid emerald hues. Viewed at ground level this new landscape has a certain abstract beauty, like a Mark Rothko painting in shades of green—darker for a field of soybeans, a lighter tone touched with gold for corn. The only feature it shares with the vast prairies it replaced is its uniformity. In Nebraska, 84 percent of the cropland is devoted to just two commodities—corn and soybeans. In Illinois the number is 91 percent, and in Iowa it is 94 percent. The sweet little farmsteads that once lined the roads—a picturesque red barn, a few grazing Holsteins, a wooden corn crib or chicken coop—are mostly gone, replaced at much longer intervals by huge, round steel grain bins that are industrial in their scale and corrugated sheen. Dropped blindly into corn country, you wouldn't know what state you were in: Ohio looks remarkably like southern Minnesota, which looks like Iowa, which looks like eastern Nebraska—endless miles of corn and soybeans, some patches of wheat, all of it etched with the deep straight ditches that carry water away into ever larger channels to the Mississippi River. In the thirty-one states that make up the Mississippi River basin, about half of all the land is devoted to row crops, bare of grass and flowers, underlaid by a vast capillary system of tile lines, and reliant on artificial nutrients and pesticides.

During the last 150 years, the march of colonialism, the sanctity of private land ownership, and the Industrial Revolution and all its inventions have created a midwestern landscape now largely defined by one human activity—farming. Nearly four in ten ears of corn grown globally come from the Mississippi basin, and more than a third of the world's soybeans. Decatur, Illinois, calls itself the Soybean Capital of the World, and Iowa is the Food Capital. In 2016, the Mississippi River basin produced crops worth $131 billion and accounted for some 90 percent of the nation's crop production. All the forces that built that empire are still at work, along with new tools like genetic engineering and increasingly sophisticated, data-driven technologies that farmers can control with a touch of an app on their phones.

Agriculture, in short, is modeled on industry—simple, efficient, and predictable—and the logical result of the industrial transformations that began in the nineteenth century. But that's problematic, according to Iowa State University agronomist Richard Cruse. "The difference is that unlike industry, agriculture is based on biology," he said. "And biology doesn't do well in simple systems." The evidence of that is everywhere. Agriculture has altered—and damaged—the very biological and chemical cycles that created the extraordinary prairie in the first place, from the creation of soil to the flow and purity of water, to the web of wildlife and the circulation of elements in and out of the atmosphere. It's created intractable pollution problems that endanger human health and cripple other ecosystems, from its damage to insects to the Dead Zone in the Gulf of Mexico. The tallgrass prairie once held a complex web of life and millennia of evolution in every square meter. But since Europeans arrived, it has been valued largely for the crops it can produce instead—an economic distortion that is driving agriculture onto the nation's remaining grasslands in the West, the most threatened ecosystem of all.

# PART TWO

A thing is right when it tends to preserve the integrity, stability and beauty of the biotic community. It is wrong when it tends otherwise.

—ALDO LEOPOLD

CHAPTER FIVE

## *River*

THE MISSISSIPPI RIVER BEGINS AS A TINY STREAM IN NORTH CENTRAL Minnesota that flows over a small stone dam from its source, Lake Itasca. Now the heart of a state park of the same name, the lake is shimmering blue in summer, snow-covered in winter, and surrounded by the dark pine trees that make up the southern edge of the boreal forest. The stream's channel hasn't changed much since 1832, when a local tribal leader led explorer and geologist Henry Rowe Schoolcraft to the lake the Ojibwe called Omushkos. It is just twenty feet wide and eighteen inches deep, and thousands of visitors wade across it every year, stepping from one side of the Mississippi to the other so they can boast that they walked across the country's mightiest river. The sandy bottom is visible through the clear, cold water. Tamarack trees, cattails, and wildflowers line the edge of the stream as it winds its way out of Itasca State Park. A carved wooden marker dating back to the 1930s says the river's total length to the Gulf of Mexico is 2,552 miles, though since Europeans arrived the Mississippi has lost more than 200 miles from engineering projects that smoothed out some of its curves. After leaving the park, the stream meanders through grassy wetlands and pine forests, slowly building momentum as it carves a giant question mark through the famous lake country of central Minnesota, getting wider and faster as it collects what is still clean water from hundreds of streams and small rivers on its way south.

After traveling about a hundred miles to central Minnesota, the Mississippi meets the Crow River, the first major tributary that drains farmland. The northern half of the roughly 2,800-square-mile Crow River watershed was once mostly forest, with rolling hills and gravelly soils that dry out quickly. The lower half, drained by the South Fork Crow River, was covered by the Des Moines Lobe during the last Ice Age. The glacier flattened out the landscape and created soggy clay soils that hold water and the wet prairies that bedeviled early European colonists. This watershed, though, became one of the places in the Midwest where the great inventions of nineteenth-century agriculture—the plow, tile drainage systems, and nitrogen fertilizer—came together to create generations of prosperous farmers. At first, they grew the wheat that supplied the region's flour mills and fed the nation and the "wheat eaters" in Europe. The mills gave birth to the city of Minneapolis, which became the incubator for giant corporations such as Pillsbury, Cargill, and General Mills. Today, the Crow River watershed is still home to prosperous farmers, but they plant mostly corn, soybeans, and sugar beets.

Like much of the tallgrass prairie, central Minnesota was wet before the farmers arrived around 1850. A fifth of the thirteen-hundred-square-mile South Fork watershed consisted of wet prairies, bogs, lakes, and river bottoms. Most of that is gone now—the small town of Buffalo Lake, for example, no longer has a lake. About 80 percent of the land in the watershed is farmed, and half of that is underlain with tile lines. Along with the rich soils left by the glacier, tile drainage is one of the primary reasons that today it produces bumper crops. But this small, agriculturally intensive watershed is also the first place that the nation's greatest river begins to collect the fertilizers—nitrogen and phosphorus—that build up along its 2,300-mile route. Counting its full length, the Mississippi drains a watershed that covers 41 percent of the United States and includes some of the nation's most chemically intensive farmland before flowing into the Gulf of Mexico at its wide, muddy mouth south of New Orleans. There the transformation of the American grasslands is destroying another ecosystem altogether: the warm ocean waters along the southern coasts of Mississippi and Louisiana.

In Minnesota, Renville County lies about a hundred miles west of the spot where the Crow River joins the Mississippi. This is where Minnesota's rolling terrain levels out into a checkered grid of corn and soybeans, and the fields stretch as flat as a tabletop to the western horizon. Randy Kramer farms here, running an operation that's been in his family for seventy years. His land is in the watershed of Buffalo Creek, a tributary that flows into South Fork Crow River, which puts him near the top of the vast watershed of the Mississippi River.

To visit Kramer's farm is to understand how dramatically agriculture changed in the twentieth century and, at the same time, why it is so hard for farmers to change their practices in a way that would clean up the nation's waters. A compact and energetic man even in his sixties, Kramer has spent most of his life in western Minnesota, a place dotted with small-town water towers and crisscrossed by a grid of gravel roads. When Kramer's parents started out here in the 1950s they farmed 120 acres, raising corn and soybeans for the market and alfalfa and small grains to feed their livestock. They kept about three hundred hogs and fifty head of beef cattle to diversify a revenue stream based mainly on grain. His dad sowed the fields with a six-row planter—meaning he could plant six rows in each pass across the field—and harvested his crop with a four-row combine, a machine that simultaneously cuts the stalk and threshes the grain from the stems. Today

Kramer pilots a twenty-four-row planter, meaning that with each pass across a field he can cover four times as much land as his parents did. His cab is outfitted with a computer that reads GPS mapping and knows every square foot of his farm; tiny feelers on the planting arms brush the ground and help the machine make precision turns at the end of each pass. In the fall he takes out the corn with a twelve-row head on his combine; he can harvest two hundred acres in a single day. With these massive machines, Kramer and his partners work far more land than their parents did; they farm 2,500 acres, part of it owned and part of it rented from retired neighbors. They grow corn and soybeans, like most midwestern farmers, as well as sugar beets, a lucrative crop that thrives in this part of the Midwest, and a small crop of sweet corn and peas that they sell to Seneca Foods/Green Giant for the supermarket freezer aisle.

On a blustery November day, decked out in blue jeans and comfortable hiking shoes, Kramer is giving a tour of the farm and explaining how the scale of farming has changed in a generation. His machine shed is the size of a hockey arena—space for three immense tractors and a combine worth $700,000. Outside, five steel grain bins, each the size of a house, stand on the concrete pad. These huge silvery structures have mostly replaced the quaint red silos once visible from every country road and allow farmers to store much more grain on site rather than depending on the local grain elevator. Behind the shed are two small concrete ornaments left over from his mother's garden—a cow and pig. "That's all the livestock I've got left," Kramer says in a wry nod to the wave of specialization that swept the Midwest in the late twentieth century. Today grain farmers like Kramer don't raise animals. It's grueling year-round work, and a field of corn or soybeans turns a higher profit than a field left in grass. Cows and hogs are raised in giant confinement barns by specialist farmers who contract directly with dairies and meat processors such as Hormel or Cargill; typical dairy farms now milk an average of one thousand cows, and many hog operations raise five thousand or more.

Kramer is a talker who waves his hands in the air when he gets going, and he says with a shrug, "I'm an open book." He's the kind of guy who shows up—at church, the county fair, local government boards, and farm fests. But his extroverted personality masks a mind like a calculator. He can rattle off the corn yield of every acre of every field and the loss he would take if wet weather delayed his soybean planting by a week. But then it takes a mind like that to survive in agriculture today. Every spring he takes out an operating loan of at least $1 million to buy seed, fertilizer, diesel, and other supplies. "That's why you lie awake nights wondering if you're going to have a crop," he says. He watches world commodity markets on a computer in the cab of his tractor, and if corn hits an attractive price in, say, early July, he will call his broker and lock in a contract to sell ten thousand or twenty thousand bushels. He can tell you how much of his crop was already contracted before harvest (70 percent in 2023) and the risks of shorting the commodities market. He's also a skilled mechanic, able to maintain and repair his huge equipment and manage storage tanks of diesel fuel, and he understands the hydrology beneath his fields—the pattern tiles, pumps, and drainage ditches that are key to his productivity.

Even in the era of so-called corporate farming, most midwestern farms are still essentially family operations, and Kramer's is no exception. He formed a limited liability partnership with his father and two brothers when he came home from college in the 1970s, and although his father retired years ago, it's still basically a family concern, soon to be passed on to two cousins. During the planting and harvest seasons, Kramer and his relatives work ninety-hour weeks. He and his wife take two weeks off in the winter to go to Hawaii. Yet farms like his lie squarely at the nexus of industrial agriculture. The seed and fertilizer he buys each spring are produced by one of the handful of giant multinationals that dominate global agrochemical production: Four huge companies, including Bayer Crop Science and Syngenta, produce 76 percent of the world's soybean seeds and 85 percent of the corn seed; four massive grain merchants, including Cargill, buy 82 percent of the nation's soybeans and 84 percent of the corn crop.

Like many farmers in the Midwest, Kramer is still haunted by the farm crisis that began in the late 1970s, when commodity prices crashed and borrowing costs soared after the Federal Reserve Board raised interest rates in a historic campaign to tame inflation. With the rural economy in collapse, protesting farmers drove tractorcades to Washington, D.C., and celebrities such as Willie Nelson, Bonnie Raitt, and B. B. King organized Farm Aid benefit concerts. Farmers began committing suicide at the rate of one per week. In Kramer's part of Minnesota, interest rates on an operating loan hit 21 percent and land prices collapsed, from $3,200 an acre to $800. "What banker was going to lend you money with that as collateral?" he asks. To this day, the rural Midwest still bears the scars of the period. At any Home Depot or Main Street café, you still encounter aging men who stare out the windows with wistful expressions. They may have taken jobs mowing lawns or repairing factory machinery because they lost the family farm in the '80s. Even today, farming has the highest suicide rate of any American occupation.

That sort of tension colors Kramer's attitude toward conservation. He knows exactly what fertilizer does to water and what happens to the water that runs off his fields into Buffalo Creek, then into the South Fork Crow River, and ultimately the Mississippi River and the Gulf of Mexico. "We're environmentalists," he says of his family. "We live here. We know our neighbors and we want our grandkids to live on this land." But he's also pragmatic about financial risks, margins, and the cost of making big changes. Take the conservation practice known as cover crops like wheat and winter rye. These crops, widely advocated as a solution to farm pollution, are planted after the fall harvest and stay on the land through the winter, preventing erosion and soaking up excess fertilizers and pesticides that would otherwise flow into the creek. Like a growing number of farmers, Kramer plants cover crops on a portion of his land for exactly these reasons, but he can't do it on all of his fields. Sugar beets, for one, are harvested late in the fall—too late, this far north, to plant another crop after they are out of the ground. He also observes that they are "just one more crop to worry about. It costs you money to put it in, and then you worry about it all winter." As an eleven-year member of the Renville County board, he's helped draft a water management plan to clean up the farm pollutants that flow into the South Fork Crow River. At the same time, he says farmers can't afford to give up the field drainage systems that deliver farm pollutants into the creeks and rivers, because today's large-scale operations require perfect soil conditions to generate the highest yields. If even twenty acres in a big field are too wet to plant, the lost yield can cost a farmer $30,000. "I can't go out and plant 80 percent of a field," Kramer says.

The consequences of these decisions, however, are evident where the Crow River enters the Mississippi above Minneapolis. North of that confluence, the Mississippi is pretty clean. It flows through some small cities, where it picks up stormwater runoff, and it can carry bacteria after rainstorms. But the Crow River carries so much fertilizer, mostly from the intensely farmed watershed of the South Fork, that nutrient concentrations at that point in the Mississippi River immediately skyrocket.

It takes about three months for water from the Crow to flow down the Mississippi before it reaches the Gulf of Mexico. Only the Congo in Africa and the Amazon in South America drain larger watersheds. Along its way south, the river gets bigger and bigger as thousands of tributaries collect water and runoff from the 1.2-million-square-mile Mississippi basin. In this part of the country, which contains the Corn Belt, more than half of the land is planted in row crops that consume 65 percent of the fertilizer used in the United States. By the time the Mississippi powers through New Orleans on its way to the ocean, it looks nothing like the tiny clear stream in Itasca State Park. It's brown and muddy and a mile wide, and carries tons of nitrogen and phosphorus from farms across the basin. When it finally flows into the Gulf, it carries so many nutrients that for much of the year, the polluted water obliterates the sea life across thousands of square miles along the Gulf coast. Scientifically, it's called "the hypoxic zone," because the excess nutrients launch a biological cascade that robs the water of oxygen. More commonly, and more descriptively, it's called the Dead Zone because almost nothing can live there.

The warm Gulf waters off Louisiana were once one of the most productive fishing grounds in the world, generating billions of dollars for the state's economy—much of it from shrimp that live near the ocean bottom. The Mississippi and other rivers slow down as they near the ocean, dropping sediment that over centuries formed the coastal estuaries and wetlands that became incubators for sea life of all kinds. But starting in the 1970s, shrimpers and other fishing operators began reporting that they were finding "dead waters." A young researcher at Louisiana State University named Gene Turner hitched a ride on a research vessel out of Mobile Bay, Alabama, operated by the National Oceanic and Atmospheric Administration, to investigate. The trip started him on a nearly fifty-year scientific journey to understand how a century of farming on the American grasslands had depleted oxygen in coastal waters in the Gulf hundreds of miles away. Initially, the annual cruises identified only spots of low oxygen, which he documented every summer for ten years. But eventually Turner and his wife, Nancy Rabalais, also a Gulf scientist at LSU, began to understand the breadth of the ecological disaster. Every year melting snow and spring rains wash tons of nutrients out of the soil across the Mississippi River basin. The river's fresh water pours into the Gulf, sliding out on top of the heavier salt water. Fueled by sunlight and fertilizers, massive algae blooms explode on the surface. When they die, they sink to the ocean floor, where bacteria decompose them, consuming all the oxygen in the lower layers of water as part of the process. When the oxygen levels drop to two milligrams per liter or less, the water can't support all the small organisms that live in sediments, an important food source for fish and other species, or other sea life up the food chain. In areas that are sixty feet deep, up to 80 percent of the water column above the ocean floor can be devoid of oxygen, or hypoxic. Fish, shrimp, crabs, turtles, and dolphins are mobile enough to escape the dead waters. They abandon the area, but have to travel miles before they reach water with enough oxygen—and food—to survive. Those that are less mobile, like starfish, clams, and oysters, try to move higher in the water column in their search for oxygen. Some brittle starfish have been seen standing on their points to escape suffocation. The hypoxic condition persists until hurricanes or other big storms sweep across the Gulf and mix the surface with lower, saltier water, but that often doesn't occur until September or later. Then, in springtime, the whole process starts again.

In 1989, Turner and Rabalais found that the Gulf Dead Zone covered nine thousand square kilometers west from New Orleans as prevailing ocean currents carried the polluted river water. Then in 1993, one of the wettest years on record, the river dumped almost twice its usual volume of water into the Gulf, and the Dead Zone doubled in size, reaching all the way to Texas. Since then, the pair have conducted annual measurements, which show that the size of the hypoxic areas changes every year, growing and shrinking depending on the amount of fresh water and nutrients that pour out of the nation's farmland. But every summer, at the height of the shrimping season, it is there.

Over time the scientists at LSU and other research groups found that the history of the Dead Zone precisely tracked the history of agriculture in the Midwest. Studies of sediment core samples on the bottom of the Gulf showed that nitrogen levels began to rise in the 1950s, just as commercial nitrogen fertilizer came on the market. As the use of both fertilizer and agricultural tile drainage rose, so did the concentration of nitrogen in the Mississippi. Flows from urban areas and municipal wastewater treatment plants made a contribution, but much less than the volume coming off farmland. Between 1955 and 1970, the annual flux of nitrogen into the Gulf tripled, along with the intensification of agriculture, before peaking in the early 1980s. More than half of the fertilizer load enters the river north of its confluence with the Ohio River at Cairo, Illinois. In short, it's coming from a handful of Corn Belt states such as Iowa, Minnesota, and Illinois. As the Dead Zone grew, the fishermen followed their catch. The shrimpers moved out into deeper waters eight to ten miles offshore where there was still oxygen below their hulls. Their boats had to get bigger, the trips longer. They added refrigeration and spent more money on fuel. They learned to find shrimp along the edge of the hypoxic zone. The shrimp became smaller and brought less money at market. As costs rose, the number of shrimpers dropped.

The hypoxic zone isn't the only cause of the decline of the Gulf shrimp industry. The 2010 BP Gulf of Mexico oil spill hurt fishing for years, and the competition of low-cost shrimp from Asia and South America has been even more devastating. Today around 90 percent of the shrimp consumed in the United States is imported from other countries, and most of that is farmed. Twenty thousand shrimpers once plied the waters off the Louisiana coast; now it's closer to five thousand and still falling.

At the same time, fertilizer isn't the only thing that's increased in the Mississippi channel. The river now carries much more water—a 30 to 40 percent rise in overall volume since 1940. Higher rainfall driven by climate change is almost certainly responsible for a good share of the increase. But other great river basins in the United States that have also seen changes in precipitation—the Columbia, the Rio Grande, and the Savannah—have not shown a comparable increase in water volume during that same period.

It's an extraordinarily complex undertaking to tease out how weather patterns and ongoing changes on the land have altered the movement of water across the Mississippi basin. But it has become increasingly clear that farming has changed the hydrology across more than a million square miles of the country. Since the 1970s, the amount of land planted in corn and soybeans increased sharply, along with the chemical fertilizers and pesticides commonly used with them. Those two crops now cover a majority of the land in the Mississippi's giant watershed, and up to three-fourths of the area in its upper region. Small grain crops that were once planted throughout the year, such as oats and rye, used more water more consistently throughout the growing season—from spring to fall and, in the case of wheat, even in winter. In contrast, corn and soybeans are planted in spring and use water for only six months a year, and mostly in the midsummer, when precipitation is at its low point. For the rest of the year, most of the farmland is fallow, with no living plants to soak up the heavy rains that come in spring or late fall. The deep-rooted, long-lived perennial grasses that once grew across most of the Mississippi River basin soaked up nearly twice as much water as annual crops.

But all the water that falls from the sky has to go somewhere, and whatever isn't absorbed by crops either flows off the surface into ditches and streams or seeps into the ground. Water running off the surface of farm fields has been reduced since the Dust Bowl as farmers adopted land management practices such as hillside terraces and grass buffer strips along fields and ditches. As a result, the amount of dirt and phosphorus flowing off the surface of farm fields has declined significantly since the 1930s. But that also leaves even more precipitation to soak into the ground, creating a parallel problem: Once in the soil it picks up nitrogen from fertilizer before collecting in the intricate web of drainage tiles that lie below the surface of tens of millions of acres in the Mississippi watershed. Altogether, the combination of shallow-rooted crops, the elaborate drainage systems, and months of bare soil have created a perfect delivery system for nitrogen pollution and a lot more water in the Mississippi.

Like other farm technology, tile drainage keeps getting better and better, and midwestern farmers are continually upgrading this network because corn doesn't do well with wet feet. In recent decades many have turned to "pattern tiling" their fields to permanently lower the underground water table in a systematic way. Pattern tiling systems are precisely designed to match the rainfall

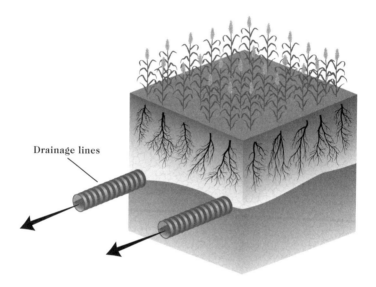

In tile drainage systems, farmers install long perforated plastic tubes below their crops, especially in wet areas, to reduce and manage water in the soil around the roots. When it rains, the tile lines collect water as it flows through the soil, preventing it from saturating the roots, and promoting growth and crop yield. The water in the tile lines eventually drains to a ditch or stream, much like stormwater sewers in urban areas. *Design credit: Sara Bereta*

patterns and soil types of the field above them—tighter spacing for wet areas, wider for dry. These advanced grids can remove all the water from the upper few feet of ground in just a day or two, keeping the surface and the root zones dry. Drained fields increase crop yields dramatically by providing just the right amount of water and preventing the buildup of salts in the soil. In spring, they help the fields dry out much faster, allowing farmers to bring out their heavy equipment as soon as the soil warms up. The same is true in fall. Farmers can harvest precisely when the crop is ready instead of waiting for fields to dry out if it rains. It can take just three seasons for a farmer to recover the investment in pattern tiling through increased yields, making drain tile as critical to a successful farm as fertilizer and plows. Without it, farmers in many parts of the Midwest would be planting only the hilltops.

IN 1927, LONG before the advent of today's sophisticated tile drainage and intensive agriculture, a year of relentless rainfall created a flood that brought ruin to the lower Mississippi River Valley. Water broke through the levees, flooding 27,000 square miles and destroying the lives of a million people. That historic disaster prompted passage of the Flood Control Act of 1928, which created a federal program designed to prevent the worst Mississippi River floods conceivable. It has resulted in a massive reengineering of the Mississippi River by the Army Corps of Engineers. The water control systems they constructed not only manage flooding, they also prevent the river from following its natural inclination to change course from time to time. In earlier centuries, as the river deposited sediment in one spot the land would rise there, forcing the channel to shift toward lower ground. Over time this process created a two-hundred-mile-wide swath of marshy land along the Gulf coast. Much of southern Louisiana would not exist if it weren't for the restless river that wouldn't stay in one channel. In the 1950s it appeared ready to shift again, this time to merge with the Atchafalaya River, which splits off from the Mississippi in northern Louisiana, and make its own way to the Gulf farther west. But the reengineering of the Mississippi in the last century has artificially fixed the Mississippi in one place. It is a defiance of nature on a scale equal to the reengineering of the prairies. Today, the giant river is imprisoned by thousands of miles of levees, dams, and spillways from Cairo, Illinois, to its mouth in the Gulf. These structures release floodwaters at crucial times, protecting the cities, factories, refineries, and farms that have grown up along the Mississippi River's floodplain over the last two centuries—as well as some 4.5 million people. As a result, the Mississippi has become a sort of immense shipping channel that carries the enormous agricultural output of the Corn Belt on the massive barges that float south to the ports in New Orleans for export around the world. But it also has become a highly efficient carrier of the environmental consequences of that wealth in the form of pollution and too much water.

A central piece of this engineering marvel is the Bonnet Carré Spillway, a 1.5-mile concrete weir cut into the east bank of the Mississippi just above New Orleans. It was completed in 1931 to divert Mississippi floodwaters away from the Crescent City and into Lake Pontchartrain, a 630-square-mile saltwater estuary that sits just north of the city. The Army Corps opens the spillway when a river gauge operated by the United States Geological Survey at Carrollton, Louisiana, measures a flow of 1.25 million cubic feet per second. The structure, which hasn't changed since 1931, consists of 350 bays that hold 7,000 "needles," or long vertical timbers that fit tightly together, creating a wall to hold the water back. Closed, the spillway acts like a levee, guiding milk-chocolate-colored water toward New Orleans and the Gulf. When the Army Corps gives the order to open the spillway, a crew of men with a crane work their way slowly across the top, pulling up the timbers one by one. It's always an event—people gather in lawn chairs to watch the river water surge through the gates to a floodway that runs six straight miles to Lake Pontchartrain.

When the spillway was built, the Army Corps envisioned a structure that would be opened "infrequently and for short periods of time." And for most of its existence, that was true. It has been opened just fifteen times in eighty-nine years. But the last two decades of higher precipitation from climate change and increasing water flows off the land have made a mockery of that original intent. It's been opened six times in the last fifteen years, and four times between 2018 and 2020. Each time, it dumps a massive load of fresh water and fertilizer into the naturally brackish Lake Pontchartrain.

The year 2019 was one of the wettest on record. As always, the Army Corps could see the flood coming long before it arrived—a slow-motion disaster. Nearly thirty-eight inches of precipitation had fallen on the vast watershed of the Mississippi, almost eight inches more than average. The flooding began in earnest in March, as heavy rains in the Midwest fell on frozen ground covered with a deep layer of snow. Communities in South Dakota, Nebraska, and Iowa were quickly overwhelmed. As waters from the Ohio and Missouri rivers flowed into the Mississippi, the flood became biblical. By the end of June, the flooding was so bad that a dozen states sought federal disaster relief for four hundred inundated counties. Mississippi River gauges operated by the United States Geological Survey measured more water than at any time in the previous twenty years. That year the Army Corps opened the Bonnet Carré Spillway in February to protect the city of New Orleans, and then again in May. As a result, the spillway was open for a total of 118 days—an extraordinary span that solved one disaster, but inflicted another.

The Lake Pontchartrain estuary receives fresh water from half a dozen little rivers, and, at forty miles long and just twelve feet deep, it is one of the largest wetlands along the Gulf. Tides push salt water into the lake from two inlets on its east side, the Rigolets strait and Chef Menteur Pass. Both connect to Lake Borgne, which also isn't a lake—it's a saltwater lagoon open to the Gulf on its east side right at the Mississippi state border. Pontchartrain is a popular place for anglers, home to countless species of fish and other animals like oysters that need the mix of fresh and salt water, and is an incubator for sea life of all kinds. Dolphins and sharks travel through the connecting straits so they can use it as a nursery for their young. When the polluted fresh water from the Mississippi flowed down the spillway and hit the lake in 2019, satellite images showed massive plumes of brown water slowly encroaching on the blue. It created immense algae blooms like the big one in the Gulf, depleting the water of oxygen and killing much of the sea life that couldn't escape. The muddy, polluted water moved through Lake Pontchartrain, down the straits, and into Lake Borgne, and from there flowed east along a stretch of Mississippi and Alabama coastline known as the Mississippi Sound.

The long opening of the spillway in 2019 was just one in a series of environmental catastrophes that have hit the Mississippi coast in recent decades, starting with Hurricane Katrina in 2005, then the BP Gulf oil spill five years later. The hurricane vaporized the infrastructure—the road that hugs the coast, the marinas, fishing docks, harbors, and the Victorian mansions that faced the sea. The oil spill wiped out the sea life, the seafood industry, and the tourists, as did the chemical dispersant sprayed by flying tanker planes to sink the oil floating on the surface. But the amount of fresh, polluted water that poured into the Gulf in 2019 was so massive and the flow lasted so long that it turned the ocean water brown all the way east to Mobile, Alabama, and from the coastal beaches out to the federally protected barrier islands twelve miles offshore.

Louis Skrmetta runs tour boats from Biloxi out to the islands through his family business, Ship Island Excursions. He advertises the emerald-green waters and sparkling white sands as the best swimming and nature experience in the sound. But throughout the summer of 2019 the water was the color of a butterscotch milkshake. "They made us look like liars," he says. That season, he had only eighteen thousand customers—the same number he had the year of the Gulf oil spill. If he doesn't get at least fifty thousand during his eight-month season, he loses money.

All of Mississippi's beaches closed, each one marked by a red sign warning of toxic algae blooms. The swarms of college students who represent the financial lifeblood of the Mississippi coast flocked instead to Florida for spring break. The algae blooms sucked the oxygen out of the shallow Gulf waters, and, like the permanent Dead Zone in Louisiana waters, forced the shrimp and fish—and the fishing boats—out to the edges of the depleted waters. The crabs scrambled onto shore by the thousands to get air.

The fresh water, however, was even more destructive than the algae. The salinity in the sound crashed to near zero, a death sentence for dolphins because fresh water destroys the film that protects their skin. Instead of swimming out of the polluted area like fish and other mobile wildlife, they refused to leave their territory—or they didn't realize that the water had become toxic until it was too late. Fishermen saw the dolphins leaping from the water, their sides opened with bloody sores from infections. It was, the fishermen say, like watching them rot from the outside in. That year, Mississippi lost nearly 150 dolphins, 188 sea turtles, and some $215 million in revenue from the decimated fishing and tourism industries.

It took four years for the sea turtles to return to nesting sites on the Mississippi coast. The dolphins are back, as are Skrmetta's customers and the college students during their spring breaks. The shrimpers have also returned, but there are fewer of them and their annual take is down by one-third since the spillway opened. What has not come back, and may never, are the oysters.

THE TOWNS OF Bay St. Louis and Pass Christian, Mississippi, just east of the Louisiana border, occupy opposite sides of a small bay and are connected by a bridge. This is the start of the Mississippi coastal recreation and tourist area, and both towns are replete with marinas, restaurants, boat rentals, and beaches. All the restaurants advertise oysters—fresh, fried, in po'boys, you name it, the briny morsels are a tourist draw and the signature dish of the region. Oysters were the major industry along this section of the Mississippi coast, and once upon a time were a primary reason that Biloxi was crowned the Seafood Capital of the World. In the early 1900s, "white-winged" Biloxi schooners collected millions of pounds of oysters and shrimp from the sound for processing at local canneries. The white road that ran along the coast was made from oyster shells.

The shallow, warm waters off Pass Christian and Bay St. Louis were a perfect place for oysters. A number of small freshwater rivers flow into the Gulf there and mix with water from the Atlantic to create the perfect level of salinity. Over the centuries, they grew up in ever deeper layers, each new generation taking root on the shells of the previous ones below, creating the oyster reefs that are their permanent homes. At one time there were approximately twelve thousand acres of oyster reefs in the bays and shallow waters along the Mississippi coast, most of them near Pass Christian.

There is no question that oysters in the Mississippi Sound have been over-harvested, and in recent decades the demand for fresh oysters on the half shell has deprived oyster reefs everywhere of the material they need to sustain themselves. Additionally, Eric Powell, director of the Science Center for Marine Fisheries at the University of Southern Mississippi, says that scientists have only recently begun to understand that the old shells making up the base of the reefs degrade much more quickly than was previously known. Then, starting in 2000, there was a major "regime" shift, an acceleration in the pace of climate change and an increase in water temperatures that affected sea life in the entire North Atlantic. It had a particularly big impact on oysters from the Gulf coast to the Chesapeake Bay, for reasons that are not well understood. In other words, Powell says, oysters in the Gulf came into the twenty-first century in a weakened condition, something that the state natural resource managers, who set harvest limits, had no way of knowing. Their reefs were not being replenished by new shells, they were no longer as resilient as they used to be, and they did not reproduce as well. After Hurricane Katrina and the Gulf oil spill, they started to recover, and the oyster boats would have been back on the sound—if not for the opening of the Bonnet Carré Spillway.

Oysters can withstand low salinity and low oxygen for quite a while if the water is cold; they clamp shut and wait it out. Powell believes that in 2019 they would have been fine if the spillway had been open only between February and April. But the water in the river didn't recede that spring, and the spillway was opened a second time, from May until July. With the long exposure to warm waters, the oysters were forced to open to get food and oxygen, and they all died. In all likelihood, the reefs could have recovered again if there had been enough oysters left to create another generation, but with 100 percent mortality, there were no larvae and no young spats to start anew. "The Bonnet Carré Spillway was something that no oyster population could survive," Powell says.

Just a decade ago the sight of oyster boats on the water at the start of the winter season was a reason for celebration in Pass Christian. Darlene Kimball, whose family has been buying and selling seafood on the waterfront in Pass Christian for four generations, says that customers who saw the boats would swing by the dock in the morning to ask if she would have oysters later in the day. And she always did. Fish and shrimp could be unreliable—the fishermen never knew what they would get or what they could sell to her. But the oysters were always there. They were like insurance for her business. "I used to unload two hundred sacks a day," she says. She points to a white metal boom that hangs over the edge of the cement pier where she runs her business from a turquoise metal shipping container that's been turned into a cluttered office. "Now look—my boom is rusting." She pulls out a green magic marker and begins scribbling on her steel table, checking off a list of calamities that hit her business—the hurricane, the oil spill, the oil dispersant, and, the final blow, the flush of Mississippi River water that brought catfish and alligators by the dock. She will be the last of her family in the business, and "it hurts my heart," she says.

The state of Mississippi isn't going to try to rebuild the reefs around Pass Christian. Over the years, it's already spent millions of dollars to promote the oyster industry. Now it would cost millions more, and it could all be wiped out again the next time a flood comes down the Mississippi River and the Army Corps opens the spillway to protect New Orleans. The state just can't afford that anymore. It is moving its investment in oyster reefs farther east, away from the threat of Mississippi River water, and will concentrate on restoring them in the estuaries where the oysters can do their part to clean the water. It's also adopting the same strategy as Louisiana, Texas, and other states by privatizing what has always been a public resource. Commercial oyster operators can buy leases from the state to work what's left of the natural reefs and build artificial ones, and they will be responsible for taking care of them. The public reefs open to small fishing boats will shrink to 20 percent of the total.

In the wake of the Bonnet Carré Spillway disaster, Pass Christian, Biloxi, and other cities and counties along the sound, as well as the state Lodging and Hotel Association and the commercial fishing association, sued the Army Corps of Engineers for damages. But they have lost their case all the way up to the Louisiana Court of Appeals. Federal law is clear—the Army Corps has the authority to manage the Mississippi River, and only Congress can change it.

Ryan Bradley is executive director of Mississippi Commercial Fisheries United, one of the entities that sued the Army Corps for its operation of the spillway. Not too long ago, he was just a fisherman. He got started at the age of six when his grandfather took him shrimping for the first time. They filled the hold and when they got back to shore, his grandfather paid him $300 for his work. He ran shrimp and oyster boats throughout his high school years, and when his first son was born he quit college and started fishing full time. When his grandfather retired in 2014 and fuel prices went through the roof, he switched to deepwater fishing for high-priced red snapper. These days he represents 275 oyster, shrimp, and fishing operators who make up the state commercial fisheries group. When the wind is too high and he can't head out, he studies the workings of Washington, D.C., agriculture, and the national farm lobby. He's brought oysters to farmers in Wisconsin, and joined forces with a nonprofit that lobbies in Washington, D.C., for independent food producers—community fishers, ranchers, and small family farmers. He's learned that farmers and fishermen have a lot in common—they're both independent, self-employed, entrepreneurial, and they provide food for the nation. But he's keenly aware of the difference in power. Farmers have that long-standing piece of federal legislation—the Farm Bill—renewed every five years, and which enshrines billions in subsidies and protects them from bad weather, low prices, and the other risks inherent in working with nature. With a raised eyebrow, Bradley says, "We'd like to have a fish bill." Coastal fishing around all of the United States is under pressure from climate change, rising costs, and environmental problems, and his industry is an increasingly smaller cog in the national political arena. But the fishing industry along the Gulf coast is also up against something more—a hundred years of changes on the land. "How are we going to come up with a solution for a watershed that covers thirty-one states?" he asks.

THOSE STATES, THE ones that generate the pollutants that create the Dead Zone, have been asking themselves that question for nearly a quarter of a century. And they haven't found a solution either. It took Gene Turner, Nancy Rabalais, and other scientists who studied the Dead Zone two decades to get the attention of the federal government, and they only succeeded in part because they started issuing annual press releases and holding conferences about its size and impact. Finally, in 1996, during the first Clinton administration, the Environmental Protection Agency convened the Mississippi River/Gulf of Mexico Hypoxia Task Force. That launched a frenzy of scientific debate over whether farming and fertilizer was responsible for the Dead Zone, and led to pushback from agricultural interests. Finally, in 2000, the task force devised an action plan. The goal: to reduce the size of the Dead Zone to an average of three thousand square miles by 2015 by reducing the amount of pollution in the Mississippi by 30 percent. For all its ambition, the action plan had a serious flaw: Everything would have to be voluntary. There would be no regulation of farming, even as it generated 75 percent of the nitrogen and 56 percent of the phosphorus delivered to the Gulf of Mexico.

By almost every measure, nothing has changed—except for more water in the Mississippi. Nitrogen and phosphorus levels have remained consistently high, and, in many areas of the basin, they have increased. The annual size of the Dead Zone far exceeds the three-thousand-square-mile goal set in 2000. The problem has proven to be so intractable that in 2015 the Hypoxia Task Force was forced to move the goalposts. Now the plan is to reduce the Dead Zone by 20 percent by 2025, and 30 percent by 2035.

It's also true that scientists have only recently begun to understand just how much excess nitrogen and phosphorus there is on the landscape. Nutrient concentrations in the Mississippi River rose steadily from the 1950s to the 1980s, when they peaked. Since then, taxpayers have spent billions to help farmers reduce their fertilizer use, reduce tillage, improve soil health, create grass buffers, plant cover crops, construct wetlands, and find other strategies to hold back fertilizer, soil, and water. And while that has improved things in some watersheds, others have gotten worse. At the same time, the amount of corn and other row crops on the land has increased, as has urban development—constant changes that make measurements extraordinarily difficult to track. Climate change is only adding to the complexity. It turns out that nutrients can take a decade or more to move from the land to the river, depending on the terrain, the types of soil, and, most important, the movement of water above- and belowground. Now rainfall is increasing, and it could be mobilizing fertilizer that's been stored in soil or in groundwater for up to fifty years. Scientists refer to it as "legacy nutrients." What that means, however, is that even if farmers, golf courses, and gardeners all stopped using fertilizer today, it could be decades before everything stored across 1.2 million square miles washes out into the Mississippi River, and decades before the Dead Zone starts to shrink.

THE SOUTH FORK Crow River may be thousands of miles from the Dead Zone, but it's one of the many small places where the future of the Gulf is being made. Minnesota, one of the twelve states that belong to the Hypoxia Task Force, has a goal to reduce its share of the nutrients that enter the Mississippi River by 45 percent. Phosphorus levels have come down over the years, but there has been no progress in reducing nitrogen, which makes up the vast majority of the problem in the Gulf. It's not just the Mississippi that suffers. Nitrogen pollution is a serious water problem across the agricultural regions of the state, polluting private wells and municipal water sources. It's also enormously destructive to lake and river ecosystems. Water bugs, fish, clams, and birds have even less tolerance for nitrate than humans do. The state's environmental scientists have determined that in order to protect wildlife, surface waters should have no more than eight milligrams of nitrate per liter. But regulating farmers the way that industries and water treatment plants are regulated by the Clean Water Act has, so far, been politically out of the question in Minnesota and every other state along the Mississippi River, even when human health is threatened.

Like every other state that sends nitrogen pollution into the

Mississippi, Minnesota is instead trying to tackle the problem locally, enticing farmers to make voluntary changes on the land with public money, watershed by watershed. In 2023 the South Fork Crow River was one of them. A committee with members representing the local government agencies that manage watersheds, state officials, and commissioners from three counties—including farmer Randy Kramer—devised a plan.

Mark Zabel, a committee member who ran a nearby watershed district, described himself as a conservationist at heart. He grew up on a small Minnesota farm that his family lost during the farm crisis of the 1980s. He knows the history of the land in central Minnesota. His fellow committee members are mostly businesspeople and farmers. His best hope, he says, is that somehow they can all find "that sweet spot" and agree that their ultimate goal is to improve water quality in the river. That would require more wetlands, which catch and filter water, and more cover crops in the fall and winter to take up fertilizer before it runs off the surface when the snow melts in spring, or flows through the tile lines belowground. But even before the first meeting, he was concerned that the priority for local people was to preserve and improve farm drainage. That's a mistake, Zabel says. "We think we are gaining something by overengineering the landscape," he says. But if managing the land is not done in concert with nature "we will fail," he says. "Nature bats last."

The final, ten-year water management plan for the South Fork Crow River, completed in the summer of 2023, predicts that not much will change in the coming decade. Land use will remain largely the same—that is, in agriculture. It also points out right at the top that farm drainage needs to be improved to manage the larger rainfalls driven by climate change, and because many tile lines are old and collapsing. Altogether, the plan expects $13.6 million in public funding to help farmers plant cover crops, improve livestock operations, add wetlands or "water storage sites," improve soil health, and make other changes on the land. And if all that was implemented, the plan estimates, in ten years phosphorus and nitrogen in the South Fork Crow River would be reduced by 1 or 2 percent.

Perhaps, if multiplied across every watershed in Minnesota and the rest of the Mississippi River basin, that would be enough to change the future of the Gulf of Mexico. It's hard to say. But what's clear is that the fate of the Dead Zone, the Gulf tourist and fishing industries, the oysters in the sound, and the health of the Mississippi River from top to bottom will not be decided by the larger community of people who are touched by the continent's greatest river. Nor, it appears for now, will it be determined by the federal regulators who have the satellite-high view of the altered landscape that surrounds it. The fate of the Gulf will be decided by people like Randy Kramer—the farmers, county commissioners, drainage officials, and watershed supervisors—who make decisions largely in obscurity in kitchens, county courthouses, and village halls scattered across the small towns of the upper Midwest. Most of them know that farm chemicals are polluting their local streams, that the Mississippi River is under assault, and that parts of the Gulf of Mexico are dying a slow death. But they also know that generations of American farmers have come to depend on the chemicals and drainage engineering that made the Midwest so extraordinarily productive, that no one has the power to dismantle economic and political systems that took decades to build, and that their local communities would collapse without the prosperous farmers who support them. For the great majority of these people, the Dead Zone is a long way away and the default position in their little watershed is that agriculture has to come first.

CHAPTER SIX

# *Dirt*

Floyd and Dale Coen were playing outside their parents' farmhouse in southwestern Kansas on a spring day in 1932 when they saw an ugly cloud forming in the west. As the sky darkened, the family horses began whinnying and a few hens scurried for the chicken coop. Fine particles of grit began stinging the boys' faces as the black cloud swept toward them.

"It looked like you were going to be enveloped," Dale Coen told the filmmaker Ken Burns decades later. "It was scary, no doubt about that, it was really scary."

In minutes the storm blew into Morton County at full force. It slammed into a nearby schoolhouse so hard that it blew out the windows. Drivers caught on local roads had to pull over until the darkness passed. The boys' mother called them inside, where they sheltered in a corner of the kitchen as the wind howled outside.

The dust storm lasted for twenty-one days. When it cleared, families emerged, stunned, to survey the damage. Chickens lay dead in their yards, cows wandered the pastures in a daze, and dirt was piled waist-high against barns and houses.

At first, many Kansans tried to shrug off the April storm. Dust and wind were facts of life on the southern Great Plains; people told themselves that the weather would soon go back to normal. They were wrong. That summer thirteen more blistering storms ripped across Kansas and Oklahoma. A year later the number climbed to thirty-eight. The United States Weather Bureau began classifying storms by their severity—the worst were those that cut visibility to a quarter mile or less—and counted 350 major dust storms over the next decade. In the first four months of 1935, the wind blew so incessantly that health officials in Dodge City, Kansas, recorded only thirteen days when the air was safe to breathe.

The blowing dirt buried tractors, killed livestock, and turned Kansas's golden fields of wheat into rolling sand dunes. When a storm hit, it was like "midnight at noon, except without the stars," one rancher remembered. In the hours before a storm arrived, the air became so charged with static electricity that children's scalps tingled, car radios shorted out, and telephones went dead. Neighbors avoided shaking hands because the static charge could knock them off their feet. When a storm arrived, the dust cloud was so dense that a man could get lost walking from his barn to his house. Families tied ropes between buildings so they wouldn't lose their way while doing chores. Children caught outdoors without dampened masks would come inside and cough up clumps of mud. Even in homes sealed against the wind, some families cooked under umbrellas to keep dust off the food and set the table with dishes upside down until everyone was ready to eat. Historian Donald Worster has called it "the most dramatic ecological disaster in American history."

The photographer Margaret Bourke-White, visiting the plains to document the calamity, got caught in a dust storm and sought shelter in a nearby farmhouse. She wrote, "Three men and a woman are seated around a dust-caked lamp, on their faces are grotesque masks of wet cloth. Their children have been put to bed with towels tucked over their heads. My host greets us: 'It takes grit to live in this country.'"

For livestock the fate was worse. Cattle ground their teeth to the gums trying to chew grass coated with grit. Thousands of cows perished, suffocated by the dust in their lungs or lost in the fields, their eyes glued shut by mud made of dirt and tears.

When the winds ceased, the heat could be nearly as deadly. Ann Marie Low, a farm girl in southeastern North Dakota, wrote in her diary, "Yesterday was so hot the men didn't go in the fields in the afternoon for fear of killing the horses. The thermometer registered 114 degrees. Dad didn't believe it, so he put out another thermometer. It registered 116 degrees."

Meridel Le Sueur, the writer and activist, described an almost spectral scene during a visit to the Oklahoma panhandle in 1934: "We got in a car and drove slowly through the sizzling countryside. Not a soul in sight. It was like a funeral. The houses were closed up tight, the blinds drawn, the windows and doors closed. . . . Through all these windows eyes were watching—watching the wheat go, the rye go, the corn, peas, potatoes go."

At first, other Americans didn't grasp the scale of the disaster. Then, in May 1934, a bigger storm came raging out of the northern Great Plains. Winds blowing at sixty miles an hour raised a dust cloud two miles high and six hundred miles wide. It passed over the Dakotas, southern Minnesota, and Wisconsin and then, on the afternoon of May 9, swept into Chicago. The sky went dark at midday and streetlights came on early. Drivers turned on their headlights and shopkeepers had to sweep South Dakota dirt off their sidewalks along Michigan Avenue. Some twelve million tons of dust fell on Chicago that day. Two days later the storm reached the East Coast. In Manhattan, the sky went dark for five hours. Tourists atop the Empire State Building could scarcely detect Central Park or the Statue of Liberty. Along Fifth Avenue, shoppers stood coughing and wiping tears from their eyes. Meteorologists said the storm stretched from Albany to Nashville; ship captains more than one hundred miles out at sea reported poor visibility and a scrim of dust on their decks.

For a nation already devastated by the Great Depression, it seemed like a biblical curse had descended. In 1934, drought parched forty-six of the nation's forty-eight states. But nature wasn't finished.

The morning of April 14, 1935, dawned warm and sunny in the Oklahoma panhandle. Families put their washing out to dry and set off on picnics. About 7 A.M., forecasters in the Dakotas detected a Canadian cold front moving south at high speed. The winds picked up dust as they tore south across Nebraska and Kansas. In western Oklahoma the temperature dropped by thirty degrees in a few hours and the sky began to darken. By midafternoon birds were chattering and flitting nervously from tree to tree. At about 4 P.M., a dust cloud two thousand feet high and four hundred miles wide swept into the Oklahoma panhandle, carried by winds at up to fifty miles an hour. Families raced for shelter; a traveling journalist said he had to open his car door and watch the edge of the pavement to avoid driving off the road. In Pampa, Texas, a young folk singer named Woody Guthrie recalled that "you couldn't see your hand in front of your face." The storm cloud was so big that it cloaked the region in darkness for two hours until it finally dissipated near the Mexican border. Even people who had sheltered indoors emerged coughing up dirt and cleaning grit from their nostrils. The hospitals in Meade County, Kansas, would later report that half their admissions for the month of April were suffering acute respiratory illness; thirty-three patients died. Robert Geiger, an Associated Press reporter who happened to be visiting the Oklahoma panhandle, wrote a dispatch describing what he had seen in the nation's "dust bowl" that day. The phrase stuck, for both a place and a calamity.

By 1934 two million farmers were on government relief. Caroline Henderson, a Mount Holyoke graduate who had moved to Oklahoma as a young woman to stake a claim, wrote to a friend in

Maryland about the accelerating pace of farm foreclosures. "These sales take place just outside the window of the assessor's office and . . . have become . . . a matter of routine. No one tries to redeem the property in question; no one ever makes a bid on it; in fact, no one appears but the sheriff and the lawyer representing the plaintiff."

By the time the rains returned, in 1939, a region that had aspired to be America's Eden had collapsed into destitution. One

Farm families devastated by the Dust Bowl in Kansas and Oklahoma were captured by Dorothea Lange in a series of acclaimed photographs for the Farm Security Administration.

quarter of the southern Great Plains—23 million acres—had lost most of its topsoil. One in four residents had simply moved away, their heartbreak chronicled by John Steinbeck and Woody Guthrie.

The Dust Bowl had many causes, and environmental historians

still debate the relative weight of each. One cause was certainly the astonishing drought that settled on the region from 1931 to 1939. As crops shriveled up and blew away, they left nothing to shade the baking ground or hold the crumbling soil.

But, plainly, people made a natural disaster worse. Kansas, Nebraska, and Oklahoma had fallen to the plow a bit later than states such as Illinois, Missouri, and Iowa. Settlers who reached this flatter, windier part of the Midwest around 1850 recognized that it lacked the plentiful rainfall of states farther east. It had even less timber and water than regions where the tallgrass prairie grew. But frontier fever had gripped the nation, stoked by cheap land and glowing tales of the soil's fertility. Kansas's population grew tenfold within two decades after it achieved statehood in 1861, and wheat farms multiplied faster than grasshoppers. By 1930 the state's population had doubled again: Its endless wheat fields had established it as a prairie powerhouse and its fertile farms gave rise to a bizarre scientific theory that held that plowing open the land would cause more rain to fall.

Between 1910 and 1930, farmers plowed up fifteen million acres of grassland on the southern plains, replacing a thick mat of mixed- and shortgrass prairie with wheat fields where the soil was exposed to wind and rain for months each year. And who could blame them? When World War I devastated grain harvests in Europe and Russia, Congress urged farmers to answer the challenge and guaranteed $2 a bushel for wheat—roughly twice what the market was paying. Then, too, the 1920s were unusually wet, prompting an additional influx of farmers who planted more land on the assumption that the rains would never go away.

The ecologist and essayist Wendell Berry summed it up this way: "We plowed up the prairie and never knew what we were doing—because we didn't know what we were undoing."

To anyone taking the long view, the disaster shouldn't have come as a surprise: Throughout recorded history, societies that overtaxed their land have paid a bitter price. The prehistoric city

Windstorms blew more than eight hundred million tons of topsoil off prairie farms during the Dust Bowl, destroying crops and burying farm implements. *Photograph courtesy Library of Congress*

of Ur, the biblical birthplace of Abraham and once a vital Sumerian port near the Persian Gulf, today stands many miles inland, its surrounding plain silted in by soil from the Euphrates River. Much the same happened to ancient Miletus, a powerful Greek port city in Anatolia until its harbor filled up with silt from the Maeander River; it, too, wound up standing some distance from its own coastline. The geologist and historian David Montgomery says that Plato noticed erosion silting up the rivers near Athens in ancient Greece. "The rich, soft soil has all run away, leaving the land nothing but skin and bone."

Colonial Americans also understood that soil was a fragile thing. Southern tobacco farmers routinely moved to fresh land when their demanding crop had exhausted the soil where they started. Writing to Alexander Hamilton in 1796, George Washington observed, "A few years more of increased sterility will drive the Inhabitants of the Atlantic States westward for support; whereas if they were taught how to improve the old instead of going in pursuit of new and productive soils, they would make these acres which now scarcely yield them anything, turn out beneficial to themselves."

But Americans in the twentieth century were bent on making history rather than studying it, and the Dust Bowl landed like the wrath of a spiteful god. It also delivered an indelible lesson: Soil isn't just valuable, it is vulnerable—a lesson whose implications have only grown clearer and clearer in the subsequent decades.

In Washington, D.C., President Franklin Roosevelt recognized that he had to expand the scope of his New Deal to include the rescue of dust bowl farmers. He turned to a pioneering soil scientist named Hugh Hammond Bennett at the U.S. Department of Agriculture. Bennett was a cosmopolitan man with a graduate degree in chemistry, but he loved getting dirt under his fingernails. After graduating from the University of North Carolina, he traveled the country by car, camping out by night and taking soil samples by day, and he soon completed what was then the most comprehensive survey of the nation's soils. Bennett was among the first to recognize that erosion could sap the fertility of a farmer's field by washing away the rich topsoil layer. The top five to ten inches of soil generally hold the richest stores of nitrogen, carbon, and other nutrients because they are constantly being replenished by decaying leaves, stems, and animal droppings—and because they have enough oxygen to support the aerobic bugs and microorganisms that unlock nutrients from dead plants. In what became a famous paper for the U.S. Department of Agriculture, *Soil Erosion a National Menace,* he wrote, "To visualize the full enormity of land impairment and devastation brought about by this ruthless agent is beyond the possibility of the mind." If Bennett's prose seems overheated, his prediction does not: "An era of land wreckage destined to weigh heavily upon . . . the next generation is at hand." It was published in 1928.

By the time he went to work for Roosevelt, Bennett believed that the Dust Bowl wasn't merely one more natural disaster, but reflected a series of catastrophic blunders by human beings. In a speech given toward the start of the Dust Bowl, Bennett said, "Of all the countries in the world, we Americans have been the greatest destroyers of land of any race of people, barbaric or civilized." The Dust Bowl, he said, was a sign of "our stupendous ignorance."

After a long swing through the Great Plains, Bennett returned to Washington and asked Roosevelt for money to establish a national network of experiment stations where he could test soil conservation techniques. When that money began to run out in 1935, Bennett proposed something more ambitious: a permanent branch of the Department of Agriculture that would study soil conservation and promote the new techniques among farmers. At the same time, he knew that selling Congress on another expensive New Deal project would be difficult. On a spring morning of that year, Bennett began making his case before a congressional subcommittee that oversaw agriculture policy. He knew from regional newspaper accounts that another big dust storm was heading for the East Coast, and he asked his regional field offices to keep him apprised of its progress by telegram. He resumed testifying the next morning and, knowing the storm was on its way, began stalling the committee. By midmorning the sky over Washington took on an orange glow amid swirling winds. Federal workers stepped out of their offices and walked onto the National Mall to stare upward. By Bennett's account, when the dust cloud finally hit Capitol Hill, members of the committee stopped the hearing, went to the windows, and stared at the dark sky in disbelief. "Gentlemen," he said, gesturing at the dust cloud outside the windows, "there goes Oklahoma." Five weeks later, Congress passed Public Law 74-46, creating a new soil conservation service within the USDA and declaring erosion a "menace to the national welfare."

In the next three years, Bennett and his new agency planted

nearly forty million saplings on the Great Plains to create windbreaks that Roosevelt called prairie shelterbelts. Nearly one million acres were taken out of crop production and replanted with drought-hardy grasses. In 1937, the Roosevelt administration sent governors a set of guidelines to establish soil conservation districts in their states. These volunteer planning bodies, generally run by farmers and local landowners, monitored local soil and water conditions, promoted soil conservation techniques, and spread the idea that landowners had to act in common to stop erosion, much as they had joined forces to sponsor drainage projects. Soon Caroline Henderson was writing her Maryland friend with a novel development: Her neighbors were experimenting with shallower plows known as "listers" and with contour plowing—plowing that followed the natural curves and ridges of a field rather than straight up and down its slopes. By 1939, roughly twenty million acres were enrolled in soil conservation districts. Over time, Bennett's infant agency grew into an important branch of the U.S. Department of Agriculture, now called the Natural Resources Conservation Service. It operates with a $5 billion budget and runs more than three thousand locally operated soil conservation districts that remain influential in rural communities across the United States even today.

By the middle of the twentieth century, American farmers had adopted Bennett's conservation practices on millions of acres, soil erosion had fallen sharply, and the Dust Bowl was beginning to fade from the nation's memory.

Perhaps too soon.

ALTHOUGH HUMANS HAVE been tilling the land for millennia, only recently have they begun to comprehend the hidden ecosystem beneath their feet. Until the 1500s, for example, Europeans thought plants obtained nutrients by consuming the soil around their roots. Then a Belgian chemist named Jan Baptist van Helmont performed a simple experiment. He planted a willow sapling in a pot of soil, first weighing both the soil and the sapling. When he weighed them five years later, the willow had gained 164 pounds and the soil had lost almost nothing. Clearly, plants obtain nutrients from a source other than soil. Almost two hundred years later, a Dutch English physician named Jan Ingenhousz performed another breakthrough experiment. Taking two sets of plants exposed to plenty of sunlight, he submerged one in water and left the other in the open air. He noticed that the plants submerged in water gave off tiny bubbles. The bubbles turned out to be oxygen—Ingenhousz had discovered photosynthesis, in which plants exhale oxygen and inhale carbon dioxide, a process that ultimately deposits carbon in their roots and the surrounding soil.

In just the last decade or two, soil scientists at universities across the Midwest have extended this scientific legacy with a series of remarkable discoveries. In 2010 researchers at Kansas State University began a small-scale, carefully controlled experiment in grassland restoration. At the Konza Prairie Biological Station, an 8,600-acre expanse of protected grasslands near Manhattan, Kansas, they fenced off a field that had been planted in conventional cash crops for decades. They divided it into a series of small plots, each twenty meters square, then began replanting them, one by one, with native prairie plants. In the first year, they planted a block of four plots. Two years later they planted the next block, and so on every two years for the next decade. This "sequential" experiment would answer a series of questions about soils and prairie plants: How do grasses fare in soil that has been exhausted by decades of crop production? If a block is planted in a dry spring, will its grasses eventually catch up to blocks planted in a wet year? And, crucially, would the depleted soil heal itself as the native plants took root?

On a warm spring day twelve years later, John Blair and Charles Rice are taking a walk around the sequential plots. Blair is an ecosystem ecologist at Kansas State and director of the Konza Prairie

Biological Station. Rice is a Kansas State agronomist and geographer who shared the 2007 Nobel Peace Prize for his work on soils and climate change. Standing at the foot of one plot, Blair kneels down and scoops up a handful of dirt. This section had been planted in cash crops until a few months earlier; the soil comes up gray-black, powdery dry, and finer than sand. It blows off Blair's palm the minute a gust of wind comes out of the east. One hundred yards down the road, in a test plot where prairie grasses have been growing for a decade, a scoop of dirt looks utterly different. The soil comes up moist and black. It has the texture that farmers liken to chocolate cake and emits the intoxicating earthy smell well known to gardeners. Even a stiff wind can't blow it away. Examining similar samples in the lab, Kansas State researchers found that the soil itself changed dramatically once planted with native species. Within three to five years it had richer networks of roots and beneficial fungi, increased enzyme activity, and a burgeoning underground population of microbes.

Walking around Konza Prairie, it's easy to assume that all the action is aboveground in the rich jumble of prairie flowers, grasses, sumac, and sedges. But if the surface were transparent and you could peer through with a microscope to the layers underground, you would see a lively community as busy as any human city.

"It's not just dirt," Rice says. "It's full of life."

And like any community aboveground, soil is full of players that rely on one another. Bacteria known as actinomycetes, for example, break down dead stems and leaf fragments—even tougher compounds such as cellulose—releasing nitrogen in a form useful to plants. Burrowing arthropods, dining on leaf litter and dead roots, create tunnels that allow air and water to penetrate to depths where they can feed the plant's lower roots. Certain fungi nestle just inside the plant's roots, extending tiny tendrils known as hyphae that extract water from the soil and break down minerals, then carry them back to the plant. The hyphae also secrete a sticky protein called glomalin, which binds soil particles together into tiny clumps known as aggregates. These miniature capsules, in turn, store carbon and water, provide protective homes for beneficial microbes, and give soil a porous, spongy structure that allows it to store water and oxygen.

Many of these microbes perform highly specialized tasks, such as the remarkable class of bacteria known as symbiotic rhizobium. These microbes are attracted to the roots of legumes, where they form nodules that become tiny nitrogen factories. Using the enzyme nitrogenase, they break down molecules of atmospheric nitrogen inhaled by the plant and recombine them with other elements to make nitrates and nitrites, the nutrients usable by plants. Other beneficial microbes, gathered around a plant's roots, can reach such a density that they form a protective coating, shielding the roots from parasites and consuming carbon that would otherwise feed hostile pathogens. Other prairie plants nourish a community known as arbuscular mycorrhizal fungi, which have an extraordinary ability to help soil store carbon, stimulate the growth of helpful bacteria, sequester water, and prevent erosion.

TAKEN TOGETHER, THESE residents of the root zone, or "rhizosphere," form extraordinarily dense communities. A plot of healthy soil the size of a kitchen stovetop can contain as many as 1,000 earthworms, 60,000 mites, 100,000 arthropods, millions of the minuscule worms called nematodes, fully one pound of bacteria, and more than 100 miles of fungal filaments. David Montgomery notes that just a tablespoon of soil contains more living things than the human population on Earth. And when all those living things die—plant roots, worms, insects, fungal threads—they add more carbon to the soil. Repeat that process thousands of times over hundreds of years, and you understand why prairies have some of the richest soil on the planet. The famed black soil of the Midwest is black for a reason—it's loaded with carbon.

The new appreciation for this underground world has produced one discovery after another in recent years. To understand the importance of arbuscular mycorrhizal fungi, an Oklahoma State University professor named Gail Wilson applied fungicide to a test plot during her doctoral research in Kansas. She found that where the fungal population was suppressed by the fungicide, the soil was less able to store carbon and hold water, and it broke down and washed away much faster when doused with water. Charles Rice himself has conducted a series of experiments to see how soil might be encouraged to sequester more carbon, pulled from the atmosphere and stored in tiny underground particles, a project with huge implications for climate change. Rice and a colleague, Ganga Hettiarachchi, showed that treating soil with complex carbon inputs—a rich variety of organic sources such as manure—can enhance its ability to store carbon, apparently through increasing the diversity of microbes and adding minerals that bond strongly with carbon. Other researchers have shown that increasing the diversity of grasses and flowers on a parcel of land—that is, mimicking the original prairie—can also increase the soil's ability to absorb and store carbon.

It turns out that this hidden ecosystem also provides hugely valuable benefits to the human community above it, collectively known as ecosystem services. Healthy soil filters rainwater so that people can drink it: Microbes suppress pathogens and the soil acts as a sieve that captures contaminants as water seeps down into aquifers and wells. Healthy soil provides natural pest control because beneficial bacteria, arthropods, and fungi tend to outnumber the parasites, either devouring them or suppressing their growth. While the first antibiotic, penicillin, was discovered in mold, at least half of subsequent antibiotics used in medicine today were derived from soil microbes. Roots and their hyphal tendrils also create thousands of tiny air pockets that hold water and reduce flooding after heavy rains. And in a feat that humans are just coming to prize, grasslands have an amazing ability to absorb greenhouse gases through photosynthesis and lock carbon in underground soil particles that create deep, durable storehouses.

An agronomist named Jason Hill at the University of Minnesota has been studying soil and its ecology for decades. Asked if soil can be considered a valuable natural resource, like petroleum or gold, he paused, as if puzzling over the comparison. "It's more valuable than oil or gold," he said finally. "Humans can survive without oil or gold."

ON A COOL day in April, a brisk wind whips across the John Redmond Reservoir in southeastern Kansas, sending whitecaps against the riprap of its southern shore. The reservoir lies four hundred miles east of the county where Floyd and Dale Coen grew up, but the landscape is not altogether different: dry and flat, with amber fields of grass and wheat that roll off under the blue sky.

The reservoir, created by damming the Neosho River and maintained by the U.S. Army Corps of Engineers, is one of the biggest lakes in Kansas and a popular recreation spot. It provides a crucial source of drinking water in a state where two-thirds of the households rely on reservoirs to supply their taps. It also provides essential cooling water for the massive Wolf Creek nuclear power plant, which rises from the plains just a mile or two from the reservoir's southeastern shore.

So when state water authorities announced in 2012 that erosion along the Neosho was filling John Redmond with sediment, threatening its capacity to hold water, the news ran through Kansas like a high-voltage shock. That year was extremely dry on the Great Plains, and the reservoir's water volume fell by 70 percent between June and December. State forecasters predicted that, if nothing changed, within two years the reservoir would be unable to meet the state's drinking water obligations and supply the nuclear plant. After a series of tense hearings at the state capitol, legislators authorized $20 million to scour the sediment out of

John Redmond, the largest reservoir dredging project in Kansas history.

The dredging worked, the drought eased, and Kansas averted a disaster that year. But the sediment problem hasn't gone away. Kansas has spent hundreds of thousands of dollars to reduce stream-bank erosion along the Neosho and its tributaries, and many thousands more to help farmers reduce runoff and erosion from their fields. Even so, the Redmond Reservoir is projected to lose half its storage capacity by 2070. It's not alone. State water officials expect the rest of Kansas's two dozen drinking-water reservoirs to lose 40 percent of their water capacity by then as well. "The amount of water able to be retained in reservoir storage will be insufficient to meet the demands of multiple user groups," according to the state's 2020 water plan, "and puts the state in the position of being unable to supply adequate amounts of water."

The unfolding crisis in Kansas's reservoirs is just one sign that, despite the grim lessons of the Dust Bowl, the United States has not solved the problem of soil erosion. Even though farmers adopted a variety of soil conservation measures pioneered by Hugh Bennett, millions of acres are still plowed open in the spring and fall, then left unplanted—bare and unprotected—for months at a time, allowing wind and rain to wash away the exposed soil. In Iowa, a roadside exhibit in Adair County shows that the ground has dropped by more than a foot since the nineteenth century because of lost topsoil—roughly half of that since 1950. In South Dakota, the summer of 2022 brought a sequence of dust storms so severe that they closed roads, caused traffic accidents, and evoked images of the Dust Bowl. And in Kansas, in the dry autumn and snowless winter of 1995–96, the state lost as much as thirty tons of topsoil from every acre of cropland, a number that ranks with the worst of the 1930s. Windbreaks, shelterbelts, and other New Deal practices have certainly reduced wind erosion, but they have done little to reduce erosion caused by heavy rainfall. Even in a modest storm, individual raindrops strike the ground with explosive force, and when they land on bare soil they can break soil clumps into fine particles that quickly wash away into nearby ditches, marshes, and streams. Counting both wind and water, researchers at the Union of Concerned Scientists estimate that between 1982 and 2015, erosion cost the United States the equivalent of 25 million acres of prime farmland—an area the size of Ohio—by washing away the upper layers of soil.

In 1972, determined to gauge the nation's continuing erosion challenge, Congress directed the secretary of agriculture to conduct a periodic assessment of the nation's soil, farmland, and surface water, a report known today as the National Resources Inventory. The agency found that wind erosion has dropped sharply since the Dust Bowl, as farmers planted windbreaks and embraced contour plowing. And after Congress created the Conservation Reserve Program in 1985, which pays farmers to take millions of erodible acres out of crop production and plant them in grasses and trees that protect the soil, erosion dropped a stunning 25 percent.

But even so, the agency estimates that, between wind and water erosion, the nation's farms still lose 1.7 billion tons of soil annually—enough to fill 100 million dump trucks every year.

Of course nature is constantly replacing lost topsoil. Leaves, stems, roots, and dead grass decompose into rich humus; mineral fragments such as silica break down into tiny grains; animals large and small die, contributing their own organic matter to the mix. The question is: Can the rate of regeneration keep up with the rate of erosion on any given farm? To find out, the Department of Agriculture estimates erosion rates in locations all over the country, extrapolating from test plots and adjusting for local factors such as rainfall, soil type, and topography. Then it sets a "T factor," or tolerance factor, which is the amount of erosion a given locale can tolerate, given the depth of existing soil, without losing agricultural productivity. In the 2020 National Resources Inventory the government estimated that in much of the Farm Belt, the erosion rate is well below the T factors, meaning that American agriculture is roughly in equilibrium with nature and erosion is not ruinous.

Starting in the 1980s, however, geologists and agronomists began to challenge the government's T factor, arguing that the Department of Agriculture was underestimating erosion and overestimating the speed at which soil replenishes itself. In 1995 an agricultural ecologist at Cornell University named David Pimentel published a landmark paper with several colleagues concluding that the United States has been losing soil at least ten times faster than nature can replenish it. Taking their own measure of soil regeneration, Pimentel and his colleagues concluded that 90 percent of American farmland is still losing soil at a rate that is not sustainable in the long run. In 2007, David Montgomey at the University of Washington tried a more systematic approach to the logic of the T factor. A geologist by training, Montgomery knew how to measure change over long time scales. He reasoned that landscapes generally achieve some sort of natural equilibrium over long periods, so that over many decades, the Earth should regenerate soil about as fast as it erodes. Surveying a dozen geological studies, Montgomery concluded that soil rebuilds itself under natural conditions at just fractions of a millimeter per year. Doing the math, he concluded, much like Pimentel, that the nation is losing soil ten to one hundred times faster than nature can replace it. To put it another way, nature will require centuries to replace the soil that American farmers have lost since John Deere introduced his polished-steel plow.

In 2020, a young geologist at the University of Massachusetts made a second breakthrough. Evan Thaler wanted to get beyond annual rates of erosion and find a way to calculate cumulative soil loss over decades. Thaler had come across a research paper that noted that hilltops in Iowa fields have a lighter shade of soil than low-lying ground, presumably because the rich, black topsoil has eroded off the hilltops, washed away into gullies or nearby streams and rivers, and exposed the gravelly, tan-colored dirt underneath. He realized that if you could take a giant overhead picture of the Corn Belt, you could count the light spots and measure with some precision the places that, over time, had lost all their topsoil. Thaler and his colleagues compared satellite images from eight midwestern states to a USDA gradient of soil colors used to identify levels of fertility. They concluded that a shocking one-third of the Corn Belt—an area the size of New York State—had completely lost the darkest, most fertile soil.

Erosion on this scale causes damage that runs into billions of dollars annually. Dirt washed off farm fields, like the silt in Kansas's reservoirs, winds up in the ponds, streams, and rivers of the Midwest. In the bluff country of southeastern Minnesota, for example, a scenic masterpiece known as Lake Pepin—actually a wide stretch of the Mississippi River—receives so much sediment from farmland upstream that its upper third will essentially become a marsh by the end of the century. In northeastern Nebraska, a beloved recreation spot known as Lewis and Clark Lake is silting up so fast from the Missouri River that it will be half full of sediment by 2045. The Mississippi River silts up so often—not all runoff makes it downstream to the Gulf of Mexico—that barges run aground and the U.S. Army Corps of Engineers spends millions of dollars annually to maintain a shipping channel deep enough for commercial traffic.

IT'S NOT JUST dirt that pours off farmers' fields. Erosion from runoff carries the full cocktail of herbicides, insecticides, and fertilizers that American farmers apply every year—the very combination that has polluted long stretches of the Mississippi River and contributed to the Gulf of Mexico Dead Zone mapped by Nancy Rabalais and Gene Turner. Across the United States, roughly

400,000 miles of rivers and 1.6 million acres of lakes are too polluted to support aquatic life, mostly because of agricultural chemicals. Across the upper Midwest, thousands of private and municipal wells are so tainted by nitrate, a byproduct of nitrogen fertilizer, that their water is unsafe to drink. Some parts of Minnesota—the Land of 10,000 Lakes—have entire counties where not one lake is safe for children to swim. The U.S. Environmental Protection Agency has concluded that agriculture has become the nation's leading source of water pollution in rivers.

In addition to polluting lakes and rivers, modern erosion almost certainly contributes to the devastating floods that have struck the Midwest in recent decades. Because healthy soil is full of tiny air pockets and spongy organic matter, it has an extraordinary ability to absorb water during a heavy rainfall. Andrea Basche, a soil scientist at the University of Nebraska, demonstrates the point in a delightful YouTube video using what is known as a rainfall simulator and two beakers of soil. The first beaker holds dirt from a farmer who has used soil-conservation techniques that keep the soil moist and spongy; the second holds soil from a field where repeated trips by plows and heavy tractors over years have left it badly compacted. When Basche pours water into the first beaker, it disappears quickly, seeping deep into the dirt. In the second beaker, the water instantly pools on the surface, forming a tiny lake just waiting to escape. Basche hoists an umbrella, glances at the camera, and steps to a second table, which holds a toy farm complete with a fence and tiny cows. She opens the tap on an overhead shower to show the same effect on a larger scale. As the simulated rainstorm erupts, water pools on the compacted fields and then gushes off in a torrent that washes out the adjacent highway and sets toy cars afloat in the town across the road.

In a recent study, Basche reviewed the devastating midwestern floods of 1993, 2008, and 2011—when surging rivers crushed grain bins, carried off tractors, destroyed highways, and deluged cornfields, causing tens of billions of dollars of damage in Illinois, Iowa, and Missouri. She concluded that flooding was almost certainly made worse because the surrounding region had lost much of its spongy topsoil. During the flood of 2011, for example, she estimated that runoff into the Missouri River could have been cut by 20 percent if the most erodible land had been planted with cover crops or grass.

Perhaps most troubling, erosion also threatens the soil function that humans prize most: food production. In 2015, the United Nations Food and Agriculture Organization published a report concluding that one-third of the world's farmland is already badly degraded by erosion and overuse, and that only sixty years remain before harvests begin to collapse. Already, the authors noted, per capita production of grains around the world has been falling since 1984. In a sobering conclusion the authors wrote, its "loss and degradation is not recoverable within a human lifetime."

It seems unlikely that American harvests will collapse anytime soon because of soil degradation. But it's clear that the combination of erosion and industrial-scale farming is quietly ravaging the nation's soils. Every ton of topsoil that washes off a farmer's field carries away one to six kilograms of nitrogen and one to three kilograms of phosphorus, nutrients that are essential to plant growth. David Pimentel calculated that soil carried away by erosion—the top layer of soil—has up to five times more organic matter, the rich decaying humus that gives soil its structure and fertility, than the dirt left behind, leaving a field that is 50 percent less productive for grain crops. Agronomists who measured this effect in several midwestern states found that it reduced corn yields by 9 percent to 18 percent in Indiana, and by as much as 24 percent in Iowa. Evan Thaler calculated that Corn Belt farmers now lose $2.8 billion annually in soil productivity due to erosion.

As a result, farmers are spending huge sums to replace what nature once gave them for free. As recently as 1940, despite the ravages of the Dust Bowl, American farmers spent almost nothing on commercial fertilizer—even though the Haber-Bosch breakthrough was beginning to make it available—because those ancient prairie soils were still full of carbon and nitrogen. In 2020, decades after the widespread adoption of commercial nitrogen, they spent more than $24 billion—a sum passed on to consumers in the form of higher grocery bills. As David Montgomery has noted, "This puts us in the odd position of consuming fossil fuels [the feedstock for synthetic fertilizer]—geologically one of the rarest and most useful resources ever discovered—to provide a substitute for dirt, the cheapest and most widely available agricultural input imaginable."

Industrial-scale agriculture, which leads farmers to specialize in nitrogen-heavy commodity crops such as corn and wheat, compounds the loss of soil fertility. When planted on the same ground year after year, they gradually exhaust the soil's native stocks of nitrate. This is why for centuries farmers rotated their crops, resting a field by letting it lie fallow for a year or planting it with crops such as alfalfa, soybeans, and peas to replenish nitrogen. And it explains why some Native Americans planted beans and squash with their corn—the "Three Sisters"; they shaded the soil and added nitrogen. As recently as the 1960s many farmers, like Randy Kramer's parents, had livestock, so they planted oats and alfalfa as livestock feed—crops that diversified their rotations and rebuilt the soil. Today, thousands of farms across the Midwest plant nothing but corn and soybeans, and diverse rotations have virtually disappeared. It's not unusual to find a field that's "corn on corn" for two or three years in a row, with only the occasional soybean harvest. These practices have been drawing down soil nitrogen stocks that took centuries to accumulate, a process known as "nitrogen mining." Despite regular additions of synthetic fertilizer, by one estimate soils of the Great Plains have lost more than 40 percent of their original fertility, as measured by soil nitrogen, since the start of the twentieth century.

"Agriculture has become an extractive industry," says Tim

Crews, chief scientist at The Land Institute, a Kansas research nonprofit that is trying to develop a more sustainable form of agriculture. "We are living off natural capital that took centuries to accumulate—and exhausting it in a matter of decades."

GAIL FULLER DIDN'T need academic studies to understand the Midwest's erosion problem. Fuller grew up on a family farm just upstream from Kansas's John Redmond Reservoir, on bottomland of the Neosho River, and took over part of the operation after graduating from high school in the 1970s. One spring afternoon in 1993 he was plowing a big cornfield when a thunderstorm appeared on the horizon. He quit for the day, leaving half the field unplowed, and reckoned he would finish after the storm blew over. But that night the Neosho overran its banks and came pouring across Fuller's land. When he went out a few days later he found that the river had washed away eight inches of topsoil from the field he had just plowed—a disaster. Then he noticed something else: On the half where he hadn't finished plowing, where last year's corn stalks and soybean roots still held the ground in place, topsoil loss was negligible.

Over the next few years Fuller began experimenting with a suite of soil-conservation practices that were well known to conservationists but unusual on most midwestern farms. He adopted no-till planting: Rather than plowing a fresh furrow in the spring, which aggravates erosion and shreds the vital underground microbial community, the farmer drills seeds directly into the soil amid the previous year's crop stubble. He abandoned the region's conventional, narrow corn-soybean-wheat rotation and began raising as many as eleven cash crops, including sorghum, barley, triticale, oats, corn, and sunflowers, rotating them among fields in a way that rebuilds soil, diversifies soil nutrients, and suppresses pests. He also kept his land protected twelve months a year with cover crops such as alfalfa. Looking back today, he laughs at all the mistakes he made, but before long he became a respected conservation evangelist recognized by the Kansas branch of the Natural Resources Conservation Service. Today Fuller and his partner, Lynette Miller, raise sheep, chicken, and ducks in addition to various crops. Even in his relatively dry part of Kansas, his pastures are a dense, lush carpet of grass, clover, and alfalfa. He stresses that sheep and poultry aren't the only critters on their land. Walking across his farm one spring day, he stoops to inspect a handful of dirt that is full of bacteria, fungi, and bugs. "I'm looking after my underground livestock," he says with an insider's smile.

The growing appreciation for healthy soil has spawned an expanding community of small conservation groups composed of farmers, academics, and environmentalists: the Kansas Soil Health Alliance, No-till on the Plains, the Nebraska Soil Health Coalition, and the South Dakota Soil Health Coalition, to name a few. The U.S. Department of Agriculture's Natural Resources Conservation Service also created a new branch, Sustainable Agriculture Research and Education, which provides grants to support and spread research on soil health and regenerative farming. Even so, farmers like Gail Fuller remain outliers across the Midwest. The experience of Brett Niebling is probably more typical. Niebling farms with his father and uncle in northeastern Kansas, about 150 miles north of Gail Fuller, raising corn, soybeans, and cattle. The land is hilly in their part of the state, and not far from the Missouri River, so they have placed a priority on soil conservation for more than a generation. They use no-till cultivation and they've constructed terraces on the slopes to reduce erosion. But they first tried cover crops only in 2021, and the experience left them frustrated. They hired a local contractor to scatter a mix of rye and turnip seeds from the air on fifty acres. It cost roughly $1,500 and produced only patches of green here and there. They were able to graze cattle on the shoots that emerged, but for only a few days. "It was a pretty spotty crop," he says. The Nieblings haven't given up on cover crops—they planned to try it on 150 acres the next year—but their experience explains why farmers across the Midwest are skeptical of cover crops and similar practices. As Randy Kramer observes, conservation techniques often require extra money upfront for seeds and special equipment. For another, erosion can be all but invisible on a typical farm, so that even careful farmers might not realize just how much valuable soil they're losing each year. One ton of topsoil, spread across one acre, has roughly the thickness of a sheet of paper, so a farmer could lose tons of it and scarcely notice. Finally, using popular chemicals and profitable crop rotations is the path of least resistance, endorsed by crop consultants, corporate sales reps, the manager at their local co-op, and sometimes even state university extension agents. Farmers who experiment with soil-conservation techniques, by contrast, find that they have to teach themselves, learning painful lessons by trial and error. One big error, financed with an expensive bank loan and combined with a poor year in commodity markets, can put a farmer out of business. It's no wonder that farmers hesitate to stray from the safe and tested path.

As a result, adoption of modern soil-conservation techniques has been limited in the heart of the Corn Belt. No-till planting is expanding but still accounts for only about 15 percent of cropland in the upper Mississippi River basin. Acreage planted with cover crops has risen nearly 50 percent since 2012 but still represents only about 5 percent of harvested cropland. "For a lot of farmers, erosion is a 'next-year' problem," says Marcia DeLonge of the Union of Concerned Scientists. "They have something more urgent to worry about most of the time."

And so for months at a time, in the period between fall harvest and spring planting, the soils of the Corn Belt remain mostly naked and unprotected. To drive across the Midwest in these months is to observe an endless expanse of black—the fat, dark furrows of precious soil exposed to wind, rain, and blizzards. For many Midwesterners, this is subconsciously a comforting sight—

the tidy and weed-free domain of diligent farmers. But it is an anomaly in the natural world, where land is protected by continuous blankets of living plants and sturdy roots that nourish the soil. "Nature," says The Land Institute's Tim Crews, "does a better job of protecting its soil."

CHAPTER SEVEN

# *Bugs*

THE BLUE DASHER FARM LIES IN EASTERN SOUTH DAKOTA, HALFWAY up the state to the border of North Dakota, a tiny oasis in the sea of corn and soybeans in the upper Midwest. Here there are piebald pigs, sheep in grassy pastures, a few llama, a collection of barn cats, hives of honeybees, a small vineyard, and a vegetable garden. The farmyard has an old-fashioned feel to it—chickens, ducks, and a few peacocks run free around a gravel parking area outside a long white shed. Blue Dasher is the home of Jonathan Lundgren, who was once an entomologist for the U.S. Department of Agriculture. He was an expert on the most common pests that plagued farmers in the Corn Belt—corn borers, rootworm, soybean aphids, wire worm—that in a bad year could cost them billions of dollars in damaged crops. He was also an expert in all the chemicals and genetically modified seed technologies designed to defeat those bugs, and which generate billions in revenue for the companies that make them. Today Blue Dasher Farm, named after the sapphire-colored dragonfly common throughout the country, is also the home of the nonprofit Ecdysis Foundation that Lundgren established after he quit his job with the agriculture department in 2014. After decades as a scientist who thrived alongside the agrochemical industry, he came to recognize the futility of its war against insects, the failure of its default strategy of using toxins for pest control, and the enormous damage it caused to the rest of the natural world. His own research showed how generation after generation of insecticides were silently killing off the bees and butterflies as well as the many unseen insects that form the base of the natural world.

The heart of Blue Dasher Farm, and Ecdysis, is a long white shed that once held farm equipment. Today it's filled with a dozen or more young scientists sitting shoulder to shoulder along long tables as they peer through microscopes or count tiny bugs. A flattened cardboard box hung at one end of the room lists goals and ideas for the coming year. In one corner is a walled-off section full of shelves and boxes that house a growing insect collection—up to a million species so far—that is the foundation of a public database for entomologists around the world. Some of the boxes hold insects that have long been categorized as pests, usually defined as the kinds of bugs that eat crops. Others are beneficial insects—they pollinate flowers, convert dung to soil, and prey on the pests.

The goal of the Ecdysis Foundation is to find a way to balance the good and bad bugs, to use them with and against each other in a different kind of farming popularly described as regenerative agriculture. It uses far fewer chemicals, insecticides, fertilizers, and herbicides, and, instead, leverages natural systems like biodiversity and richer soils to solve one of the fundamental paradoxes in modern agriculture: Its reliance on toxins to control pests is driving its own never-ending crises with insects. In truth, Lundgren says, "pests are something we created." And that is something entomologists have recognized for decades.

First, early colonists plowed up the grasslands and wetlands, the naturally diverse biological systems that supported thousands of insects that were in balance with one another and plant life. Then, they replaced them with crops—small monocultures that were a banquet for only the insects that ate them. In speeches and in the influential agricultural journals published during the mid-1800s, entomologists begged farmers to diversify their crops and to stop growing wheat or corn altogether during outbreaks of pests. But their advice to diversify, to manage insects by understanding their life cycles, and by using modest natural controls like tobacco extract and chrysanthemum dust were impractical for the increasingly large farms that grew up on the prairies. Nor did it fit with the powerful economics of growing monoculture crops for world markets. Managing insect problems on individual farms eventually evolved into a government-driven national quest to solve the insect problem for society at large. The development and manufacture of insecticides first became a booming industry during World War I in the belief that they would increase food production to feed the army and protect soldiers from scourges like lice and malaria. After the war, and encouraged by the policies and advice of the federal government, farmers began using poisonous chemicals like arsenates, cyanides, and rotenone in the millions of pounds.

Ever since, toxins to control pests have remained a fixture in agriculture, and the historic paradox that plagued early colonialists continues on a much larger scale. Farmers eventually lose the war against their insect enemies because the pests that thrive on monoculture cropping systems are able to adapt to any weapon used against them. Evolution always wins. At the same time, that loss of plant diversity across an agricultural landscape, combined with the ubiquitous use of insecticides, is causing massive collateral damage to the helpful bugs that our food system—and the rest of the natural world—needs to survive.

E. O. WILSON, the pioneering evolutionary biologist, pointed out that if humans disappeared, the insect world would carry on and even thrive without us. Humans, however, could not survive without them. Though insects may be one of the least appreciated and most invisible forms of life on Earth, they form the structural and functional base of all living systems. They predate mammals and other vertebrates on the planet by 100 million years or so, and, at more than 5 million different species, have become the most successful and diverse of all organisms. They drive most plant reproduction through pollination—carrying pollen from male plants to female plants to create seeds. The evolution of most plants, their extraordinary diversity in scents, shapes, and colors, was driven by the competition among them to attract the insects that play such a vital role in their existence. By eating plants, and by being eaten themselves, bugs are like livestock—they turn sunshine into protein. They are the base of the food chain, providing sustenance for innumerable species, including fish and most birds, that are in turn eaten by other animals, including humans. They keep the world clean by decomposing organic matter, including dung and dead creatures. Dung beetles that process manure keep pastures healthy, provide nitrogen to plants, and reduce the number of problem bugs—flies and parasites. By one estimate, they are worth an estimated $380 million in services to the U.S. cattle industry. Insects create rich soils by cycling nutrients from organic matter through their waste, and keep their numbers in check by eating one another. Without them, humans wouldn't last long because the crops we eat—fruit, vegetables, nuts—would crash. Populations of amphibians, fish, birds, and mammals would collapse, followed by forests, flowering plants, and grasslands. After many decades of global rot and decomposition, life would be reduced to bacteria and algae. Wilson described that catastrophic cascade in an essay, "The Little Things That Run the World," published in 1987. In the years since, scientists have begun to recognize that the demise of those little things is accelerating at a pace that may exceed the loss of all the other living species that are disappearing around the world in what scientists call the Earth's sixth mass extinction.

Compared to concerns about the loss of "charismatic megafauna" species like the polar bear, the chimpanzee, and the condor, the scientific world came late to the realization that many insects are also in decline. So even now there are huge gaps in the knowledge of which of the five million species are in trouble and where, and which ones are increasing in number. There are many that have never been identified, and may have disappeared before we knew they existed. One of the studies that shocked the entomology world was published in 2017. A twenty-seven-year project to monitor insect populations showed a 76 percent decline in flying bugs at several of Germany's protected parks and wild areas. That's an average 2.8 percent loss of insects by weight per year—and that was in natural places largely undisturbed by people. Though many scientists questioned whether it reflected larger trends, it sparked a slew of new monitoring studies. More than most other animals, insect populations are extraordinarily difficult to track. Their numbers fluctuate naturally depending on weather and other conditions, and they can swiftly rebound. Long-term counts can be unreliable for species as a whole because they tend to be conducted in natural areas, like the German study, where insects are abundant rather than cities or farms where they are fewer in number. After all, nobody goes to downtown Chicago or the middle of a cornfield to count butterflies. But the global trend is clearly downward, though there are exceptions. One widely respected analysis of 166 global reports found that aquatic insects are faring surprisingly well, in large part because legal protections like the Clean Water Act have likely improved water quality. Terrestrial insects, however, appear to be declining by an average of 1 percent per year—meaning that in the last three decades one-third of insects have disappeared. Declines in the American Midwest were among the highest.

Bees are among the most studied because hive-dwelling honeybees and their wild cousins provide such a vital service in pollinating plants, especially food crops like fruit, nuts, and vegetables.

They are also easy to see as they go about their business. As a result, they are often considered a sentinel species that reflects the fate of many other flying bugs. And they are not doing well. England has already lost twenty-three different kinds of bees. A survey of wild bees in North America in the early 1970s found a 32 percent reduction in species compared to seventy-five years earlier. Only fifty-nine of the seventy-three prairie bee species remained. Wild bees were in decline in one-fourth of the nation between 2008 and 2013, primarily in the Midwest and Great Plains and the Mississippi Valley, where corn production almost doubled during the same period. In 2023 researchers with the Xerces Society for Invertebrate Conservation published a study that found the western bumblebee, once a common sight in the western half of the United States, has become increasingly rare. It estimated that the species has declined by up to 83 percent in some parts of its range, and at the current rate, it will experience an overall decline of 93 percent by the 2050s.

While climate change, urban sprawl, and industrial development all threaten insect populations, scientists believe that the combined effects of the post-1950s intensive agriculture have had the largest impact. The widespread use of herbicides and pesticides and wetland drainage, and the rapid expansion of land devoted exclusively to monocultures that support only a few species of insects all contribute to the loss. That's true globally, and even more so in the former tallgrass prairie, where corn and soybeans now occupy half or more of the landscape. Little can survive in those carefully managed fields. Like soil erosion and contaminated in water, the impacts of agriculture on insects spread far beyond the farm fields. The fertilizer that generates algae blooms and depleted oxygen in water from Minnesota to the Gulf of Mexico also kills off aquatic insects. Herbicides that drift in the air or move in the water kill many of the wild plants that bugs need. While little is known about the number and species of insects that lived in the tallgrass prairies before Europeans arrived, it's clear that the populations and types of many once-common insects have declined precipitously.

The insects that do survive are rarely the iconic ones such as the monarch and Karner blue butterflies, which are both on the federal endangered species list, or the clouds of fireflies that once lit up warm summer nights, or the pollinators like domesticated honeybees and wild bees that are critical to growing fruits and vegetables. Many pests, however, which have been part of the landscape since long before Europeans arrived, have managed to evolve along with the expansion of farming and the seismic shift after World War I in how chemicals are used to control them.

The Colorado potato beetle, for example, historically ate a type of wild potato in the West. It adapted to domesticated potatoes so quickly that its eastward expansion starting around 1850 was compared to Sherman's march to the sea in the Civil War. In just fifty years, the bug ate its way through potato fields from Colorado to the Chesapeake Bay. The chinch bug originally lived in native grasses, and its numbers were naturally controlled by a fungus. By the late 1800s, however, it had adapted to eating wheat and potatoes, and in dry years, when the fungus was scarce, it caused tens of millions of dollars in crop damage. The Rocky Mountain locust blew out of the West and covered the mid-continent in swarms, devouring whole counties and destroying farms and families. Native plants had evolved to resist the insect, but new cultivars of wheat, corn, and barley had no such defenses.

The process accelerated after World War II, when many of the chemicals invented for the war effort were adapted for farming, DDT in particular. The Allied armies used the compound to kill lice that spread typhus among their soldiers throughout Europe. The adoption of DDT by farmers coincided with the use of synthetic nitrogen fertilizer from the Haber-Bosch process. The combination of chemical breakthroughs, synthetic fertilizer, and insecticides allowed farmers to abandon the ancient practice of rotating crops to control pests and build soil health. Soon farmers were growing corn season after season in the same fields, using fertilizer for fertility and insecticides to control the pests that inevitably came with monoculture.

Widespread use of DDT came to a halt when the biologist Rachel Carson exposed the extraordinary damage it caused to other wildlife and the risk to people in her landmark book *Silent Spring*. From the 1960s through the 1980s, DDT and other insecticides gave way to the far more environmentally friendly practice known as Integrated Pest Management (IPM), which was very much like the practices farmers used in the early to mid-1800s: diversity through crop rotations, pest monitoring, and the judicious use of insecticides only when necessary. The goal was to protect crops—and farmers—while doing minimal damage to the land, the water, and the bugs. In the 1980s, that meant only a third of cornfields and 5 percent of soybean fields were treated with insecticides.

But IPM quickly lost favor when agrochemical companies such as Syngenta and Bayer Crop Science developed the technology of genetic engineering. In contrast to the standard practice of spraying insecticides or drenching the soil in infested fields, one of the great novelties of the technology was that it functioned as insurance. It armed every seed and every plant against a particular pest—whether it was a problem or not. It was also one of the first commercial applications of manipulating the DNA of one species to use against another. It was based on extracting a gene from a natural bacterium that was widely known to be toxic to insects, and inserting it into the DNA of the crop seed. The first genetically manipulated seed, introduced in 1996, was Bt corn, which carried a toxic protein produced by the common bacterium *Bacillus thuringiensis*—or Bt—commonly found in soil, aquatic, and agricultural environments. The bacteria was first identified in 1901 and had for decades been used in insecticidal sprays, but they were generally ineffective. The toxin broke down quickly when exposed to air and sunlight. But it had intriguing possibilities: The different varieties of Bt bacteria carry different toxins that work against different kinds of bugs—caterpillars, moths, beetles, or mosquitos. When the targeted bug eats the bacteria, the toxic protein punches a hole in its digestive system and it dies. These genetically modified crop seeds with their built-in toxins aimed at specific pests eventually became the standard in most farming, and are common in corn, soybeans, cotton, and canola. In corn, Bt protein targets the European corn borer, one of the most common and damaging pests known to farmers. It starts out as a caterpillar that eats all parts of the plant before transforming into a small brown moth. Bt corn was—and still is—a great invention. Unlike the earlier insecticides that were broadly sprayed in and around fields, killed all kinds of insects indiscriminately, and were a serious health threat to the farmers who used them, Bt just kills the corn borer.

Decades of experience with insecticides, however, has shown that through natural selection, pests eventually become resistant to toxins. In any population, a few individual insects by chance tightening and avoiding redundancy have a fortuitous mix of genes that allow them to survive exposure to the toxin. Those are the bugs that survive, and they produce the next generation of insects, which all carry the genes that make them resistant. It's the same process that makes human bacteria grow resistant to antibiotics. When a class of insecticides becomes ineffective because of resistance, manufacturers introduce new ones that farmers swiftly adopt, and the same thing happens again. It also occurs with herbicides, like Roundup, that kill weeds. In farming, it's called the pesticide treadmill.

In order to slow down that evolutionary process, farmers who use Bt corn are supposed to plant what is called a "refuge"—a plot of ordinary, non–genetically engineered corn in a section of their fields where some insects are not exposed to the toxin. The idea is that the corn borers born inside the refuge will fly off when they become adults and mate with the few bugs outside the refuge that carry the rare genes that give them immunity. That dilutes resistance in the entire population because fewer offspring with the protective genes are born. In many crops and in many parts of the world, Bt-engineered seeds have been a remarkably effective tool against some destructive pests. They've been instrumental, for example, in nearly eradicating the pink bollworm from U.S. cotton crops. Bt corn has been credited with reducing the density of destructive corn borer pests throughout much of the Midwest, and in some places their numbers are the lowest they've been in three quarters of a century. The toxin in the seed is highly effective, killing 99 percent of the insects that are exposed to it. Also critical to that success, however, were the refuges that farmers are required to plant to slow adaptation. Ten years after it was introduced, however, resistant populations of corn borer were found in Canada. Around the world, at least two dozen other crop pests have become resistant to Bt seeds, and seventeen more are starting to show early warning signs. But so far no significant problems have occurred in the United States.

That is not true, however, for the yellow-and-black-striped beetle called the western corn rootworm, aka the billion-dollar pest. It has successfully used natural selection to defeat a century's worth of insecticides, including genetically engineered corn. The corn rootworm arrived in what is now the southwest United States more than a thousand years ago. At some point it became a bug exclusive to the corn grown by Native Americans, though it proved less problematic in those smaller, more diverse farming systems. The first documentation of the little striped beetle after Europeans arrived was in Colorado in 1867. Their life cycle is tied to that of their host plant, corn. Adults mate on corn, then lay their eggs in the soil at the base of the plant. As larvae (the worm) they eat a plant's roots down to little nubs, and the adult beetles eat all parts of the corn plant.

Corn rootworm was first recognized as a pest to corn farmers in 1909, and as corn spread across the country, so did the bug.

Because the females lay eggs almost exclusively in cornfields, and the tiny larvae feed on roots to survive, farmers were advised to control their number by rotating their crops on a regular basis with another crop that the larvae didn't eat. In the 1940s, farmers began increasing drainage systems in their fields and the use of chemical fertilizers on their corn crops. That made corn yields so abundant and profitable that instead of rotating crops, farmers began to grow corn in the same fields season after season, a practice called "corn on corn." With more and more corn on the landscape, outbreaks of rootworm accelerated and insecticides like DDT became the weapon of choice against rootworm. Then, in 1959, scientists realized they were seeing adaptation in action when the first report of rootworms developing resistance to a commonly used insecticide was documented in Nebraska. Though using a wide variety of insecticides allowed the continual planting of corn for farmers who needed it for livestock feed, rotating crops became a more common defense. Without as much corn around to fuel the population of bugs, the problem diminished. But after a few decades, the bugs even adapted to crop rotation by evolving in a way that changed their very biology.

In the late 1980s, entomologists realized the adult beetles were hedging their bets by indiscriminately laying their eggs in both corn and soybean fields at the same time. In the two-crop system, no matter where corn was planted, there would always be rootworm eggs waiting for it, even though there was only a fifty-fifty chance that there would be something for them to eat when they hatched. Once again evolution helped the bugs outsmart farmers and the chemical industry.

When entomologist Joe Spencer arrived at the University of Illinois in 1996, the rootworms were winning the war. Illinois alone had thirteen counties where farmers were losing half their corn crops to swarms of the yellow beetles, a devastating loss. In response, they were drenching the soil with insecticides in amounts that alarmed the Environmental Protection Agency. Spencer's job was to study the problem—and beetle behavior—to find a way for farmers to stay ahead of the insects.

In 2003, the agrochemical industry devised a new kind of Bt corn designed to kill the rootworm as well as the corn borer. The two toxins were "stacked" together in a single seed, a combination that appealed to farmers because they could deal with both pests at the same time. But the strategy did not work as well on rootworm. They aren't like corn borers, which become moths that fly away as adults and mate far from where they hatched. Instead, rootworms become beetles and don't travel as far. That made the strategy of including refuges in the fields less effective because they mate close to home, and are more likely to produce resistant offspring. Moreover, the new Bt toxin for rootworms wasn't as lethal as the one that targeted corn borers. About 10 percent of the beetles could survive exposure, compared to less than 1 percent of corn borers. That meant the rootworm beetles' adaptation could occur much faster. Spencer, who studies how insects behave, remembers sitting in a meeting where the new Bt variety was presented, and thinking, *Why are we letting the life cycle of one bug determine the way we fight another?*

Nonetheless, farmers across the country quickly adopted the new Bt corn for rootworm. And it took the bugs only six years to develop a defense. The first reports of resistance came from Iowa in 2009. The seed companies ultimately responded by introducing new varieties of Bt toxins and stacking multiple layers in a single hybrid seed. Today, virtually all corn planted in the United States expresses multiple toxins against rootworm. They still have the benefit of leaving other insects unscathed, but the pests have managed to develop resistance against all of them. Resistant populations are widespread in the heart of the nation's Corn Belt—Iowa, Illinois, Nebraska, Minnesota, and North Dakota. Farmers helped drive that swift evolution for two reasons. Many overused Bt corn by planting it in fields even where they didn't need to. Farmers saw it as a form of insurance against the possibility of larger infestations, or because it was just easier to plant the same variety of corn in all their fields, regardless of whether the bugs were around in large numbers or not. But it was enough to create resistance in the small populations that were present. Perhaps more important, some farmers—an estimated 30 percent—failed to create enough non-Bt refuges in their fields to thwart the spread of resistant genes. Spencer described it as a "tragedy of the commons." A minority of farmers was enough to sink the technology for all of them.

The agrochemical industry has recently introduced yet another new genetic weapon against rootworm. But it's not a toxin, it's a whole new strategy called RNA interference (RNAi). The corn seed contains genetic material from the rootworm itself, an RNA molecule that gives instructions to the genes contained in its DNA. Only in this case, the RNA molecule acts like a tiny Trojan horse. It shuts down the production of a critical protein in the cells of the insect that eats the corn plant. It's still not the perfect weapon; laboratory studies have shown that the bugs will develop resistance to it as well. Still, seed companies are combining the new RNA molecule with a pair of Bt toxins in the hope that the rootworm will succumb to at least one of the three.

In a 2022 YouTube video directed at Illinois farmers, Spencer explains that the rootworms have "a formidable capacity to overcome whatever management tactic we put in front of them." In the video, he stands in a sandy field with a wall of perfect corn behind him, the leaves clacking gently in a breeze while he implores farmers to use the latest generation of genetically manipulated seeds with care. With every subsequent growing season in Illinois, rootworms are surviving at higher and higher rates. It's imperative, he said, that farmers scout their fields to look for the insects, that they use the genetically engineered seeds only in fields where the pests are a problem, that they rotate their crops, and that they plant the prescribed refuges to slow the development of resistance. With the new RNAi seed, farmers now have a window of time to get out ahead of the corn rootworm. It is, he said, "a golden opportunity to do the right thing."

THE SEEDS WITH genetic manipulations like the Bt toxins and the RNAi molecule were popular in part because they were weaponized against specific, so-called primary pests like the rootworm and corn borer, leaving the rest of the insect world untouched. In 1991, shortly before Bt corn came to market, the chemical industry also produced a different kind of weapon. It used an old insecticide delivery method—seed coatings—for a new class of toxins that worked indiscriminately against all kinds of insects, not just a targeted few. In the 1800s, farmers tried brining wheat seeds to prevent contamination with a fungus. In the early 1900s, farmers tried coating seeds with coal tar, tobacco, and heavy metals to prevent fungi and pests. Though those efforts were mostly unsuccessful, the use of mercury on seeds continued for decades. By and large, however, seed coatings were not nearly as effective as spraying insecticides onto fields. But that changed dramatically starting in the 1990s with a new class of toxins called neonicotinoids.

Neonics, as they are often called, come in a variety of candy-colored coatings on the outside of the seed, arming it against not just one kind of insect but all the so-called secondary crop pests. That includes the largely earth-dwelling insects, such as maggots, white grubs, and leaf miners, that are hard for farmers to see but nibble on the seed and the tiny seedling. The toxins, named after the nicotine molecule they resemble, act by continuously stimulating neurons in the central nervous system of insects, leading to epileptic-type seizures and death. Just a few parts per billion, like a pinch of salt in ten tons of potato chips, is enough to kill. Their appeal was undeniable. While highly toxic to pests, they are benign to the people who grow the crops, and, unlike most of the other broad-spectrum insecticides on the market, bugs had no resistance to them. They are also supremely easy to use, and came precoated on the same seeds stacked with Bt traits. Farmers no longer had to worry about scouting for insects or investing in expensive spraying equipment to manage them when they showed up. With or without pests, the insecticide was already in place on the seed. Every seed and plant was armed with multiple defenses against multiple pests that functioned as an insurance policy against infestations, just like a vaccine acts as insurance against an infectious disease—you may not need it, but it's there when you do.

Neonicotinoids achieved instant popularity among farmers, taking over the market in just a few years. They are now the most widely used insecticide in the world for wheat, rice, cotton, corn, and soybeans. Over time, farmers had increasingly less choice in whether to buy neonicotinoid-coated seeds. Companies like Bayer Crop Science that engineered the seeds had patents on the Bt genetic technologies farmers had adopted, which meant they also had near-total control of the seed markets. That made it easy to routinely apply neonicotinoid seed coatings on everything they sold. A wave of mergers in the agrochemical industry further reduced the number of companies competing for a farmer's business, giving them even fewer choices in seeds. Eventually, federal law even compelled farmers to use them. Federal subsidy rules restrict the amount of crop insurance farmers can claim from insect damage if they don't use neonicotinoid-treated seeds. The argument is that they shouldn't collect an insurance payment if they aren't using every tool available to prevent crop damage.

Today, about half of all soybean seeds and virtually 100 percent of nonorganic corn seeds are coated with these pesticides. The amount of insecticide used on each acre is far less than it was with earlier chemicals that were sprayed, and the risk to humans is far lower. But the use of the coated seeds has become so prevalent that the number of acres across the United States that are now treated with insecticides has exploded, reaching 150 million or more every year—an area about the size of Texas. It's not just farmers who use neonicotinoids—nurseries use them to grow geraniums and marigolds, and they are routinely injected into ash trees to protect against emerald ash borers. Pet owners buy them in flea and tick collars, and they are widely used on seeds that grow food crops. In short, they're everywhere. In just a few decades they have spread well beyond the farm fields where the seeds are planted every year, and they harm far more than agricultural pests.

Beekeepers were among the first to sound the alarm. They began noticing troubling developments within their hives and questioning the widespread use of neonicotinoids across the landscape where their bees foraged in wildflowers, and around the orchards and vegetable fields where they are used to pollinate food crops. Their warnings launched an avalanche of research on the unintended consequences of the compounds. For a while, the insecticides were blamed for the terrifyingly named "colony collapse disorder," in which honeybee hives would inexplicably die over the winter. It turned out that neonicotinoids were only one hostile element among many in the complex interactions of insecticides, diseases, parasites, and a shortage of flowers in barren agricultural landscapes. More profoundly, however, researchers have found neonicotinoids do cause a variety of "sublethal effects" in honeybees, including impaired reproduction, flight, navigation, and immune response, all of which weaken populations over time. Honeybees are fortunate to have beekeepers who watch over them and document the threats against them. The rest of the insect world is not so lucky.

Study after study has found that the insecticide does not stay on the seed after it's planted—only about 2 percent stays with the plant as it grows, moving through its circulatory system. In corn, some of the seed coating is blown into the air during planting to drift across the landscape. The rest of the toxin on the seeds—more than half—is absorbed into the soil, where it can stay for months, or dissolved in water, spreading hundreds of feet from the planted crop. It moves through underground drainage tile lines into streams and wetlands and is absorbed by wild plants. And because farmers buy and plant new insecticide-treated seeds every season, the amount in the environment accumulates year after year. Earthworms and other soil insects that are decomposers are exposed to it. The toxins have been found in surface waters across the Midwest, and many aquatic insects—including in the water fleas and other tiny insects and in larger ones like the larvae of dragonflies—are especially sensitive to them. And the effects can move through the ecosystem. One study in Japan found that when neonicotinoids were introduced into a rice crop, the runoff into the nearby lake killed all the zooplankton and aquatic insects at the bottom of the food chain, resulting in the collapse of the fishery at the top.

It's impossible to know the total impact that neonicotinoids have on the universe of insects, or their impact on insect life across the Midwest. Just as problematic for prairie insects is the sheer loss of grasslands, wetlands, and their other natural habitats. It is clear, however, that the impact of neonicotinoids on pollinators—the domesticated honeybee, bumblebees, solitary wild bees, and butterflies—can be profound. As a result of the risk to the insect world, Europe has banned the outdoor use of neonicotinoids. In the United States, the Environmental Protection Agency is only now beginning to evaluate their impact on all the species considered endangered after legal action brought by environmental groups forced its hand. As of 2022, the EPA had reviewed three of the neonicotinoids used in cotton, corn, soybeans, and wheat, and found that they all could pose a risk to many endangered insects. That launched another round of federal biological reviews to determine where the agricultural use of neonicotinoids would jeopardize their survival. But no matter what regulators find, neonicotinoids are unlikely to be banned the way they are in Europe. Instead, the EPA will issue a multitude of new rules that farmers have to follow. They might be allowed in some counties and not others, depending on where the endangered species or their habitats are found. There could be new restrictions on how to use them that would minimize exposure to specific species. But one way or another, it will be up to the farmers, who have little or no choice in seeds, or what goes on them, to protect the natural world from their impact. In the meantime, there is increasing evidence of the pesticide treadmill—that pests are, not surprisingly, becoming immune to neonicotinoids. The first report was published in 1996, and by 2013 there were some twenty species of crop pests that had developed immunities to the toxins, including the Colorado potato beetle.

ONE PLACE WHERE these forces all converge—the farmers, the pollinators, the pests, the insecticides, the monocultures, the treadmill, the scientists, and the food system—is in the watermelon fields of southwest Indiana. Here a small region of sandy soils is perfect for watermelons, making the Hoosier State the fifth-largest producer in the country. Watermelons are a crop that cannot grow without pollinators. And this region of watermelons, cantaloupe, pumpkins, and other specialty crops is an island in a sea of insecticide-treated corn and soybeans that stretches from Indiana north to Michigan and west to Iowa. That makes it a perfect place for scientists to study the life and death of both good and bad insects and their broken relationship with farming.

Walking through a watermelon field is an acquired skill, and Zeus Mateos-Fierro spent a summer in southwest Indiana learning to do it as efficiently as possible. Three mornings a week, starting as soon as the dew dried, he would stride across dozens of green fields with steps wide enough to cross over the rows of giant fruit, or agile enough to slide his foot in the space between them. The melons are hard to see. Their own wide leaves, weeds, and vines protect them from the sun and hide them from view. Stepping on top of a hard green rind means stumbling, falling, or, worse, sinking ankle deep into an overripe, pink stinking mess. But on a rare, cool, gray day in mid-August, he moves effortlessly. The voices of distant farmworkers singing along with Spanish music from a portable radio drift across a hill as they cut row after row of ripe watermelons. Every dozen steps or so Mateos-Fierro squats down, turns over a leaf, or peers into a yellow flower in a never-ending hunt for bugs. Every time he sees one, he notes it on the chart on his clipboard—an aphid (pest), a wasp (predator), or a bee (pollinator). Each has a role to play in the success or failure of this year's watermelon crop, and it's Mateos-Fierro's job, in his first year as a research entomologist for Purdue University, to keep score. It's been a struggle at times. Summer temperatures reach almost one hundred degrees, and the humidity is just as bad. After weeks of drought and heat, biblical midsummer thunderstorms turned the fields into thick mud. As one watermelon grower said, "We prayed for rain, and we got it." Farmers had to abandon some fields, leaving swaths of rotting melons on the ground. Mateos-Fierro is from Spain, and his accent, along with his tattoos and cluster of ear jewelry, make him stand out in this small Indiana community around Vincennes. Still, he loves the science and the bugs. He is most fond of wild pollinators, like the tiny iridescent green hoverfly that floated next to him with a load of pollen hanging like yellow pantaloons on its back legs. "Isn't he cute," he said. But, he added sadly, waving at the fields around him, there aren't many of those native insects left.

Mateos-Fierro and other scientists at Purdue have spent years studying insects in and around the fields in Vincennes, Indiana. They have watched farmers fight pests with generations of pesticides, only to see the chemicals fail over time—or kill the good insects along with the bad. Their studies have often challenged the claims of pesticide companies about the effectiveness of their products. Every summer students like Mateos-Fierro sit vigil in farm fields across the state, watching, counting, and trying to understand the secret lives of bugs. Long ago, these fields were part of America's eastern forests. The tallgrass prairie once extended into Indiana north of Vincennes. But today, the former prairie and forest landscapes look pretty much the same—a continuous expanse of farm fields, and like most of the Midwest, it's mostly corn and soybeans, which, critically, don't need insects for pollination. Corn is pollinated by the wind and soybeans are self-pollinating, meaning that the pollen in their flowers fertilizes the pistil of the same flower. Corn and soybean farmers are free to make use of chemicals to control the pests that are inevitable, including neonicotinoid-coated seeds. But watermelons, like most edible crops, do need pollinators, and in the Midwest they have to grow in what many describe as a "toxic landscape" for bugs.

Each year more than $500 billion worth of food, pretty much every fruit, vegetable, and nut, is touched by a pollinator before it becomes something we can eat. That doesn't include the 80 percent of wild plants that need them as well. Decades ago, watermelon growers didn't worry much about how their crop was pollinated. The wild insects did the job just fine. But starting in the 1970s and 1980s, wild insects declined as the land became more intensively farmed, and that was a problem for watermelon growers. If not pollinated correctly, fewer flowers turn into melons in the first place, or the fruit inside the rind splits into three separate sections, a condition poetically known as "hollow heart," and it can't be sold. In other words, watermelons don't just need pollinators—they need enough of them to produce a decent crop. Growers started bringing in domestic honeybees and commercially raised bumblebees to fill the niche in the natural world that had been vacated by wild pollinators.

That has made agricultural pollination an industry in itself, one that now produces more revenue for beekeepers than honey. Beekeepers truck their tiny workers across the country, moving them from state to state and crop to crop, depending on the season. Every summer watermelon fields around Vincennes are marked with a stack of white beehives. Many fields also have large white cardboard boxes that hold nests of wild bumblebees. Like honeybees, they are grown commercially, but are still wild. With the exception of the queen bumblebee that hibernates over winter to start a new nest in the spring, bumblebee colonies live for only one season.

Dennis Mouzin, a third-generation farmer who grows melons and other specialty crops on ten thousand acres in Indiana and elsewhere, started renting honeybee hives for pollination ten or fifteen years ago. It costs him $100,000 or so every year, but "we are convinced we can't do without them," he said. That's especially true now that seedless watermelons have become so popular among consumers, making them the biggest and most profitable crop for growers. But seedless watermelons are more difficult to pollinate, and need a lot more insects to do it right. The old-fashioned watermelons that come full of those dark seeds perfect for summer spitting contests grow on vines with both male and female flowers. Pollinators simply have to hop from one flower to the next along the vine, depositing their load of pollen as they go. Seedless varieties, however, were developed with only female flowers. In order to set their fruit, farmers also have to plant a row of a seeded variety, with their male flowers, nearby. Instead of just randomly flitting from flower to flower, pollinators must first go to a male flower on another vine, and then visit a female flower on a seedless vine farther away in order to create the fruit. And they have to transfer enough good quality pollen to fertilize enough of the many ovules in the female pistil to set it right—to prevent hollow heart. And that means a lot of flights by a lot of insects.

Watermelons also have their own destructive pests. They include tiny mites and aphids that eat the leaves and hurt the plant's ability to process sunlight; nematodes and maggots that eat the roots of young plants; and cucumber beetles that are related to corn borers eat the roots of seedlings and damage the rinds of mature fruit. These pests force watermelon farmers, and all the others who grow the food we eat, into a delicate balancing act that their corn- and soybean-growing neighbors never have to consider. They have to kill the pests that harm their crop without also killing the pollinators they need to make it grow in the first place. Or at least not too many.

Mike Horrall is another producer who has been farming near Vincennes his whole life. His dad started using bees on the family farm in the 1980s. Now Horrall, who grows a variety of crops on 2,500 acres with his three kids, relies on the honeybees, and his farm is a regular research site for the scientists from Purdue. Horrall's business, Melon Acres, operates out of a low white building set back from the main road that heads south from Indianapolis to Vincennes. On the door is an invitation to the annual Hispanic Thanksgiving celebration for the farm's Spanish-speaking workers. Horrall is a big man with a warm Indiana lilt to his voice and is quick with a joke. He's learned a lot about bees from his beekeeper, Curtis Simpson, who runs a multistate pollinating business out of Kentucky. Every spring Simpson brings up a load of bees to pollinate Horrall's and other growers' watermelon crops around Vincennes. They have a pact to do everything they can to protect the bees from the insecticides Horrall uses. For instance, thanks to the advice of Purdue entomologists he doesn't spray insecticides on a routine schedule, which used to be his practice. The regular applications killed pollinators, which were always around, but it wasn't always necessary to control the pests, which only showed up from time to time.

Now, he hires scouts to walk the rows looking for bug outbreaks, and he sprays only when they find enough to cause problems in the yield or in the appearance of the fruit. He's mindful of when the bees are flying—during the day—and so sprays at night, when they are in the hive. Simpson, who was a mechanic before he turned to beekeeping ten years ago, has also learned to navigate the ever present risk of insecticides in the area. If he sets the hives into the watermelon fields too soon—before the yellow flowers appear in spring—the bees will find someplace else to forage. It might be on wildflowers contaminated by the toxic dust that floats across the land when farmers plant corn seeds coated with neonicotinoids and other insecticides. The dust comes from the talc powder, or other lubricant, that is added to planting equipment to separate the sticky seeds so they flow smoothly through the machine. When the planters move across the fields in the springtime the talc dust, contaminated by the seed coatings, flies up in clouds that drift through the air and settle on wildflowers nearby. To avoid that risk, Simpson sets his hives in the watermelons only when Horrall tells him that the vines are in bloom, so the bees stay put. And he places the hives as far from the corn and soybean fields as he can.

Horrall used to buy melon seedlings grown from neonicotinoid-coated seeds—similar to the ones used by corn growers—to protect against cucumber beetles. But Purdue researchers found that when the seed grew into a flowering watermelon plant, there was still enough toxin in it to affect the pollinating bees. They became disoriented, one of the "sublethal" effects documented by researchers in different field environments across the country, and couldn't find their way home again. Now Horrall tries to avoid using seedlings grown from treated seeds. But the global fruit and vegetable seed industry is also dominated by three companies, and, just as is the case with corn and soybean farmers, they dictate what most farmers can buy. The same is true for other commercial fruit and vegetable seeds, as well as those grown in the nurseries that sell ornamental plants and vegetables to home gardeners. Sometimes Horrall doesn't have a choice and has to plant the ones that are pre-poisoned.

Simpson's bees, like all the bees that are used to pollinate the nation's food supply, face a multitude of health risks beyond insecticides: contagious diseases, infections, and parasites are a constant threat among apiaries across the country. But the widely used insecticides like neonicotinoids weaken the bees' ability to resist them, so Simpson does his best to baby his bees throughout the summer. He checks the hives once every two weeks, and if a hive doesn't have enough bees or if they look sick, he'll boost their food supply, replace an ailing queen to replenish the hive, or give them an antibiotic. It's his mission to make sure that the growers who contract with him get all the bees they paid for—their crops depend on it. And he only works with growers like Horrall who are mindful of his bees; it's a symbiotic relationship.

As soon as the watermelon flowers are pollinated, Horrall releases the hives for the year, and Simpson hustles them back to Kentucky. He puts them in a place far from corn and soybean fields where the bees can forage on a variety of uncontaminated flowers—goldenrod and asters are ideal—so they can get healthy again. Then he moves them to Mississippi for the first part of winter, where the bees can conserve energy in a warmer climate. Still, like all beekeepers, every winter he loses about a fourth of his hives—sometimes more. Starting in January or February, he and his hives join the nationwide migration of bees and beekeepers to California, where a majority of the country's bees are deployed to pollinate almonds, and the annual cycle of pollinating food crops around the United States begins again. Simpson says that it is his job to help his bees survive in an environmental war zone. When someone sprays at the wrong time or in the wrong place, or the wind blows the wrong way, "it really hurts us," he says. So whenever he and his son open a hive and see it swarming with healthy bees, they joke that it's like opening a Christmas present. Domesticated honeybees are fortunate to have keepers like the Simpsons, whose own economic survival depends on the survival of their bees—another symbiotic relationship.

AS EVIDENCE AGAINST the widespread use of insecticides continued to mount, scientists at Purdue wanted to find the answer to a fundamental question for growers. They knew insecticides were bad for pollinators, and watermelon needs a healthy population of pollinators. But if growers minimized their use of insecticides—both sprays and neonicotinoid-coated seeds—and managed their fields to improve the lives of all insects, would nature create a healthy balance between pests and pollinators? And how would that affect yields? In other words, could growers make more money without insecticides than with them? As simple as it sounds, it was a novel approach. Most other research on pollinator-dependent crops examined pests and pollinators separately, even though they coexist in the same fields. And getting the answers wasn't easy—it took a total of six growing seasons, and a lot of bug watching.

Jacob Pecenka, a former Purdue PhD student who'd spent his life in the Midwest and studied insects in South Dakota, ran one study. To re-create a real-world agricultural environment, he planted neonicotinoid-treated corn seeds on five of Purdue's research fields, and, for comparison, paired each with an equivalent plot about five kilometers away planted with insecticide-free corn seeds. The following year, mimicking the kind of crop rotations common around Vincennes, he transplanted watermelon seedlings into a portion of all the fields, and again divided them by insecticide use. In the fields that had previously been planted with treated corn seeds, he adopted the common practice of spraying insecticides on a regular schedule to combat striped cucumber beetles and any other pests. In the others he used an Integrated Pest Management system, meaning he scouted for cucumber beetles and sprayed them only when they reached a population level that would affect growth and yield—five or more per plant. And once a week, on the warmest, sunniest days, he and the students who were helping sat in the fields between rows of yellow flowers, and immersed themselves in the world of pollinator behavior while counting all the wild insects and honeybees that came by.

He expected to find that the pesticide-treated corn seeds did little to boost yields for farmers—which he did—because the same thing had been found in previous studies on corn and soybeans. He also predicted that minimizing insecticides on watermelons would increase the number of pollinators without hurting yields. In fact, over three seasons he had to spray only four times to control beetles in the IPM fields, compared to seventy-seven times in the fields on a routine schedule. Most likely, he said, the pests in the IPM fields were managed by nature; either heavy rains knocked them back, or they had greater numbers of predator insects that thrived in the lower-toxin environment. But he also found something he wasn't looking for, and it was completely surprising.

In the watermelon fields with minimal insecticides, it was the wild bees, not the domesticated honeybees, that were the pollinating champions. In fact, the honeybees were lollygaggers that buzzed around one flower for a long time before wandering off to the next. Many of the wild insects were all business as they zipped from flower to flower, and they showed up in droves—twice as many as compared to the fields that had been heavily sprayed. It was as if, even in the Midwest's ocean of insecticide-treated corn, every wild bee and hoverfly in the region had somehow found its way to the little islands of pesticide-free watermelon fields. The pollinators in the IPM fields visited 129 percent more flowers per minute than in the conventionally sprayed fields. The bottom line: Between money saved on insecticides and increased yields from better pollination, growers could earn $1,800 more per acre if they didn't spray.

Ashley Leach, a researcher in the same lab group as Pecenka, led a second study that seemed to indicate farmers would be better off if they didn't focus on pests at all. She grew up in a cherry orchard and has spent most of her life trying to help the people who grow food—unsung heroes in her mind. As she sees it, every growing season they have to balance all the forces of the natural world against an unforgiving marketplace. Her goal with the watermelon study was an extension of the research she often does in food crops—figuring out the lowest thresholds needed for agrochemicals. It's good for the environment, and good for the farmer.

Leach wanted to know the answer to two main questions: What would happen in a watermelon field where striped cucumber beetles, honeybees, and wild pollinators were allowed to coexist with little or no insecticides? And what kind of pollination produced the best watermelons? How many visits by a honeybee, how many visits by a wild bee, pollinating by hand, or random pollination by whatever bugs happened to come by? Her field trial used pesticides just to protect against aphids and weeds.

The study was one of the most boring things she ever did, sitting on a bucket in the hot sun along with seven other students for hours at a time, staring at the flowers in front of them. Like Pecenka, she found in the end that the most important factor was the native bees, especially the somewhat obscure *Melissodes bimaculatus,* an all-black solitary bee that did about half of the pollinating among all wild insects. It had a body built for moving pollen, Leach marveled, with little yellow hairs that coated its back legs. She also found that more visits from wild pollinators resulted in more fruit and bigger fruit than with the honeybees. The only thing that performed better was pollination by human hand—an unrealistic strategy in agricultural fields. As for cucumber beetles, they had very little impact on the crop, and spraying against them did more harm than good. Whatever damage they inflicted was vastly outweighed by a healthy population of bees and other pollinators.

Leach, now an assistant professor of crop entomology at Ohio State University, concluded that growers might need to turn their insect strategy upside down. She doesn't believe that growers who raise fruit, nuts, and vegetables will be able to do without pesticides altogether. But rather than trying to find an impossible balance between spraying for pests and protecting pollinators, growers might be far better off by managing their fields simply to attract pollinators. Wild bee conservation—planting flowers and other habitat for wild pollinators and protecting them against insecticides—might give the farmers far more and better quality produce, and higher profits. For free.

The two research projects are small, covering just a handful of farm fields in Indiana. And they focus on just one crop, watermelons, with its own unique requirements for pollination and growth. Still, they show how the food system has created its own cascade of problems—and that letting nature do its job may be a solution. But adopting that in the real world of farmers and the food business is extraordinarily difficult. Horrall has learned a lot from the Purdue researchers who walk his fields every year, and he's adopted some of their advice, including using less insecticide. But he also has to sell his watermelons by the truckload to companies like Costco, Walmart, and Target. They expect the fruit to be perfect. Because that's what consumers have learned to expect—perfect, unblemished produce in their grocery stores—without understanding what it takes to get it that way. Cucumber beetles, for example, have a way of making watermelons look imperfect, even though they don't affect the quality of what's inside. The larvae chew on the bottom of the growing melons, leaving scars, and the adults eat the green rind on the top. Quality standards set by the U.S. Department of Agriculture allow buyers to reject whole truckloads of watermelons if just a few have damage on the rind. Horrall said he loses about 3 percent of his crop from such rejections every year. Growers would rather not spray against beetles, he said, but you would hate to lose truckloads of watermelons because you missed a spray. That's what everybody fears.

The National Watermelon Association has asked the USDA to update the rules that govern appearance. Horrall says growers will start changing how they grow fruit if more research substantiates what Purdue has found about the futility of insecticides and the value of wild pollinators and it proves out in real growing conditions on real farms. In the meantime, he's experimenting with other ideas. This year he's trying a new kind of plant, one that is hand-grafted onto the root of a type of wild watermelon that is naturally impervious to root pests. It costs three times as much as the other seedlings he uses, but it needs less fertilizer and less water, and he'll use them in a field that's had a lot of bugs and doesn't produce well. Maybe it will all pencil out and produce perfect fruit.

BUT NO MATTER what they do, Horrall and his farmer neighbors still have to survive in the larger agricultural landscape with its enduring use of insecticides and other agricultural chemicals. Changing that is as large a task as plowing up and draining the prairies was in the first place. The strategy of using what is now called "ecological intensification," or leveraging natural systems like pollinators to feed the world, would also require a landscape-sized approach in building a kind of agriculture that is both sustainable and sustained by nature. To date, there is very little research or funding in figuring out how to do that, but that is the mission of Jonathan Lundgren and his colleagues at the tiny Blue Dasher Farm in eastern South Dakota. They are studying the differences between conventional farms that rely on intensive soil tillage, chemicals, and insecticides and those that, like Purdue's insecticide-free watermelon fields, leverage nature to do the same things. They are recruiting growers from a thousand farms across the country, from neighbors in South Dakota to almond growers in California, who are interested in experimenting with different methods. They want to know what works, and what doesn't, and to help farmers switch from one kind of agriculture to another. When completed, Lundgren expects to have a database built on experience that can be extended to farmers nationally.

The work has produced some convincing results in regard to insecticides and pests. For instance, one of Lundgren's students found that corn grown conventionally in fields with chemical inputs had ten times more pests than fields that had never seen insecticides. That fits in with what others have found as well. Tim Crews, of The Land Institute in Kansas, pointed out in one of his papers that despite a tenfold increase in the use of insecticides between 1945 and 1989, the amount of crops lost to pests nearly doubled to 13 percent. What that shows, Lundgren says, is that we need a paradigm shift in the way we see insects. We compartmentalize them into "the good, the bad, and the ugly," he says. Treating many as a problem has simply made the problem worse. He sees a way to rely on evolution—survival of the fittest—as a useful force in farming, rather than the endless treadmill it is now. He expects to build ecologically diverse farms, from the soil to the crops to the insects, that keep nature in balance. That's how pollinators can do the job that evolution gave them, how insects can keep one another in check, and how pests can become just another bug.

CHAPTER EIGHT

## *Water*

THE RACCOON RIVER MAKES A LAZY ENTRANCE INTO DES MOINES, completing its journey from northwest Iowa in a series of graceful switchbacks through an urban park, its olive surface shaded by locust trees and towering cottonwoods. At the edge of downtown it turns north and, on the southern bank, passes a squat concrete culvert: the main intake for the Des Moines Water Works, which supplies drinking water to the city. The scene has the pastoral calm of an English landscape painting and gives the impression of a river scarcely touched by progress or tainted by the twentieth century.

But sometime in the 1980s, biochemists working for the waterworks began to detect something wrong with the river. At a chemistry lab inside a stately brick building not far from the intake culvert, they tested river water samples and found rising levels of nitrate, the nutrient that forms as part of the nitrogen cycle. The chemical conversion, which happens naturally in soil and water, changes the harmless element $N_2$ into a compound that, in high concentrations, has been linked to premature births and several cancers.

They continued to monitor the river, and in 1989 found that its nitrate levels had shot above ten milligrams per liter, the federal government's threshold for safe drinking water. The Iowa Department of Health and Human Services ordered Des Moines to treat the water. The utility's trustees, a board of five appointed by the mayor, went to ratepayers, raised $4.1 million, and in 1992 christened a state-of-the-art facility consisting of eight massive steel tanks that could purify ten million gallons of water a day and would kick in whenever the river's nitrate content reached unsafe levels. On the day it began operation, it was the largest nitrate-removal facility in the world.

But nitrate levels in the Raccoon River kept on rising.

In 2013 peak nitrate loads in the lab's samples of river water hit twenty-four milligrams per liter, more than twice the federal safety standard. Forced to reduce their draw from the Raccoon River that summer, the waterworks trustees issued a plea to the people of Des Moines to conserve water and cut back on lawn sprinkling. The nitrate-removal plant operated for a record seventy-four days that fall.

By this time, the trustees had good reason to be alarmed. The federal government had begun regulating nitrate in 1997 after a series of public health studies linked it to several medical conditions, including a phenomenon known as "blue baby syndrome," in which an infant's blood loses its ability to carry sufficient oxygen to the body's organs. A bad case can cause a baby's skin to turn blue; a severe case can cause cell damage and even death. After federal regulations took effect, cases of blue baby syndrome became quite rare. But the scare caused medical researchers to probe for other health threats from nitrate. And before long they had found several. A study of Missouri infants born between 2004 and 2008 found an increased risk of limb abnormalities, heart defects, and neural tube defects such as spina bifida in babies whose mothers were exposed to drinking water with nitrate levels at half the federal safety ceiling. Subsequent research from a project known as the Iowa Women's Health Study found higher risk of bladder cancer and thyroid disease in women with more than four years of exposure to drinking water with nitrate levels above five milligrams per liter. In 2019 researchers writing in the journal *Environmental Research* estimated that across the country, elevated nitrate levels in drinking water caused more than 2,900 low-birthweight babies annually, 1,725 premature births, and roughly 6,500 excess cancers of the bladder, ovaries, colon, and thyroid. Together, these cost the U.S. healthcare system at least $250 million a year. But even collectively, the research found only correlations that might implicate nitrate—not the hard proof of causation required to trigger the difficult and controversial process of tightening federal health regulations.

The waterworks had a backup water source, the Des Moines River. But it faced an even more frightening problem: heavy blooms of blue-green algae in a reservoir fed by the river, caused by warm weather and excess nitrate from fertilizer. The algae grows in slow-moving water, and produces a primitive microbe called cyanobacteria, which is highly toxic to humans and animals. Unlike nitrate buildup, which can take weeks or months, cyanotoxins can appear in lakes or slow-moving streams overnight and make the water instantly unsafe to drink. In 2014, on Ohio's Maumee River near Toledo, a combination of fertilizer runoff and unusual weather sparked a massive algae bloom that turned the water a putrid green and sent cyanotoxin levels to twice the maximum allowed by federal law. The city ordered an emergency shutdown of its waterworks. Stores sold out of bottled water within hours and the governor sent in the National Guard to deliver water and ready-to-eat meals to homes across the city.

Facing these threats, the waterworks trustees were left with two choices: massively increase their spending to treat Raccoon River water or expose the people of Des Moines to grim health risks.

Des Moines's predicament was not unusual. The twin triumphs

of synthetic nitrogen and highly efficient tile drainage didn't merely foul the Mississippi River and the Gulf of Mexico, they sent large volumes of nitrate into the drinking water sources of the upper Midwest. In Iowa alone, some sixty other towns and cities had to treat their drinking water for elevated nitrates; across the Midwest thousands of private wells, mostly unregulated by state health authorities, were contaminated too. In state after state, thousands of people were discovering that their tap water might be unsafe to drink.

What set Des Moines apart was a public servant named Bill Stowe, the general manager of the city's waterworks and a guy who wasn't afraid to challenge the status quo. Stowe was alarmed by the nitrate levels and he was pretty sure he knew where it was coming from: the heavily fertilized farm fields around the river's source in northwestern Iowa and the drainage tiles that carried excess nitrogen off the fields and into local waterways.

In the spring of 2014, Stowe dispatched a team of employees to test his hunch. They drove 150 miles north and west to the upper reaches of the Raccoon River, amid the network of drainage ditches and straightened streams that crisscross the farms of the Des Moines Lobe. Parking on county roads that provided public access to private land, they held their collection bottles beneath the drain pipes that poked out of fields alongside creeks and ditches. Stopping at twenty-seven sites in three counties, they filled scores of sample flasks. Back at the lab, they found what Stowe had expected: Dozens of the water samples showed extremely high levels of nitrate—many exceeding ten milligrams and some exceeding twenty.

It was dry that year and Raccoon River ran unusually low, concentrating its contaminants as the summer weeks rolled by. The waterworks's nitrate-removal plant, which had never been used for more than a few months annually, ran for 177 days that year, costing ratepayers more than $2 million. Stowe warned his trustees that if the nitrate didn't subside, it would soon overwhelm the

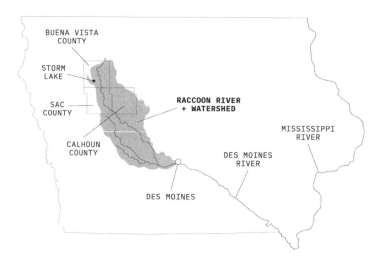

The city of Des Moines, Iowa, gets its drinking water from the intensively farmed Raccoon River watershed, where the vast majority of land is planted in corn and soybeans. The Des Moines Water Works unsuccessfully sued three county drainage authorities in the upper reaches of the watershed for failing to stop farm chemicals from polluting the city's water. *Graphic by Alexander Hage*

plant. An upgrade could cost as much as $80 million and force a double-digit rate increase for the citizens of Des Moines.

Art Cullen, editor of the *Storm Lake Times Pilot* in west-central Iowa, remembers that year well. Cullen grew up not far from Raccoon River, near the lovely lake that gave his town its name, and he treasured the little country sloughs and prairie potholes that throbbed with wildlife. But he had also seen pristine small lakes turn into muddy swamps because of soil erosion, and he was known for fearless editorials on agriculture and water pollution—work that eventually earned him a Pulitzer Prize. That year Cullen appeared with Stowe at a panel discussion organized by the Iowa Environmental Council. When their presentation ended and the crowd was thinning out, Stowe leaned over to Cullen and said, "I'm going to sue your county." Cullen knew that Stowe was giving him a big story. He remembers thinking, *This guy has no fear*.

The waterworks's lawsuit, announced in January 2015, had a

simple premise: People who pollute a river should pay to clean it up. It asked the court to order three counties along the upper reaches of the Raccoon River—Calhoun, Sac, and Buena Vista—to reduce nitrate levels running into the river through the farm drainage ditches they supervised.

Stowe and his trustees conceived the legal strategy carefully. They knew that suing individual landowners was a fool's errand. Farmers command enormous public sympathy in Iowa and, in any case, no one farmer could reengineer the hydrology of the entire Des Moines Lobe. Instead, the lawsuit targeted that obscure but powerful unit of local government: the drainage district. These descendants of Alonzo Thornton's vision had supervised the epic replumbing of Iowa farmland in the nineteenth century and continued to wield substantial power into the twentieth. They could commission dredging projects costing hundreds of thousands of dollars, levy assessments on any landowner who benefited from drainage, and even use eminent domain to acquire land they desired. They were so valuable to Iowa agriculture that their powers were enumerated in state statutes and protected by a special provision of the state constitution. Taken together, the state's three thousand drainage district boards controlled the water that fell on more than a quarter of Iowa's farmland. They replaced nature's tools for managing excess water—sloughs, bogs, and meandering streams—with local officials whose mission was to dry out the land as fast as possible.

If they had that much power to move water, Stowe reasoned, they had the power and the responsibility to clean it up.

The lawsuit struck Iowa like a thunderclap. In a state where one in six jobs depends on agriculture, where almost everyone grew up in the country or knows someone who did, no one had ever tried to make farmers pay for pollution. No one had even considered it. Farmers were widely regarded as stewards of the land and pillars of the community; their property taxes funded rural schools, their commerce supported small-town retailers.

They fed the world and earned billions of dollars for Iowa's economy in the process. The lawsuit not only challenged one of Iowa's cherished myths—the farmer as conservationist—it indicted one of the state's proudest historical achievements, the vast system of drainage ditches that had converted the hated mosquito bogs of the Des Moines Lobe into priceless farmland.

It also deepened the emerging cultural rift between Iowa's urban and rural communities. In a state known for corn and hogs, Des Moines is the urban heart, a metro area of 500,000 people and a regional center of banking and insurance. It has a clean and bustling downtown, a historic warehouse district with brew pubs and ethnic restaurants, and a well-educated and civic-minded population—many who are sympathetic to environmental causes. But to many rural Iowans, it was "the big city," a place of traffic congestion, liberal ideas, and arrogant urbanites.

Almost overnight the lawsuit attracted national attention in the conservation and public health communities. Which was no surprise. Across the Midwest more than three thousand streams and river segments had been declared "impaired" under federal law—unsafe for drinking, swimming, or fishing—mostly because of farm chemicals. Thousands of private wells across the upper Midwest had tested positive for excess nitrates. Along the Missouri River near St. Louis, outbreaks of poisonous blue-green algae, often caused by excess phosphorus and nitrogen, had forced several communities to close parks and beaches. In the Chesapeake Bay, runoff of nutrients and sediment from farms and suburban lawns was snuffing out aquatic life. A win for the Des Moines Water Works could set a legal precedent that would change environmental law nationally.

The lawsuit also underscored a crucial but little-known feature of federal environmental regulation. Agriculture is largely exempt from the nation's core water pollution law, the Clean Water Act. The law, passed in the early 1970s, focused on large "point source" polluters such as factories and sewage treatment plants, which are easy to identify and simple to regulate. The nation's three million farms—small, widely dispersed, and hard to regulate—mostly got a pass.

The Des Moines lawsuit promised to change that.

WHEN BILL STOWE took the job of general manager at the waterworks in 2012, no one figured him to be an insurrectionist. An imposing but congenial six-footer, Stowe had an easy charm, a striking resemblance to the actor Russell Crowe, and a mane of silver hair that cascaded to his shoulders. A lawyer by training and a public servant by accident, he was also an amateur philosopher who read Jesuit theology and spent his Sunday mornings cruising the Iowa countryside on a Harley-Davidson. Unlikely as it seems, Stowe had become something of a celebrity in his previous job, director of public works for the city of Des Moines. Whenever a blizzard threatened or flooding loomed, Stowe could be found in front of a TV camera reassuring residents that the levees would hold and the streets would be plowed. His fame reached such proportions that admirers started a Facebook fan page, a local company printed T-shirts with his likeness, and colleagues produced a *People* magazine mock-up featuring the snowy-haired Stowe as the World's Sexiest Public Works Administrator. His son, Liam, then a teenager, remembers high school teachers confiding that they had T-shirts with his dad's picture.

Bill Stowe, the charismatic director of the Des Moines Water Works, led a legal battle to make Iowa agriculture accountable for water pollution. *Photograph courtesy of the Des Moines Register/Gannett*

But behind the telegenic calm, colleagues recognized an intensity in Stowe, an almost missionary sense of stewardship. His wife, the Des Moines attorney Amy Beattie, recalled a night when a city levee broke and water poured into an important business district. When the phone rang after midnight, Stowe hauled himself out of bed and went out to lead the emergency crew in person. At sunrise the next morning, a local television crew captured Stowe walking the levee top, and for years Beattie remembered the heartbroken look on his face, as if he personally had failed the people of his city.

When the waterworks announced its lawsuit, Stowe became the board's chief spokesman and emissary. He spoke at Rotary meetings, breakfast clubs, and in church basements, explaining why his trustees had resorted to legal combat and why Iowans could not continue using the Raccoon River as a sort of agricultural sewer. "Small towns and our cities are heavily regulated on what they can discharge into the river," he said in one interview. "Agriculture essentially gets a pass, and that, for us, is a huge problem."

Beattie said the lawsuit struck Stowe as a moral imperative. "He thought: No one else is going to do anything about that river, so it's up to me."

The backlash was swift, brutal, and personal. Governor Terry Branstad, a Republican with close ties to agribusiness, accused Stowe and the waterworks board of "declaring war on rural Iowa." Small-town newspapers denounced the utility—and Stowe in particular—on their editorial pages. A powerful group representing big hog farmers called for a boycott of Des Moines businesses. The chair of the waterworks board, a popular Des Moines businessman named Graham Gillette, began hearing whispers that clients would desert his communications and marketing firm. Before long, a mysterious citizens' coalition produced a series of television commercials showing Stowe in the sort of grainy black-and-white footage used in political attack ads. One featured an ominous voice-over that accused Stowe of using legal tricks to conceal public documents from farm organizations and referred viewers to a website called What Is Bill Hiding.com. Stowe began getting hate mail at home. Beattie had to work gingerly with some colleagues at her law firm because they represented rural or agribusiness clients. The tension and Stowe's notoriety reached a point where the couple and their son stopped going out to eat at Des Moines restaurants.

Colleagues say Stowe was stunned by the personal nature of the attacks. And yet it's no mystery why big agriculture targeted him personally. His talent for explaining the science of water pollution in simple but compelling terms made him the perfect spokesman for an environmental crusade. His years as the city's trusted public works director had given him credibility with the people of Des Moines. A poll by the *Des Moines Register* found that 63 percent of Iowans—urban and rural alike—supported the waterworks.

Coordinating the agricultural industry's response was a Des Moines attorney and prominent Republican named Doug Gross. Gross had run for governor in 2002, then served as chief of staff to Branstad. At the time the lawsuit was filed, he worked for a prestigious Des Moines law firm and ran his own dairy farm. He also served as outside counsel to the Agribusiness Association of Iowa, a powerful trade group whose members included Koch Agronomic Services, a branch of the giant Koch brothers energy empire, and Bayer Crop Science, the world's largest seed and pesticide company.

When the defendant counties hired Gross's law firm, the battle escalated. Gross formed a legal defense fund to raise money from AAI members and farm advocacy groups, using a provision of the federal tax code that allowed them to make large contributions without disclosing their identities. Within months, the fund had raised tens of thousands of dollars and begun contributing toward the counties' sizable legal bills. Gross argued that Stowe had deliberately targeted counties with limited resources, and that he created the legal fund simply to ensure they had the means to defend themselves. All of Iowa had a stake in the litigation, Gross said, because a victory for the waterworks could place hundreds of farmers under crippling regulation. "You just can't shut down 40 percent of your state," he liked to say.

In a measure of agribusiness's influence in Iowa politics, the battle soon found its way to the state capitol. Politicians invoked the appealing image of the family farmer—modest, hardworking, struggling to feed the world while enduring the scorn of feckless urban elites. The Republican-controlled legislature took up a bill to dismantle the waterworks as an independent agency and place its duties under the Des Moines City Council, some of whose members had criticized the lawsuit. A year later, the legislature would strip funding from two research centers at Iowa State University that promoted conservation farming and published widely cited research on nutrient pollution in the state's rivers.

The backlash was no surprise, given agriculture's outsized role in the state's economy. Although it's not a particularly big state, Iowa dominates production of the nation's major commodity crops—it is the nation's leading corn producer and second-leading producer of soybeans. It slaughters more hogs than any other state, produces more eggs for commercial sale, and exports more grain and animal products to overseas markets. Agriculture and agribusiness account for more than a quarter of Iowa's economic output—five times the national average—or roughly $72 billion annually. Iowa is a state where people pay attention to the price of corn, where one in five state legislators is a farmer, and where hogs outnumber people by nearly twenty to one. Its fertile soil can fetch more than $10,000 an acre, so that even a modest farm can be worth more than $1 million, and successful farmers spend their winters in Arizona or Florida. Quite apart from its economic contribution, farming is deep in the state's DNA: One in three Iowans attends the state fair each year, many can tell you the difference between a Holstein and a Brown Swiss, and many have a favorite uncle, cousin, or grandparent who still runs the family farm.

But Stowe hadn't just affronted the state's farmers, he had provoked an enemy that is arguably the most powerful farm organization in America, the Iowa Farm Bureau. Farm bureaus sprang into being in the early 1900s as home-grown membership organizations to help farmers during a populist era that also saw the creation of drainage districts, co-op grain elevators, and the Farmer-Labor Party of the United States. They circulated advice on improved methods of crop production and animal husbandry, while pooling farmers' buying power for seed, fertilizer, and implements. But by a quirk of commercial history, the Iowa Farm Bureau had developed a large for-profit insurance branch, which by 2015 gave it assets of more than $10 billion and an operating budget of $89 million—more than twice that of its parent organization, the American Farm Bureau Federation. At the time of the lawsuit, it had an investment portfolio worth on the order of $100 million, with major holdings in agribusiness giants such as Monsanto and Tyson Foods. Its weekly newspaper, the *Spokesman*, reached 140,000 Iowans, more than twice the circulation of Iowa's largest daily, the *Des Moines Register*. Its officers typically hold seats on advisory boards at Iowa State University; at one time, the Iowa Farm Bureau president Craig Lang chaired the university board of regents. The farm bureau has a branch office in virtually every town in Iowa, giving the bureau eyes and ears in every county. Its reach is so broad that many describe the bureau as a fourth branch of Iowa government. Farm Bureau officers saw Stowe's lawsuit as an attack on farmers, a threat to the future of agricultural drainage, and, perhaps, as a legal precedent for greater environmental regulation of agriculture.

In recent years, with a membership that has grown increasingly affluent and increasingly Republican, the Iowa Farm Bureau had also taken on a decidedly conservative political tilt. In 2018, its political action committee gave $291,000 to state candidates, with 94 percent of the money going to Republicans. In 2018 it donated $50,000—a huge sum by Iowa campaign standards—to Kim Reynolds, the Republican candidate for governor, and $50,000 to Mike Naig, the Republican candidate for secretary of agriculture. Farm Bureau lobbyists were also hugely influential at the state capitol, especially when agriculture's priorities clashed with urban interests. In 2002, when many small Iowa communities were expressing alarm at the stench and pollution threat from large animal feedlots, it lobbied heavily and successfully for legislation to exempt livestock operators from municipal zoning rules; in 2017 it won an expansion of that law to shelter feedlots from civil lawsuits.

The Farm Bureau's role in the waterworks's legal battle might never have come to light if not for Art Cullen's *Storm Lake Times Pilot,* the tiny biweekly newspaper he founded in 1990 with his brother, John. The paper, based in Buena Vista County, operated out of a converted pole barn near the railroad tracks, an office where sleek Apple computer terminals sat on chipped wooden desks and a friendly dog often wandered through the newsroom. Operating with a tiny but talented news staff, the *Times Pilot* had the independence to take unpopular positions. When the waterworks announced its suit, Art Cullen and his son Tom, the paper's lead reporter, began asking two obvious questions: How much money would the counties have to pay in legal costs, and would taxpayers wind up footing the bill? The answer to the first question turned out to be eye-popping. Billing records obtained by the *Storm Lake Times Pilot* showed that the counties had hired three local law firms as well as a prominent Des Moines legal practice and a Washington, D.C., firm that specializes in Clean Water Act litigation. The legal bills had reached more than $300,000 within a year. The Washington firm alone, Crowell & Moring, was billing as much as $29,000 per month.

Buena Vista County officials assured Tom Cullen that local taxpayers wouldn't bear the cost of the litigation because of the legal defense fund assembled by Doug Gross. When the Cullens asked who was contributing to the fund, the story only got more interesting. The Cullens reported that officials of the Agribusiness Association of Iowa had met with executives from Koch Agronomic Services and Monsanto, the agribusiness giant that manufactures Roundup herbicide—both companies with a big stake in federal environmental regulation—about the time the legal fund came into being. When county officials refused to divulge the names of donors, the newspaper threatened to sue under Iowa's public records law, arguing that citizens deserved to know who was paying their county's legal bills. Rather than face a lawsuit that would have exposed the donors to scrutiny, the three counties withdrew from the legal defense fund and sought different financial allies. Two others quickly stepped in: the Iowa Farm Bureau and the Iowa Corn Growers Association.

Art Cullen thought it was an outrage that public officials would conceal the identities of the donors. "To me it was a question of who controls the decisions of the [county] board—agribusiness or the voters of Buena Vista County," he said later. As they pored over a batch of documents furnished under their open-records request, the Cullens found one that vindicated their suspicions. A letter from the Iowa Farm Bureau and the Corn Growers Association stipulated that in exchange for financial support, the counties would never concede in court that farmers were the source of the Raccoon River's nitrate contamination. In other words, big agriculture was staving off any attempt to link farmers to water pollution and avoid setting a precedent under the Clean Water Act—and paying three small rural counties to stand as the front line of its legal defense.

A SOLUTION TO the Midwest's drinking water problems lies no farther than a quiet stretch of the Raccoon River midway between Storm Lake and Des Moines. There, near its headwaters, the slender river runs through gently rolling farmland before cutting east underneath U.S. Highway 71, a few miles below Sac City. After a week of sleet and snow, the water runs high and fast on a gray afternoon in early winter. On the northern bank, a tidy black cornfield rolls right down to the river's edge, the dark Iowa soil cut through with neat rows of golden corn stubble. On the southern bank a broad, tangled strip of grass and brush separates the river from the nearest plowed field. This buffer strip of deep-rooted prairie grasses and shrubs, about thirty feet wide, holds the soil in place and absorbs farm chemicals before they can reach the river.

Simple buffer strips like this, which mimic the environmental benefits of native prairie, could go a long way to protecting Iowa's surface waters. Researchers at Iowa State University were conducting a series of experiments that showed that such prairie strips, even at very modest size, could dramatically reduce the amount of tainted water running off crop fields. Other research showed that re-creating the small wetlands that once dotted the tallgrass prairie could collect nitrate-contaminated water from underground tile lines and filter out most of the fertilizers and pesticides before they reached the streams. In 2011, the state legislature commissioned a team of researchers at Iowa State University to analyze the state's nitrate problem; after two years of intensive study they issued a report concluding that Iowa could cut nitrate and phosphorus pollution by 45 percent—a goal widely accepted by environmentalists—using proven strategies such as buffer strips, newly created wetlands, and conservation tillage techniques.

The report became a touchstone of Iowa's water pollution debate for the rest of the decade. But despite its encouraging recommendations, its defining feature was what it left unsaid: It proposed no mandates on landowners and no deadlines for state regulators. Cleaning up Iowa's water would be voluntary. In a way, this was a legacy of nineteenth-century settler colonialism: Land is private property and the government had scant role in telling owners how to treat it. The report's emphasis on voluntary efforts, however, made it a hit with Iowa's farm groups, who invoked it constantly in following years. Mandatory buffer strips or cover crops, they said, could cost a farmer tens of thousands of dollars and put many out of business. Constructing new wetlands on farmers' land could mean taking profitable acres out of production, another blow to rural incomes. And reengineering the tile drainage of the Des Moines Lobe, a hugely productive agricultural landscape built over decades at an investment of tens of millions of dollars, they said, was simply unrealistic.

In the years after 2011, Iowa did build dozens of new conservation wetlands after the Iowa State report—a shared public undertaking that reversed part of the state's historic drainage project, and farmers did expand their use of conservation tillage practices. But the voluntary approach proved utterly inadequate to the scale of Iowa's challenge. In a study published a few years later, the Iowa Environmental Council found that scarcely 3 percent of Iowa's cropland was protected by buffer strips and that construction of conservation wetlands had actually slowed down after the report's release. At that rate, the authors concluded, achieving clean water in Iowa would take ninety years, perhaps longer.

Bill Stowe was done waiting.

In March 2015, a Des Moines attorney named Richard Malm formally commenced the waterworks's lawsuit, laying out an argument that was both simple and ambitious. The suit argued that American agriculture had changed profoundly since the passage of the Clean Water Act, and that exempting farmers had become obsolete and irresponsible. It asked a federal judge to order the three counties and their drainage districts to reduce nitrate discharges into the Raccoon River and comply with federal law or else pay the waterworks for the cleanup costs.

Malm's case rested on an argument that environmentalists had been making since the 1970s as they watched the evolution of American farming. A generation earlier, a typical midwestern farm had resembled the operation of Randy Kramer's parents. It was small and diverse—perhaps three hundred acres of crops, a few dozen cows, a henhouse, and some pigs. To feed their livestock most farmers grew some oats and alfalfa, crops that nourish the soil and require little fertilizer. Rural Main Streets bustled with commerce and the school buses were full of farm kids. But by 2000, the little-red-barn homesteads had mostly given way to large, capital-intensive operations, much like the Kramers', typically partnerships involving siblings or cousins who had combined several smaller farms. Under this system, a single family might farm one thousand to two thousand acres, plowing their fields with half-million-dollar tractors, relying on synthetic pesticides and fertilizers, and specializing in the two crops made most profitable by federal subsidies—corn and soybeans. Between 1940 and 2015, farmers' use of nitrogen fertilizer rose more than thirtyfold, according to researchers at Iowa State University. And although fertilizer use per acre began to level off around 1990, there were more and more acres devoted to corn and soybeans, the most chemical-intensive commodities. At the same time, huge new tractors and multi-row combines made it possible—even necessary—for farmers to expand their acreage, driving a steady increase in the number of wetlands drained to produce arable land. This industrial model has produced a huge increase in the productivity of American agriculture and a supply of cheap food for the world's consumers, but it also hugely increased the volume of farm chemicals entering the nation's streams and rivers.

As a result, fifty years after the passage of the Clean Water Act, agriculture had surpassed industry to become the leading source of pollution in the nation's rivers, according to the EPA. Roughly one-third of the nation's river and stream miles had excessive levels of nitrate or phosphorus, according to one federal survey, and eighteen thousand miles of rivers and streams had excessive pesticides. More than 20 percent of rural private wells surveyed by the U.S. Geological Survey in 2010 had nitrate levels so high the water was unsafe to drink. Malm and his team argued that the lawsuit was a chance to hold agriculture accountable.

The waterworks's lawsuit made a second argument with profound implications for federal pollution law. First, they noted that the sort of agricultural drainage perfected in Iowa a century earlier carried water off farm fields to discharge pipes near open streams and ditches. At these outlets the water and its contaminants could be sampled, tested—and controlled. Crucially, researchers at Iowa State University had demonstrated that most of the nitrate flowing into Iowa's rivers was not carried by runoff—water running off the surface of a field in a rainstorm—which is specifically exempted by the Clean Water Act. Instead, it ran in water siphoned out of the soil through tile lines and discharged by collecting culverts—not unlike an urban storm sewer system. In other words, they argued, farm drainage systems had become "point source" polluters, like factories, and should be held accountable by the federal law.

The implications of that argument were staggering: The Clean Water Act would finally, after fifty years, apply to agriculture. Farmers might have to reduce the volume of water flowing through their drain tiles and cut their use of chemical fertilizers. That, in turn, might threaten a revenue stream worth more than $50 billion annually to agribusiness giants such as Bayer Crop Science and Syngenta. A nation of consumers who took cheap food for granted might confront higher bills at the supermarket as grocery prices reflected the cost of protecting water. "Had we prevailed, it would have caused a lot of changes in a lot of places," Malm would later say.

But the federal judge hearing the waterworks's suit wasn't quite ready to consider such a profound argument. Under a common judicial standard known as "redressability," U.S. District Judge Mark Bennett said the waterworks would first have to demonstrate that drainage districts had the authority under Iowa law to fix the pollution problem. That, he said, was a question for the

Iowa Supreme Court. He paused the lawsuit while asking the state's high court to issue findings of law.

Several months passed while the justices considered Bennett's request, and the makeup of the court left Stowe feeling increasingly nervous as the weeks passed. Several years earlier, in a landmark decision, the high court had granted the right of marriage to Iowa's same-sex couples, a ruling that triggered a political firestorm across the state and a well-funded campaign to recall the more liberal justices. Three were voted out of office in the next judicial election, and Governor Terry Branstad replaced them with jurists known for their conservative views. When the waterworks case landed before the seven-member panel, two of its liberal-leaning justices recused themselves for reasons the court did not disclose. Of the remaining five justices, Branstad's appointees held a 3–2 majority.

When the court finally handed down its findings, it was clear that Stowe's argument captured the sympathy of Chief Justice Mark Cady, who had been appointed by Branstad but sometimes broke with the court's more conservative members. In an emotional opinion demonstrating how close Stowe came to changing the course of Iowa history, and perhaps pollution law nationally, Cady wrote: "It is abundantly clear that Iowa law did not . . . develop with the thought that a drainage district could be a polluter. [But] one of the fundamental principles of law is for remedies to be available when we discover wrongs. Pollution of our streams is a wrong. If the drainage district has the power to build a drainage system, it necessarily has the power to build it in compliance with laws affecting public health."

And in a sad coda reminiscent of the Sauk chief Black Hawk, Cady added, "This state is blessed with fertile soil, vast expanses of teeming wilderness, and an overwhelming abundance of fresh water. The role and purpose of drainage districts in Iowa is important, but no more important than this state's enduring role of good stewardship. As every farmer knows, the work is never done."

But the chief justice did not prevail. The remaining three-justice majority rejected Malm's central argument, and to read their findings is to understand the exalted place that agriculture holds in Iowa history and the remarkable status granted to agricultural drainage. Writing for the court's majority, Justice Thomas Waterman quoted Iowa law: "The right of a landowner to place tiles in swales or ditches is necessary in order that low and swampy lands may be reclaimed, and a denial thereof would be productive of incalculable mischief." Waterman also cited nine decades of case law in which Iowa judges had granted drainage boards wide immunity from lawsuits brought by neighbors, businesses, and other plaintiffs who wanted the courts to rein in their behavior.

Then, in a crushing rebuke to the waterworks's case, Waterman continued: "The drainage district is empowered [by state law] only to restore, maintain or increase the flow of water through the drainage system. . . . Drainage districts have no other function, power, or purpose."

Two months later, in a ruling that leaned heavily on Waterman's findings, the federal court issued a swift coup de grace. The judge dismissed the waterworks's lawsuit, writing that Des Moines "may well have suffered an injury, but the drainage districts lack the ability to redress that injury."

Bill Stowe got the news at his office on the waterworks campus. Jennifer Terry, then the utility's external affairs director, remembers him walking into her office, sitting on the edge of her desk with a crestfallen look, and muttering, "Now what are we going to say?" The waterworks trustees, worn down by two years of civic warfare, voted against filing an appeal. Three years later Bill Stowe was diagnosed with pancreatic cancer and died a few months after. His battle to change environmental law and make agriculture accountable for its pollution was over.

The following few years were unusually wet in northwest Iowa, a lucky break for the people of Des Moines. Heavy rains created high flows in the Raccoon River, diluting its nitrate content to the point where it was considered safe to drink. Plans to build another costly nitrate treatment plant were temporarily shelved. Then drought struck in 2021. Water volumes in the Raccoon River dropped sharply and nitrate concentrations shot up. That summer the waterworks's nitrate-removal machinery ran constantly and Des Moines came perilously close to running out of drinking water. Neutered by the courts in their effort to clean up Iowa rivers, the waterworks trustees decided they had to find another water source and contracted to dig a huge new set of wells tapping a large regional aquifer. The new wells will cost the ratepayers of Des Moines and neighboring communities $150 million—and still will not represent a permanent solution to the problem of tainted water. Jerry Hatfield, a respected USDA scientist whose research helped link river pollution with changing farming practices, pointed out the tragedy of the Des Moines Water Works. The new wells will be a quick and expensive fix for the city of Des Moines, he noted, but they will do nothing to help Iowa's environment or the country's tainted waters. Like Des Moines, dozens of other towns up and down the length of the Mississippi River with nitrate-tainted water were left on their own in their increasingly difficult struggle to protect drinking water for their citizens. Their best chance to modernize the nation's water pollution laws and hold agriculture accountable had failed.

In addition to authorizing the new wells, the utility board also began taking steps to protect itself financially and politically. In late 2023, it formally agreed to become the Central Iowa Water Works by merging with three other neighboring water utilities, two rural water systems, and seven municipal water departments. Together, they will share infrastructure costs and the region's finite water supply from rivers and wells. Perhaps more important, it would strengthen Des Moines's political hand by enlarging the coalition of elected officials and civic leaders with a stake in sharing clean drinking water.

At the same time, pro-agriculture legislators were taking action of their own. In 2023 the Iowa legislature passed a budget that killed $500,000 in funding for a network of water quality sensors operated by the Iowa State University's Hydrological Sciences Group. The sensors had supplied critical data demonstrating the high concentrations of nitrogen and phosphorus in Iowa's rivers; they also revealed Iowa's failure to abide by a multistate compact to reduce farm pollution that pours into the Mississippi River. Republican legislators argued that Iowa would be better off putting money in the hands of farmers who could adopt conservation practices on their land. To scientists and environmental groups, it looked more like the state's leaders would rather not be inconvenienced by the facts.

THE DEFEAT OF Bill Stowe's insurrection had elements unique to Iowa, but it also offers a case study in the way big agriculture has parried conservation efforts all across the American prairie. Where legislatures have promoted buffer strips and conservation wetlands, the measures are largely voluntary. Where mandatory, the rules are poorly enforced. When states have banded together to clean up an entire watershed, such as the long-standing compact of governors along the Mississippi River, progress stops and starts every time a new governor takes office. Even progressive Democrats have failed the test. In a 2014 compromise that is infamous among environmentalists, President Barack Obama's Environmental Protection Agency and Department of Agriculture asked each state in the Mississippi River/Gulf of Mexico Hypoxia Task Force to write its own pollution-reduction strategy—all of them voluntary. All of them unenforceable.

This stalemate is all the more frustrating because, as Bill Stowe knew, saving grassland ecosystems to protect drinking water is not impossible. A 2010 compact among states in the watershed of the Chesapeake Bay showed that determined coalitions can overcome the legal barriers. There, the EPA and six eastern states united to turn back five years of legal challenges by national farm groups and agreed on specific goals and timelines to reduce farm pollution that was killing the bay. Ultimately, they prevailed at the U.S. Supreme Court and, though progress is slower than anticipated, today oysters, birds, dolphins, and other aquatic life in the Chesapeake are returning in ever greater numbers. Nor does restoring clean water have to cripple a region's economy, as many Iowans feared. Iowa State University researchers have calculated that the state could achieve its clean water goals—improving human health, cleaning up the Mississippi River, and saving money for local water utilities—using conservation wetlands for roughly $87 million annually, a sum that amounts to about one-tenth of 1 percent of the revenues that agribusiness generates in Iowa each year.

But the evolution of agriculture and the destruction of the prairie ecosystem have created an environmental paradox that America seems incapable of solving. Although science has firmly connected water pollution to the loss of grasslands and the expansion of row crops, the nation's political leaders seem unwilling to acknowledge the uncomfortable linkage. Requiring farmers to change practices on their own land is politically unpalatable in a country that reveres private property. But crafting a public solution—adequate federal funding to build buffer strips and conservation wetlands—has also proven to be beyond Congress's will. The combination of private property and technological progress that transformed Iowa in the 1800s created so much wealth that the nation seems incapable of addressing the consequences two centuries later.

As a result, hundreds of communities across the Corn Belt—many of them with far fewer resources than Des Moines—have water that is unsafe to drink. Thousands of private wells across the rural Midwest produce water tainted by fertilizers and pesticides, and most states lack any systematic way of testing private wells. Thousands of lakes and rivers are unsafe for swimming and fishing. Nutrients from a dozen states still flow into the Mississippi River by the thousands of tons each year, and the Gulf of Mexico Dead Zone remains as big as ever.

And the public pays twice, once to subsidize ever greater crop production with federal farm supports and again to clean up the pollution that results.

The Des Moines lawsuit was a chance to break that stalemate for Iowa, and perhaps nationally. But a state that had helped pioneer the transformation of the prairie could not, when asked by its own citizens, stop the destruction that resulted. Along with bountiful crops, the rich soil of Iowa has also produced powerful agricultural organizations that are now just as deeply rooted as the prairie grasses they replaced. It has created so much wealth, so much regional pride, such a national mythology of conquest, that reversing a failure that is continental in scale might not happen until Americans literally find themselves without safe water. Bill Stowe succeeded in throwing a national spotlight on problems that most Midwesterners would rather ignore. But in the end, exposing the problem wasn't enough to solve it.

CHAPTER NINE

# *Plow II*

JIM FAULSTICH CLIMBS FROM THE CAB OF HIS PICKUP TRUCK, GRABS A pair of binoculars, and makes his way down a gravel track along a row of cottonwood trees. At the end of the path a line of birdwatchers with viewing scopes are scanning a grassy meadow set aglow by the setting sun. In the near distance, they spot a flock of curlews wading in a shallow pond. Beyond them, where a line of conifers marks the back of the meadow, a family of deer grazes lazily in the golden light while looking up watchfully at the humans. Overhead, a flight of barn swallows swoops and dives, taking their dinner of insects out of the humid air. A chorus of songbirds provides the evening's soundtrack: meadowlarks, kingbirds, warblers, finches, bobolinks, and killdeers. Before darkness falls, the visitors count more than thirty bird species, erupting in *oohs* and *aahs* when someone spots a rare passerine or a flock of black-capped night herons.

The visitors have gathered this July evening for a grassland birdwatching tour, and they came to Jim and Carol Faulstich's Daybreak Ranch because it is a sort of South Dakota prairie Eden. If the word "ranch" conjures up corrals, Stetsons, and cowboys slapping dust from their chaps, it does an injustice to the Faulstiches' property: ten thousand acres of lush pastures, windbreaks of pines and cottonwoods, and meadows where Red Angus cattle graze in the company of jackrabbits, prairie dogs, whitetail deer, fox, coyotes, and an aviary of songbirds.

The abundance of wildlife on this ranch is no accident: Four decades ago Jim and Carol Faulstich decided that ranching purely for output—more cows, more acres—had no future. Stewardship of natural resources—grass, soil, water, wildlife—would replace pure production as the goal of their operation. They adopted the strategy in order to survive the devastating farm crisis of the 1980s. It turns out that working with nature, rather than fighting it, saves ranchers a lot of money—much the same way that Jonathan Lundgren and Indiana's watermelon growers demonstrated that natural approaches to pest control can save money and reduce environmental damage. At Daybreak Ranch, they began rotating their cattle frequently from pasture to pasture to give the grass periods of rest; healthier grass allowed the Faulstiches to spend less money on commercial livestock feed and chemical fertilizers. Today their pastures are a wild but lush mix of mostly native grasses and flowers: Stands of big bluestem wave above a shorter carpet of western wheatgrass; needlegrass grows in feathery tussocks, providing forage for deer and other wildlife. Patches of alfalfa restore the ground's nitrogen stocks, and native sunflowers and asters dot the fields with flecks of yellow and blue. The same techniques have enriched Daybreak's soil; it now stays moist and fertile even through dry spells like the drought of 2021, a summer when many ranchers had to buy feed or sell off cattle at disastrous prices. A flourishing population of pheasant and deer, which find habitat in the tall grass and windbreaks, has made Daybreak a destination for annual hunting parties that now furnish one-third of the ranch's revenue. And this is the ultimate goal: The ranch is surviving at a time when many are failing—and so is the grass.

But the strategy has also turned the Faulstiches' land into a festival of nature. Cattle, corn, and sunflowers still help pay the bills from one year to the next. But the foundation is a healthy grassland ecosystem, a place that echoes the riot of diversity that reigned on the Dakota prairie three centuries ago.

Jim Faulstich is seventy-three years old and lives in the ranch house where he grew up. A broad-shouldered man who looks like

Jim Faulstich, who raises cattle near Highmore, South Dakota, has become an honored leader in grassland conservation on the Northern Plains. *Photograph by Dave Hage*

he could still wrestle a bull calf to the ground, he now spends more time at a desk crowded with papers and correspondence. He gives talks on conservation ranching, hosts gatherings of ranchers and ecologists, and travels regularly to Washington, D.C., to speak with members of Congress. He also serves as vice chair of the South Dakota Grassland Coalition, a group of ranchers and academics determined to keep the grass, as they say, green side up.

Faulstich describes his mission with a conviction as tough as his weathered hands, yet he and Carol realize that they themselves have become an endangered species. In the last two decades, the same row-crop agriculture that conquered the eastern tallgrass prairie of Illinois, Iowa, and Minnesota in the nineteenth century has resumed its march west. Corn and soybeans have become more profitable than cattle, largely because generous federal subsidies and breakthroughs in hybrid seeds and herbicides have made crop farming practical in the dry Dakotas. Now the plows are slicing open the mixed- and shortgrass prairies of the western plains, a place that has been in grass for thousands of years. The result? Ecologists say the North American grasslands are among the most endangered ecosystems on Earth.

Sitting at his dining room table, finishing a lunch of chicken and vegetables produced within view of the kitchen window, Faulstich ticks off the ranches that have vanished in the last few years. A neighbor just to the north, his pastures plowed up and planted in corn. The next ranch east, now in corn and soybeans. The next ranch south and east, which fell to the plow five years ago. Down the road, the Faulstiches' daughter, Jacquie, and her husband have the last remaining livestock operation in Douglas Township, once the heart of South Dakota cattle country.

South Dakota is the epicenter of this new prairie plow-up. Between 2008 and 2016, nearly one million acres of South Dakota grass—an area bigger than Rhode Island—were broken and plowed for crops, and more grassland is vanishing every year. Geographers Christopher Wright and Michael Wimberly at South Dakota State University say the plains states have not changed so dramatically since John Deere's tractor and Cyrus McCormick's combine rumbled into town. While South Dakota has been a hot spot for grassland conversion, it is not an anomaly. The World Wildlife Fund has documented a rapid expansion of cropland from Texas in the south through Kansas and Nebraska and into Alberta and Saskatchewan—well over one million acres a year between 2015 and 2021, counting the United States and Canada. The group's annual Plowprint Report uses a slightly different methodology from that of academic researchers like Tyler Lark, but it has reached the same broad conclusions: There has been a huge expansion of corn, wheat, and soybeans into the remaining shortgrass prairie since the early 2000s.

The Faulstiches see the consequences every time they drive to town in nearby Highmore. Neighboring fields are dotted with boulder piles, a sign that they've been cleared for the plow. Halfway to town, a massive sprayer tractor sits in a field waiting to apply a pesticide, its arms folded back like the wings of a dragon. On a clear blue Saturday morning in July the song of warblers is drowned out by the buzzing engine of a bright yellow crop duster that dives and rolls while strafing a nearby cornfield with herbicide.

Jim Faulstich drives these roads noting the damage. Just west of their ranch lies a field of corn stubble where water gushed off the compacted land after a recent storm, carving ephemeral gullies in the dirt and leaving white streaks where the rain leached salt from the mineral-rich soil. The field is plowed right up to the roadside, every acre converted for profit. Faulstich says the highway department has had to close this route twice since the plows came through—once because water gushed off the field after a storm and flooded the road and once when a windstorm whipped up a blinding dust cloud from the exposed soil.

"No fences," he says, sounding like a man who just attended a friend's funeral. "That means they'll never put livestock on that ground again."

The same symptoms have begun showing up across South Dakota in the last decade. More than half the state's rivers and streams have become too polluted for swimming and nearly one-fifth are unsafe for drinking, mostly because of more chemicals for row crops and less grass to absorb tainted water—according to the U.S. Environmental Protection Agency. Dust storms that recall the 1930s are no longer rare, a result of plowed fields lying bare for months of the year. Regional flooding has become more frequent and more severe without the spongy soil of grass pastures to soak up the rainfall from big storms.

But the most disastrous consequence is invisible: The plows are destroying one of the world's last great buffers against climate change.

Like other plants, grasses inhale carbon dioxide from the air around them. But grasses sequester that carbon far underground, with roots that reach eight to twelve feet deep, and cultivate a soil chemistry that locks up the carbon for decades or centuries. (Trees can do much the same, but here on the arid prairie forests will never thrive.) Because prairies have been doing this since the last Ice Age, ecologists estimate that grasslands hold one-third of the Earth's terrestrial carbon stocks, more than forests and the atmosphere combined. The biologist Jo Handelsman estimates that the Earth's soil now stores more carbon than humans have released since the dawn of the Industrial Revolution.

Plowing this land unleashes a cascade of devastating climate effects. When broken open, the ground releases those ancient carbon stores into the atmosphere. Underground "aggregates," tiny lumps of soil that sequester carbon, crack open and release their contents. The underground web of roots and fungi, which hold soil in place and allow plants to process carbon dioxide, are shredded by the plowshare. Exposed to a new oxygen supply, tiny underground microbes that process carbohydrates go into overdrive and produce more carbon dioxide. The extra oxygen also stimulates a class of bacteria that dissolves soil aggregates, releasing more carbon. By one estimate, plowed grassland releases more than one-quarter of its stored soil carbon, most of that in the first five years. In all, midwestern grasslands and wetlands that were plowed up between 2008 and 2016 released more than fourteen million metric tons of carbon annually, according to Tyler Lark and his colleagues at the University of Wisconsin—the equivalent of building thirteen new coal-fired power plants a year. "The thing is, this is essentially irreversible damage," Lark says. "It would take ten thousand years to rebuild that lost carbon stock."

The row crops that replace grass only compound the problem. It's true that corn and soybeans also inhale carbon dioxide from the air, but they sequester much less of that carbon in underground roots than perennial plants such as prairie grasses. Moreover, much of their carbon is released again when the crops are harvested. In addition, growing these crops requires large quantities of fossil fuels. Tractors, pickups, and combines guzzle diesel fuel, and making synthetic fertilizer under the Haber-Bosch process consumes large quantities of natural gas, both as a feedstock and a heat source.

It might be a fair trade-off if these acres made an invaluable contribution to the world's food supply, but they don't. The western grasslands plowed up in the last two decades are not good farmland—rainfall is inconsistent, droughts are frequent, the growing season is short—which is why early farmers left them as pastures in the first place. Tyler Lark and the geographer Seth Spawn have estimated that the western Corn Belt acres plowed up since 2008 produce roughly 10 percent less than the average acre of American cropland, and far less than the rich soils of Iowa and southern Minnesota. They also tend to be more hilly than prime farmland and more prone to erosion. A telling indicator from federal farm statistics confirms Lark and Spawn's findings: Between 2008 and 2011 counties with the highest rates of land conversion received twice the national average in crop insurance payouts for failed harvests, $8.3 billion in total. It's as if nature were saying it was foolish to plant corn and soybeans here in the first place.

Worse, breaking these grasslands is not a one-time loss. It not only releases their stores of carbon, it removes them as a carbon sink that could capture future greenhouse gas emissions. Grass preserves 60 to 70 percent of its carbon belowground, permanently. And once plowed, prairies require a century or more to recover the original plant diversity and productivity that allow them to sequester so much carbon, according to the ecologist Forest Isbell at the University of Minnesota.

The ranchers' coalition is just one of many projects that aim to protect or restore grasslands in the Dakotas, and many of them are taking place on tribal land. South Dakota alone is home to nine Indian reservations, some of them remnants of the Great Sioux Reservation. Together they cover millions of acres, and much of that land remains in pasture and prairie. At the Pine Ridge Reservation in southwestern South Dakota, a tribal nonprofit called Thunder Valley promotes regenerative agriculture and respect for the land as part of a larger mission to establish Lakota self-sufficiency and self-determination. Its executive director, Tatewin Means, an attorney and Stanford University graduate, says that the Lakota people embraced what we call conservation for centuries before the environmental movement came along. "It's not a new idea, it's Indigenous knowledge," Means says. "Our very identity held that the natural world is not something to be measured, it doesn't exist merely for my consumption." At the Lower Brule Sioux reservation two hundred miles north and east in central South Dakota, tribal conservationists recognized more than a decade ago the crucial role that grasslands play in supporting wildlife and buffering climate change. Over fifteen years, using federal conservation money and privately raised funds, they acquired and restored eleven thousand acres of prairie, mostly by plowing up cropland and replanting it with native grasses and flowers. The project's leader, a PhD biologist and Lower Brule Sioux member named Shaun Grassel, recognized that other reservations could follow the Lower Brule example if there was an organization to coordinate projects, raise money, and provide technical assistance. In 2021, in partnership with the World Wildlife Fund, he and several other tribal biologists formed the Buffalo Nations Grasslands Alliance to increase and diversify funding for tribal conservation projects. Today the group works with more than a dozen Plains tribes and derives funding from philanthropic sources such as Ted Turner's foundation and the First Nations Development Institute.

Thunder Valley and the Buffalo Nations alliance in turn are just two examples in a growing movement of tribes using twenty-first-century techniques to protect land their people have occupied for centuries. In central Minnesota, on the eastern edge of the prairie, the White Earth band of Ojibwe have used treaty law and federal environmental regulations to impose limits on water pumping from an aquifer shared by tribal and nontribal landowners. East of there, close to Lake Superior, the Fond du Lac band of Ojibwe employs a large staff of environmental scientists and attorneys, and has filed a series of lawsuits that protect nearby wild rice waters and forced state regulators to crack down on water pollution. Grassel notes that tribes have played a disproportionate role in the efforts by federal agencies and environmental groups to protect threatened prairie species such as bison, black-footed ferrets, and the swift fox. "Tribes can control what happens on very large blocks of land," Grassel says. "And we have intimate knowledge of this land—it gave us food, medicine, clothing for centuries. For us, this is a way of giving back."

If these projects succeed they could help solve what is almost certainly the most pressing environmental problem of the twenty-first century: climate change. Many ecologists now argue that, among all the world's tools to buffer greenhouse gases, simply keeping grasslands intact is one of the simplest and most powerful. A 2020 study published in the journal *Nature* concluded that restoring 15 percent of the world's plowed grasslands in priority areas could sequester nearly one-third of the increase in atmospheric $CO_2$ that has occurred since the Industrial Revolution. Another pair of studies suggests that the United States could meet more than 20 percent of its commitments under the Paris climate accords simply by improving soil management on farmland and, especially, protecting its remaining grasslands.

But we're not.

THE 98TH MERIDIAN runs north to south down the center of the United States, neatly dividing the continent in half, and for decades it served as a sort of natural barrier to the plow. East of the 98th, in states such as Illinois, Iowa, and Missouri, farmers could count on abundant rainfall and deep black loam that produced crops instantly and reliably. Many farms had stands of oak, ash, maple, elm, or hackberry, so that settlers had sufficient wood for cabins, fences, barns, and heat. Streams and rivers coursed across the land, providing water for livestock and people.

But west of the 98th—in Kansas, Nebraska, Wyoming, and other states more in the rain shadow of the Rockies—average annual precipitation drops by half, to roughly ten or twenty inches of water per year. Temperatures soar above one hundred degrees Farenheit in the summer and plunge below zero—for days on end—in the winter. Drought is a regular feature, not an anomaly, and the wind never stops blowing. Trees are so scarce that they stand like lonely sentinels on the horizon; if you spot one you can be sure that a rare creek runs nearby.

In *The Great Plains,* his seminal 1931 account of this landscape, Walter Prescott Webb summed up the struggle of early settlers this way: "In their efforts to provide a sufficiency of water where there was none, men have resorted to every expedient from prayer to dynamite. The story of their efforts is, on the whole, one of pathos and tragedy." Webb recounted a timeworn joke: "'This would be a fine country if we just had water,' says the newcomer. 'Yes,' replied the man whose wagon tongue pointed east. 'So would hell.'"

When scientists from the U.S. Geological Survey delivered a report on frontier lands to Congress in 1902, they described states wests of the 98th meridian, such as South Dakota and Kansas, as "hopelessly nonagricultural." Of course optimists did settle west of the 98th meridian between 1860 and 1920, after Congress passed the Homestead Act of 1862 and new immigrants found that better farmland to the east was already taken. But it was a hard life, and one that baffled most other Americans. The journalist E. V. Smalley ventured onto the northern Great Plains for *The Atlantic* magazine in 1893 and returned with this dire account: "The silence of death rests on the vast landscape, save when it is swept by cruel winds that search out every chink and cranny of the buildings, and drive through each unguarded aperture the dry, powdery snow."

Like other contemporary correspondents, Smalley noticed the isolation and emotional hardship of life on the plains. "An alarming amount of insanity occurs in the new prairie States among farmers and their wives. Think for a moment how great the change must be from the white-walled, red-roofed village on a Norway fjord, with its church and schoolhouse, its fishing-boats on the blue inlet . . . to an isolated cabin on a Dakota prairie, and say if it is any wonder that so many Scandinavians lose their mental balance."

The Homestead Act of 1862, often lauded for promoting land ownership among the middle class, proved mostly a failure in this harsh landscape: It was not possible to make a living on 160 acres of land that received so little rain. The writer Richard Manning notes that most homesteaders on the Great Plains never "perfected" their claims, and just left.

Hyde County, South Dakota, where the Faulstiches live, is a good example of the region's rapid rise and steady decline. The county's population peaked in 1930, at 3,690, and has fallen by two-thirds since then. The county seat, Highmore, shows the signs of faded glory. Its courthouse, built in 1910, is a magnificent example of Classic Revival architecture, built of Indiana limestone with a spectacular stained-glass dome over a central atrium. It still houses offices for the county sheriff, treasurer, and assessor. But it looks out on signs of decay—an abandoned Dutch Colonial home with peeling paint on one side, a yard overgrown with weeds and honeysuckle on another.

A South Dakota wildlife researcher measures the tail of a Goldfinch at Jim Faulstich's cattle ranch. *Photo by Josephine Marcotty*

Sweat bee on sensitive briar at Helzer family prairie near Stockham, Nebraska. *Photo by Chris Helzer*

From prairie to cropland—a tractor prepares a field for planting near Malta, Montana, in 2018.
*Photo by Christopher Boyer/kestrelaerial.com*

A visitor at Farmfest, the annual statewide agricultural fair in Minnesota, walks by a display of modern tile lines.
*Photo by Josephine Marcotty*

A group of farmers watches as a modern trench digging machine lays underground drainage tile at the Hefty Brothers' Ag PhD Field Day on July 27, 2012, in Baltic, South Dakota.
*Photo by Renée Jones Schneider, courtesy of the Star Tribune*

Pipes carrying water from underground tile lines enter an agricultural drainage ditch in Renville County in southwest Minnesota. *Photo by Josephine Marcotty*

A sign next to an abandoned home near American Prairie in eastern Montana reflects some of the local sentiment against changes in land use by reintroducing bison to their historic home on the plains, 2022. *Photo by Josephine Marcotty*

The brown, polluted water from the Mississippi River slides over the top of the blue water of the Gulf of Mexico in this photo taken over Breton Sound, southeast of New Orleans, in June 2024. *Photo by La'Shance Perry, for* The Lens.

Workers open bays of the Bonnet Carré Spillway to divert rising water from the Mississippi River to Lake Pontchartrain, upriver from New Orleans, May 10, 2019. *AP photo/Gerald Herbert*

Bison on Dan and Jill O'Brien's Wild Idea Buffalo Company ranch in western South Dakota in June 2022. *Photo by Josephine Marcotty*

Giant windmills, shown here in southwestern Minnesota, have become a common feature amid the manicured farm fields of the Corn Belt. *Photo by Dave Hage*

American Prairie in eastern Montana under the dawn sky in November.
*Photo by Josephine Marcotty*

Leadplant extends to the horizon at Glacial Ridge National Wildlife Refuge in Minnesota. *Photo by Josephine Marcotty*

A controlled burn in May 2022 at the Konza Prairie Biological Station in the Flint Hills of northeast Kansas. *Photo by Josephine Marcotty*

A monarch at the Neal Smith National Wildlife Refuge in central Iowa. *Photo by Josephine Marcotty*

Prairie trees silhouetted by a summer sunrise near Highmore, South Dakota. *Photo by Dave Hage*

A solitary male bison roams American Prairie. *Photo by Josephine Marcotty*

Lars Anderson, field project manager for American Prairie, helps control a bison held in a squeeze chute while scientists draw blood and collect hair for research. *Photo by Josephine Marcotty*

Jerry Blanks kills a bison in the pasture for Wild Idea Buffalo Company, which developed a model meat processing system that starts with humane field slaughter. *Photo by Josephine Marcotty*

Members of the Blackfoot Confederacy in western Montana carry buffalo robes to a ceremony honoring the return of bison to tribal lands. *Photo by Josephine Marcotty*

A graveyard near Wolf Point, Montana, memorializes the hundreds of Sioux and Assiniboine people who starved to death in the winter of 1883 as a result of the U.S. government's failed reservation strategy. *Photo by Dave Hage*

Those who stayed in places like Hyde County and persisted found that they could raise a crop and feed livestock. Wheat became a staple in the Dakotas, Montana, and Kansas. In eastern North Dakota, beets grown for sugar became a profitable crop. But time and again, as they tried to press the plow just a little too far west, the settlers discovered how hostile that land could be. The agricultural historian Geoff Cunfer has noted how settlers in western Kansas kept trying crops such as wheat and oats—and generally gave up. By roughly 1870, they had settled into a durable pattern that left the western half of the state to grass and cows, even though ranching was often less profitable than raising a cash crop. As recently as the 1980s, Faulstich recalls, "you seldom saw soybeans west of Sioux Falls."

Then everything changed.

ONE DAY IN the early 1980s, a young crop scientist named Rob Fraley was pulled aside by two superiors at the Monsanto Company headquarters in St. Louis, Missouri. A decade earlier, Monsanto had introduced a revolutionary weed killer called Roundup, whose reputation for effectiveness and apparent safety had quickly made it the nation's leading herbicide. Fraley and his colleagues in Monsanto's Life Sciences Division had been working on a technique known as gene transfer—taking a gene from one organism and planting it in another. With Bt corn they had already proven genetic manipulation as a tool to fight bugs, and now Fraley's boss asked: What if we could use gene transfer to create corn or soybeans that are immune to Roundup and other weed killers? Farmers could plant a field and then safely spray it with Roundup a few weeks later, killing the weeds while leaving the crop unscathed. It would transform life for the modern farmer, who fights weeds all season. Seeds like that would command a huge price premium over conventional corn and produce even bigger sales of Roundup.

Fraley was skeptical, but quite by accident a Monsanto custodial crew in Louisiana had made a surprise discovery at about the same time. In a pond where the factory dumped residues from manufacturing Roundup, a species of bacteria had apparently developed a resistance to the herbicide. Fraley's team found a way to transplant a gene from the bacteria into a soybean seed, and after several years of tinkering and field trials, produced a soybean seed that could survive a typical dose of Roundup.

Roundup Ready soybeans debuted in 1996 and stormed the U.S. market. Within four years they represented more than half of the nation's soybean crop. Sales of Roundup soared too, reaching $2.8 billion, roughly half of Monsanto's total sales. Two years later the company followed with Roundup Ready corn.

Farmers loved the novel seeds, in part because Roundup is a highly effective herbicide and in part because it saved them time and money. As Monsanto's executives predicted, under the right conditions a farmer could grow a crop in just two passes across the field—one to plant seed and a second to spray with Roundup—instead of repeated trips to till out weeds or apply round after round of weed killers. For the same reason, the novel seed saved on diesel fuel, a major farm expense.

Roundup appeared to have additional environmental advantages. Its core ingredient, glyphosate, works by inhibiting an amino acid that is essential for growth in plants—but only in plants—and thus appeared to be harmless to humans and animals. In addition, Monsanto claimed that Roundup quickly became inert in the field: As soon as its work was done, plant and soil microbes broke it down into harmless byproducts. This made it much less likely to contaminate lakes and streams. And because it reduced the need for farmers to plow a field repeatedly to kill weeds, it hastened the adoption of no-till farming, the cultivation practice that reduces erosion and improves soil health.

Those attributes were invaluable to a corporation with a troubled reputation among environmentalists. Monsanto had been a manufacturer of Agent Orange, the defoliant that became notorious during the Vietnam War, and manufactured the industrial compounds known as PCBs, later determined to be potential carcinogens. In the early 1990s the EPA had identified eighty-nine separate Superfund cleanup sites where Monsanto was a "potentially responsible party." Inevitably, many of the safety claims for glyphosate turned out to be inaccurate, and in 2015 the World Health Organization labeled glyphosate a probable carcinogen. In 2018, a California school groundsman named Lee Johnson won a $20.5 million court judgment after a jury found that exposure to glyphosate contributed to his non-Hodgkin's lymphoma. And, much the way insects had developed immunity over time to various new pesticides, nature soon fought back against the novel herbicide. By 2006, crop scientists had identified more than a dozen weeds that could survive glyphosate, a number that tripled in the following decade.

Nevertheless, Roundup and Roundup Ready seeds remain hugely popular with American farmers. Two decades after their release, herbicide-resistant seeds represented 94 percent of the U.S. soybean crop and 91 percent of the corn crop.

Roundup didn't merely simplify weed control, it made it easier for farmers to convert grassland to row crops: You simply sprayed a pasture with Roundup, waited for the grass to die, and then plowed the ground for corn or wheat. Still, the herbicide was just one of many breakthroughs that, in combination, hastened the westward advance of row crops. In addition to genetically modified crops, big seed companies such as Syngenta and DuPont Pioneer began releasing corn and soybean seeds that could survive the drier climates and shorter growing seasons of the northern Great Plains. First came "90-day corn," a hybrid that could reach maturity in three months—two to three weeks faster than the seeds designed for warmer and wetter states east of the 98th. It was perfect for a state where the ground thaws late in the spring and frost can arrive before Halloween. Next came DT corn, a drought-tolerant hybrid suited to the hot, dry summers that are increasingly common in the West. Within a decade, DT corn had more than three-quarters of the market share in states west of the Mississippi River. The new hybrids proved their value in the summer of 2021, the worst drought since the Dust Bowl in many counties of the upper Midwest. In mid-August many cornfields stood brown and stunted; a generation earlier, one farmer said, his parents would have simply plowed the corn under as a lost crop. But that year rain arrived in September, the corn recovered, and many farmers on the dry plains had bumper crops.

By 2017, genetically engineered seeds represented 97 percent of South Dakota's corn crop and 96 percent of the soybeans. Because they produced bigger harvests per acre than their predecessors, they made every acre of land more productive and helped make crop farming more profitable than cattle ranching. Over that same period, advances in seed technology outpaced productivity gains in beef production, giving row crops another financial advantage over ranching. Economists estimate that an acre planted in row crops now pays roughly twice as much as an acre grazed by cattle. "If you break your land and make it black, you've just doubled its value," says Jack Davis, an extension economist at South Dakota State University. In addition, the improved seed technology made corn and soybeans even easier to grow than other crops—wheat, for example—hastening farmers' migration toward a corn-soybean monoculture. Between 2000 and 2017 the number of South Dakota acres planted in corn and soybeans more than doubled.

Matt Leisinger farms just a few miles up the road from Jim and Carol Faulstich, and his work has given him a front-row seat to South Dakota's transformation. Leisinger grew up helping out on his grandfather's cattle and crop operation, and at age sixteen he started working a few fields of his own. He planted Roundup Ready soybeans almost as soon as they hit the market.

"Grandpa thought I was crazy to plant soybeans in South Dakota," Leisinger recalls now. "But before long, he was planting them too."

On his own fields today, Leisinger employs a whole suite of conservation practices. He uses a diverse rotation of crops that includes wheat, oats, sunflowers, and triticale in addition to corn and soybeans, so his fields can rest and recover after the strenuous job of producing corn. He practices no-till cultivation, meaning he plants a crop directly into the previous year's stubble rather than plowing the ground open again, a technique that reduces soil erosion and holds moisture through the hot South Dakota summers. He also raises cattle, in part because they can graze for free on the stubble left after a no-till harvest, and in part because their manure provides a huge number of beneficial microbes that constantly rebuild and improve his soil.

But Leisinger also works "custom" for other farmers and neighbors, meaning that they hire him to plow their fields in the spring and harvest their crops in the fall. This practice has given him the reputation as the hardest-working farmer in Hyde County, but it also puts him at the heart of the changing landscape. By his account, he has personally broken roughly six thousand acres of grass in and around Hyde County, converting it to cropland, and has planted thousands of acres of corn and soybeans for clients and the landlords who rent him land.

The list of Leisinger's clients is long and diverse, and it explains almost the full range of forces that are driving grassland conversion. A pair of retired ranchers approached him a few years ago. Both had grown too old to handle the rigors of raising livestock. Without pensions or 401(k)s, they needed steady retirement income. They asked him to plow up two sections of their land—about thirteen hundred acres—so that a younger farmer could rent them to plant corn and soybeans. Why corn and soybeans? Because they're more lucrative than cattle and, hence, incur higher rental payments—and a higher retirement income for the ranchers.

Leisinger also has a landlord who lives on the East Coast and is essentially an investor in South Dakota land. He wanted a steady, reliable return from his asset, and in that, corn and soybeans have another advantage over cattle. Corn and soybeans are covered by generous federal crop insurance, now one of the major subsidies written into the federal Farm Bill; if hail or a drought damages a crop, the farmer still gets a check. Cattle ranching has only spotty insurance coverage in the federal program. If a drought strikes and there's no grass for the herd, ranchers take their cows to town and sell them at a loss. Studies by the U.S. Department of Agriculture have concluded that crop insurance explained only a modest share of the shift to row crops in the last two decades, but the prospect of a guaranteed return makes crops more attractive to investors who are risk-averse.

Then there is convenience. Another of Leisinger's clients, a neighbor approaching retirement, spends his winters in Florida. For a cattle rancher this would be impossible. It's a fifty-two-weeks-a-year job moving cows from one pasture to another, checking water troughs, hauling feed, repairing fences, helping pregnant cows deliver, staying on top of vaccinations and disease. Planting corn and soybeans, by contrast, has become easier and easier, Davis says. "You harvest your crop in October and you spend your winter in Florida."

Leisinger sees the changes around him as the inevitable result of markets and technology. "My grandpa put in his wheat with a two-row planter. Today I farm with a sixteen-row, computerized tractor that spaces the rows perfectly and knows which acres need nitrogen. I can plant more sunflowers in a day than my grandpa could probably do in a week."

IN THE COOL shadows of a big steel hangar outside Watertown, South Dakota, a long hopper truck full of shelled corn has pulled to a halt. The driver climbs out of the cab and attaches a crank to a pair of hinged doors at the bottom of the truck. At a signal from the control booth, he gives the crank a turn and one thousand bushels of corn pour into the vast underground dump pit of Glacial Lakes Energy. Within minutes a conveyor belt will carry the corn to a hammer mill in the next building, where giant wheels will crush it to the consistency of corn flour. Pipes will carry the ground corn to a vat where it is mixed with water and an enzyme to break down the starch, and then to giant fermenting vats where a technician will add one hundred pounds of yeast. Just as if it were a batch of home-brewed beer, this wort will ferment for several hours, eventually producing a potent grain alcohol: ethanol. As soon as the truck has dumped its corn, the driver pulls out to make way for another just like it. A dozen huge trucks form a line snaking out the back of the hangar waiting for their turn to deliver corn that is destined to produce ethanol.

If seed technology made corn possible in South Dakota, the biofuels industry has made it profitable. Lobbyists like to say that American farmers "feed the world." That's true. But it's also true that 40 percent of the nation's corn crop goes to ethanol and winds up in Americans' gas tanks, not on their dinner tables. On the same day that the hopper truck dumped its load of corn, more than a hundred trucks would roll through the Glacial Lakes compound, dumping 125,000 bushels of corn. Across its system, Glacial Lakes processes 125 million bushels of corn each year, roughly 10 percent of the corn grown in South Dakota, and it's just one of many ethanol producers in the upper Midwest.

Biofuels are actually a venerable technology—Henry Ford used a form of corn ethanol in his first Model T—but its rationale has changed regularly with world events. In the 1970s, after the OPEC oil embargoes shocked American drivers, ethanol offered a homegrown fuel for a nation seeking energy independence. During the farm crisis of the 1980s, it supplied a new revenue stream to desperate farmers. A few years later, after the fuel additive known as MTBE was banned for environmental reasons, biofuels offered an alternative way to boost the octane of gasoline and meet federal pollution standards. But it was in the twenty-first century that the biofuels industry shifted into high gear, with two landmark pieces of federal legislation, the Energy Policy Act of 2005 and the Energy Independence and Security Act of 2007, both aimed at promoting a renewable fuel that could fight climate change. The laws require the nation's petroleum refiners to use a steadily rising quantity of renewable fuel, a volume that was set to rise from 4 billion gallons in 2007 to 36 billion in 2021. They also required the EPA to certify that these biofuels be at least 20 percent less carbon-intensive than conventional gasoline—that is, in their full life cycle from production to tailpipe, they release 20 percent fewer greenhouse gases.

Because of that history, farmers came to embrace ethanol as their contribution to a cleaner planet. In addition to feeding the world, many believed they were now giving their fellow Americans a clean and sustainable alternative to petroleum. Almost immediately, however, environmentalists began challenging the science. It turns out that producing ethanol releases just as much carbon as producing gasoline, and it releases just as much carbon when it's burned in vehicles. Its claim to be climate friendly rests on its feedstock—that is, the fact that it comes from corn rather than petroleum. An acre of corn recycles carbon that was already in the atmosphere—through photosynthesis—rather than releasing carbon stored underground in the form of fossil fuels. That fact alone accounted for roughly half of the EPA's certification that ethanol is climate friendly.

But critics have pointed out this logic treats the extra acres of ethanol corn as if they were free—as if there were no other use for the land and no environmental cost to plowing it up. But if the land already was planted in corn, that grain is no longer available to feed people or livestock. And if the land was grass, plowing it for corn imposes a great environmental cost—the soil carbon that's released into the atmosphere. In a 2008 paper, the scientists Joe

Fargione, Jason Hill, and their colleagues calculated that plowing up grass releases so much soil carbon that it would exceed the benefit of burning ethanol for at least fifty years, possibly ninety years. By that time, electric vehicles might make any liquid fuel obsolete. Tyler Lark has extended this argument by drawing on his extensive modeling of land-use change in the Corn Belt. In a 2022 paper, he argued that, by raising corn prices, ethanol has accelerated the plow-up of North American grasslands. Ultimately, he says, ethanol may actually be 24 percent worse than gasoline as a source of greenhouse gases.

The environmental lawyer Tim Searchinger, who wrote one of the pioneering critiques of biofuels, makes an additional argument: If our goal is to slow climate change, we have better tools than ethanol. "If you took a hundred acres out of corn production, then put solar panels on one acre and planted trees on the other ninety-nine, you would produce just as much energy [as ethanol does] and do far more to combat climate change," Searchinger says. And if David Montgomery and Tim Crews are right about the pace at which modern farming degrades soil health and fertility, ethanol cannot be described as renewable. It becomes one more extractive industry based on a finite resource—soil.

The debate has produced volley and counter-volley—defenders of biofuels point out that farmers began expanding their corn and soybean acreage long before the federal biofuels mandate—and it heats up every time Congress revisits the issue. But there is a consensus on a few points: First, ethanol supercharged demand for corn, and that pushed grain prices higher. Corn, which seldom traded above $4 a bushel before 2007, has seldom traded below $4 since then. While this was a huge benefit for corn farmers and the economy of the rural Midwest, it almost certainly drove a wave of grassland conversion west of the 98th meridian. In South Dakota, for example, farmers had never planted more than 3.2 million acres of corn before 1997; their corn acreage rose to 4.5 million acres in 2007 and rose again to 6.1 million acres by 2021. While some of that growth reflected a shift from wheat to corn, much of it came from plowing up grass. Since the 2007 Renewable Fuel Standard, the number of acres enrolled in the Conservation Reserve Program, the government's biggest land protection program by far, has dropped by more than one-third. That means thirteen million acres came out of grass and went back into crop production. An ingenious study by the ecologist Christopher Wright and his colleagues found that the closer a farm was to an ethanol plant, the more likely its proprietor was to plow it up for row crops.

In rural South Dakota, the scientific debate is largely irrelevant. Ethanol has created so many jobs and funneled so much income to farmers that challenging it has become heresy. On a hot summer morning, Scott VanderWal answers his phone from behind the wheel of a grain truck. As president of the South Dakota Farm Bureau, VanderWal speaks for hundreds of colleagues across the state. But he's also himself a farmer, and this morning he is hauling a load of corn to the nearby ethanol plant. VanderWal points out that ethanol has provided a badly needed stimulus to the nation's rural economy at a time when it's falling behind metropolitan areas. By providing farmers a new income stream it has reduced—somewhat—the need for direct federal farm subsidies. "Without ethanol, there are a lot of farmers who couldn't make a go of raising corn," he says. And perhaps their land would never have been plowed up for crops in the first place.

TO UNDERSTAND THE recent transformation of prairie agriculture, there may be no better classroom than Ag PhD Field Day, a farm festival that takes place each summer just outside tiny Baltic, South Dakota. Driving east from Baltic on a sunny July morning, you would never guess that an agricultural spectacle is underway just over the next hill. But then you clear the rise and gaze down on something that looks like a bustling medieval fair. White peaked tents shimmer in the sun, colored pennants flutter in the breeze, and hundreds of curious visitors wander from one pavilion to the next examining the latest miracles of agroscience. On a sloping field to the south, where a cornfield has been converted to a massive parking lot, two thousand windshields glint in the hot sun.

Ag PhD Field Day is the brainchild of the Hefty Brothers, Darren and Brian, who took over a successful family seed business in the early 1990s and turned it into an agribusiness empire with fifty stores in eleven states. In addition to selling seed, pesticides, and farm equipment, they market their own line of crop additives dubbed "Hefty Naturals," operate a farm, design agronomy apps for smartphones, host a weekly radio show on Sirius FM, and sponsor the annual field day.

The event represents a celebration of the industrial-model farming that has transformed South Dakota and the agribusiness corporations behind it. Fluttering banners advertise agrochemical companies such as Belchim and Syngenta. Dow Chemical's agriculture division and Bayer Crop Science, now the world's largest seed company, sponsor huge tents where their representatives unveil their latest marvels. At one row of booths, merchants explain the latest gadgets in "autonomous farming," including remote-controlled drones and robot tractors. A few yards away, a curious farmer can climb into the cab of a $700,000 QuadTrac tractor, plug a memory stick into a USB port on the dashboard computer, and have the tractor tailor the application of seed or herbicide to every acre of a field.

Fifty yards away, an executive from Bayer Crop Science is describing the company's next generation of herbicide-resistant corn seeds—seeds bred to stay ahead of weeds that were adapting to survive glyphosate. Since acquiring Monsanto in 2018, Bayer has become the world's biggest seed and herbicide company, and is now a leader in the sort of hybrid corn seeds that allowed crop farming to spread into the arid plains. Because the corporation operates all over the world, he says, it has access to corn varieties and gene stocks from every climate zone and can conduct endless breeding experiments. Its next miracle: "short corn," a hybrid that will grow just three or four feet high so that the stalks are less prone to snap off in high winds.

Asked if Bayer's technology has abetted the plow-up of priceless grasslands, the executive replies, "We don't tell farmers what to do. We just innovate to meet the needs of farmers."

It should come as no surprise that farmers have embraced that sort of innovation. It has increased yields, provided insurance against drought, and greatly reduced the amount of labor required to raise a crop—much as mechanized implements did in John Deere's time. All across the Ag PhD fairgrounds one hears echoes of the pluck and wonder that coursed through farming communities in the nineteenth century—the rapt fascination with technology, a sense of invincibility in the face of nature's challenges, and the conviction that science and ingenuity can wrest ever more profit from an underexploited landscape.

This confident spirit finds an evangelist in Blair Karges, a farmer who has driven down this July morning from his home near Valley City, North Dakota. Karges is a family man, a successful farmer and an exceedingly friendly guy. He is a devout Christian who tithes regularly, has volunteered on church missions to Haiti and other impoverished places, and believes that people of faith have a responsibility to protect nature. He has experimented with cover crops and other conservation practices, and he argues that farmers get blamed for all the ills of rural America without getting credit for their contributions. "We feed millions of people," he says, examining the air-conditioned cab of a Case tractor. "We're using fewer chemicals, less fertilizer, and producing more food." Karges marvels at the way seed research has improved yields again and again, and he dismisses environmentalists' concerns about protecting grasslands. "People," he said, "should spend less time worshipping the Earth and more time worshipping God."

---

IT'S NOVEMBER AND Jim Faulstich is wading through the dense grass of a pasture on the south end of his ranch, looking for pheasants. He's accompanied by four men in blaze orange coveralls and two giddy hunting dogs who bound across the field sniffing for birds. The South Dakota grass has changed character since July. Bleached almost white from the sun and shed of its seed heads, the grass lies soft and pliant, as plush as the fur of a golden retriever puppy.

The fall harvest is done and Faulstich has turned his attention to the next seasonal business: fee hunters. The grass on his ranch has grown more diverse and abundant in the last two decades, the new windrows have gained height, and the land now supports a rich population of pheasants, grouse, and deer. Faulstich will host a new group of hunters every week from October until January. This additional revenue stream can be time-consuming and demanding. He meets the hunters at his machine shed around dawn, gives a genial introduction to his land, then escorts them through the fields for most of the day, building relationships with loyal customers who will return year after year. But it's an essential part of the formula that has helped Daybreak Ranch survive the transformation of South Dakota's grasslands.

When Jim and Carol Faulstich revamped their ranch in the 1980s, they adopted practices more in rhythm with nature because they believed in them, but also because working with nature saved them money and led to healthier livestock. They delayed spring calving from March to May, in cycle with deer and other local mammals. It means the calves have fewer weeks to fatten up before slaughter—a core tactic in modern cow-calf operations—but it also allows the lactating mothers to eat grass when it is most nutritious, saving the Faulstiches money on feed and supplements. They allow the cattle to live outdoors and graze all year—much as bison did decades ago—which saves on diesel fuel and transport costs in two directions. "I don't have to haul feed to the barn and I don't have to haul manure to the fields," he says. The huge diversity of plants—from bluestem to sunflowers to alfalfa—provides a form of natural pest control.

The practices have won the Faulstiches a series of awards for soil health and environmental stewardship, kept the ranch in the black, and—perhaps their proudest achievement—held the plows at bay. But now Jim fears they aren't enough to keep a cattle ranch afloat against the economic advantages of corn and soybeans. He argues that farmers and ranchers should be paid for what are called ecosystem services and other public benefits—sequestering carbon in the soil, for example, or providing habitat for threatened wildlife. He and Carol feel lucky that their daughter and her husband took up the rural life, but they worry that grassland ranching can't hold off the plows when grain markets and federal subsidies give crops such a big advantage over ranching. Will Daybreak Ranch survive long enough for one of their grandchildren to take over and work this land? He looks into the gray sky for a moment and says, "You can hope."

*PART THREE*

Plants know how to make food and medicine from light and water, and then they give it away.

—Robin Wall Kimmerer,
*Braiding Sweetgrass*

CHAPTER TEN

# *Prairie II*

MINNESOTA HIGHWAY 75 CUTS A VERTICAL LINE THROUGH ONE OF THE largest expanses of flat land in North America, the absurdly named Red River Valley, which forms the border between northwest Minnesota and eastern North Dakota. Tens of thousands of years ago this spot was the bottom of Lake Agassiz, a body of water created by a retreating glacier that was so large all five Great Lakes would easily have fit inside its shores. At its northern end, the lake was blocked by the crumbling frozen wall of the Laurentide ice sheet, two miles thick in places. As it melted at the end of the last Ice Age, it receded in front of the advancing lake, hurling icebergs into the water below as it fell apart. When the lake finally breached the last of the ice wall near the southwest side of Hudson Bay in Canada some eight thousand years ago, Lake Agassiz released a volume of icy water so large that it raised sea levels and temporarily dropped global temperatures. To the south, Lake Agassiz's long tail stretched through Minnesota and North Dakota, ending at a natural dam halfway down the western side of Minnesota. When that dam gave way, a river a mile wide surged through the opening, plowing its way southeast to the Mississippi River. The torrent carved a deep trench across the southern part of the state, creating what is now the watershed for the Minnesota River, which flows into the Mississippi in Minneapolis. What's left of the massive Lake Agassiz is the placid, muddy Red River of the north, which flows in sinuous curves up to Lake Winnipeg in Manitoba, still following the path of that long-gone glacier.

The ancient lake's imprint on the surface of the Earth is easy to see by satellite, and it's clear how it has dictated modern use of the land it left behind. Driving north on Highway 75 today, through the middle of what long ago was a diverse prairie complex of marshes, fens, and grasses, you travel through farm fields stretching uninterrupted to the horizon. Drained and plowed, the land in the Red River Valley produces the crops that feed many of America's habits—corn for livestock and ethanol fuel, sugar beets for sweets, potatoes for McDonald's french fries. Alongside the road the edges of the fields are shaved as precisely as a fade haircut. Groves of trees in the distance mark where farmhouses stand, and in the cemeteries, gravestones stand up like tiny buildings against the sky. On any given afternoon the land seems empty of people, animals, and even insects; after hours of driving on a warm summer day, the windshield is clear of dead bugs. The only remnants of the prairie are patches of wildflowers in the ditches, great layers of clouds that climb into the sky, and the big winds that sweep down from the northwest.

The road cuts through the town of Crookston in the northwest corner of Minnesota, past grain silos, endless freight trains, and a sugar beet processing plant, and turns straight east. Here the land begins to roll a bit, like swells on an ocean. On the satellite images it's obvious what these once were—beaches of the massive lake. The sandy ridges rise like giant curved steps as you drive east, stretching from far north in Manitoba to central Minnesota and South Dakota. Thousands of years ago, these were the constantly moving edges of Lake Agassiz as it grew and contracted, the place where wind and water dumped the sand and gravel.

Seven miles or so east of Crookston, the cornfields end abruptly and there is nothing but grass. A dragonfly smacks into the windshield, and overhead a flock of ducks comes in for a landing on an open stretch of shallow water. A sign at the edge of a mowed parking area provides an explanation: This is Glacial Ridge National Wildlife Refuge, the largest prairie restoration project in U.S. history.

Today, it is one of a handful of places left in the United States where you can walk for miles across a tallgrass prairie and get a taste of what the landscape once was. The silence, heightened by the sigh of wind through the grasses and the trill of birds, is immense. Green leopard frogs race ahead through the grass. In early spring, giant clacking sandhill cranes descend on the wetlands by the thousands as they stop on their way north to nesting grounds in Canada, and the ethereal booming song of prairie chickens greets the rising sun as they do their annual mating dance in the grass. From May until October wildflowers bloom in waves—pink coneflowers, false indigo, goldenrod, coreopsis, purple lead plant. In some parts of the refuge, the grass grows so high you have to stand on top of a car to see over it.

The story of Glacial Ridge explains how restoring grasslands, where it's possible, can deliver huge rewards. But it also illustrates how extraordinarily difficult it is to claw the land back from some other economic purpose, often igniting deep cultural conflicts over land use. Today the people who succeeded in restoring the prairie on Glacial Ridge—conservationists mostly—look back in amazement at what they achieved. And they recognize that it is even more difficult in places other than this part of Minnesota. Here, there was an unusual combination of poor farmland, the deep pockets of the federal government and national conservation groups, and a culture that embraces hunting, fishing, and heading "up north" for vacations in the wild outdoors. It was that mix that made re-creation of the prairie possible.

People have lived in the Red River Valley for at least thirteen thousand years, and their way of life evolved along with the climate and the changing landscape. Some eight thousand years ago, the climate warmed and the prairies pushed the forest border eastward. The people in the valley learned to use tools like flint for knives and spear-throwers that made them formidable hunters, of bison in particular. Over time, rainfall increased and the forest border moved westward again to where it is now, in central and northeast Minnesota. The Red River Valley and the parkland areas east of it—that mix of aspen trees, brush, and wetlands—became a cultural melting pot. It attracted people with forest traditions and ways of life from the east, the Great Plains nomadic people from the west, and others who lived farther south in the Mississippi River Valley. They hunted moose in the forest and bison on the plains, fished for sturgeon in the Red River and its tributaries, and grew crops around permanent villages.

Early Europeans universally described the wide Red River Valley as a patchwork of grassland, marshes, and impenetrable swaths of brush. In the winter, bison would congregate on the twisting, frozen rivers, and when the ice broke up in the spring, the animals would drown in vast numbers, their carcasses floating downstream for days on end. The land rose slowly eastward from the forested floodplain along the Red River, becoming tall grassland on the flat, former lake bottom, then rising a few hundred feet up to the grassy, rippled beach ridges that marked the eastern edge of Lake Agassiz. Water and silt collected between the ridges, creating long shallow marshes and lakes, quaking bogs, and, in places where mineral-rich groundwater welled up from below to feed them, the rare wetlands called calcareous fens.

When the first European traders arrived in the 1700s, the forest-dwelling Ojibwe occupied the northern reaches of the valley, and the bison-hunting Dakota lived in the south. When the fur traders arrived, the people's economy changed dramatically, as both Indigenous groups competed to supply the beaver, muskrats, and other furs to make the hats and coats that were in huge demand in eastern and European markets. In exchange, they received tools, blankets, guns, and other goods—the technology of another culture.

Settler colonialism in the Red River Valley began in earnest in the mid-1800s, driven by the growth of railroads and the land speculation that came with it. Steamboats moving up the Red River fueled development as they carried timber and other goods north, and buffalo hides south. Later they carried immigrants from southern Minnesota and Europe, who began to grow wheat that was carried south. Their numbers accelerated with the 1862 passage of the Homestead Act, which gave each farmer 160 acres of land, and the arrival of the railroads, which created a permanent link between Manitoba and St. Paul. The Native people were pushed to the western plains, where the fierce Lakota still reigned, or onto reservations established in the 1860s in northern Minnesota. European farmers raised barley, oats, and potatoes—and wheat. To spur colonial expansion, the federal government gave land to the railroads as well, and they sold it to settlers and land speculators.

Though productive, the clay soils left behind on the flat bottom of the ancient lake were poorly drained, and when wet were called "gumbo." They were so deep that horses would sink up to their bellies and couldn't be pulled out. Sometimes, farmers would equip them with wooden "snowshoes" to stop them from sinking into the mud.

Steam power came to the valley in the late 1800s and so did drainage, funded in part by the railroads. The Red River Valley Drainage Commission was created in 1893 to coordinate the construction of ditches and water movement across the land, which was frequently inundated with spring floods. By 1899 there were twenty main ditches totaling 135 miles, many of them draining railroad lands. Then plowing up the prairie began in earnest. As global demand for wheat increased and crop diseases devastated the harvests farther east, bonanza farms, which were up to tens of thousands of acres in size, flourished on the valley floor.

By 1890, 70 percent of the valley was plowed for crops, and a few decades later, just as in the southern and eastern grasslands, the prairie was largely gone. In the Red River Valley, with its flat landscape and gumbo soils, drainage was especially critical to the advance in agriculture. By 1920 roughly three thousand miles of ditches that dropped just a few inches for every mile on the flat land drained three million acres along the Red River.

But the ancient beach ridges left behind by Lake Agassiz, which cover thousands of square miles on either side of the Red River, were a different story. Compared to the big agricultural operations on the former lake bottom, farms on the ridges were smaller and more diversified. Farmers grew wheat and oats on the dry land and hayed the wet meadows where they could, and cattle ranching was common well into the 1970s. But it was poor farmland—the ridges dried out in the summer and the lower swales fed by groundwater stayed wet well into the late spring. Compared to the broad river valley farther west, this higher ground had more trees, more ponds, more livestock, and more grass. And in those small refuges, many of the prairie plants and animals that are now rare or endangered survived.

In the 1980s, 24,000 acres of grazing land on the beach ridges east of Crookston were purchased for $1 million by an out-of-state owner. At the time, rising prices for corn and wheat, together with a string of dry years that shrank the wetlands, made row crop agriculture possible even on the unforgiving soils of the beach ridges. Conservationists looked on in dismay as the big plows and bulldozers began tearing up the remaining prairies, and replaced them with soybeans and corn. Among them was Ron Nargang, who worked for the Minnesota Department of Natural Resources and, later, The Nature Conservancy. He figured the new owner and the local farmers who leased land from him could make money only because during bad years they could claim payments from federal crop insurance programs.

In the late 1990s the rains returned, and, as a result, so did the wetlands. As yields declined, the property came up for sale again. The Nature Conservancy staff spent hours examining maps of the region, trying to decide which scraps of land to buy, and finally someone said, "Why not buy the whole thing?" In 2000 The Nature Conservancy took the extraordinary step of putting down $9 million to buy the entire property: thousands of acres of plowed fields, drained and undrained wetlands, 165 miles of ditches, fragments of native prairie, a gravel pit, a grain elevator, and an old feedlot that came with a pile of livestock carcasses.

Across the midsection of the United States, only a few large tracts of native grasslands remain—the Flint Hills in Kansas, the Sandhills in Nebraska, and the Loess Hills in western Iowa. Northwestern Minnesota has the Agassiz National Wildlife Refuge— more than sixty thousand acres of native grass and wetlands. Glacial Ridge, however, was something else. It would become another example of the human capacity to transform landscapes, only this time it would be from cropland back to prairie.


IN ANY ECOSYSTEM, but especially a grassland, size matters. The tiny patches of tallgrass prairie scattered across the Midwest today are, in effect, islands. Their plants can't share genes with others farther away, so they can't adapt easily to changing weather and other conditions or out-compete invasive species. Butterflies and other insects are also trapped—they often can't get to the next island, and in small prairie remnants, they are far more exposed to pesticides that drift in from neighboring farms. Grassland birds and mammals, including owls, prairie chickens, and badgers, require space in order for their species to evolve and survive. Researchers have shown that over time small prairie patches in a sea of agriculture or urban development just—shrink. They lose diversity, species disappear, and the native plants become overtaken by whatever surrounds them, often invasive grasses and weeds. The 24,000 acres that The Nature Conservancy bought was a lot of land. But the real beauty of the project is that the land was adjacent to or near other native prairies owned by the state and federal governments. When restored, Glacial Ridge would connect a complex of varied prairies and wetlands that altogether covered more than fifty-five square miles.

Ron Nargang, who led the project for The Nature Conservancy, first recruited the federal agencies that might provide funding for restoration, and eventually take over management of the property. A project of that size could only succeed with the help of federal and state agencies, with their deep well of public funding dedicated to conservation and wetland protection. The U.S. Natural Resources Conservation Service agreed to pay for the wetland restoration on Glacial Ridge through federal programs, and on neighboring farms whose owners were willing to restore wetlands as well. The U.S. Fish and Wildlife Service agreed to make Glacial Ridge one of the 560 wildlife refuges it owns nationally.

Far more difficult, however, was the resistance to the project by the local community. Northwestern Minnesota, like much of the Midwest, has deep economic and cultural roots in farming. The grassland that once stretched across the beach ridges and ancient lake bottom was a thing of the past. Preserving small tracts of remaining native grassland that were left doesn't usually create much controversy. If it's never been plowed, it's probably no good for crops anyway, and those remnants are often considered wasteland. But to take 24,000 acres of land that had supported livestock and crops, no matter how poorly, and set it aside as a place that will never again produce income for a family, or revenue for a business, or taxes for local governments, can trigger a deep resistance—especially in small rural communities that struggle to maintain their populations. The hardest part of the project for Nargang wasn't finding the money to do the work. Nor was it the daunting task of re-creating a massively complicated and poorly understood natural grassland ecosystem. It was the people.

Nargang decided to approach it head on. As soon as The Nature Conservancy bought the property in 2001, he approached Warren Affeldt, chair of the Polk County board of commissioners—someone he'd opposed repeatedly, but not disagreeably, at the Minnesota legislature over wetland protection and other environmental policies. Nargang laid out his plan to create a federal wildlife refuge and told Affeldt, essentially, that he wanted to do it with him, not to him. They agreed to convene officials from the affected local governments to hash it out together—the school board, the township boards, the city councils, the watershed districts, and anyone else who cared about the county's tax base. They held a meeting at the Polk County courthouse in Crookston. The room was packed. People were tense. Nargang set up a flip chart and took notes as people stood up one by one to voice their objections. When the meeting adjourned, Nargang had a list of a couple dozen problems he had to solve. The county and other public entities would lose $97,000 in property taxes—disastrous for the small school districts and townships. The county would also lose a gravel pit that provided material for local road and construction projects, and 126 miles of ditches that, farmers believed, controlled the floods that routinely washed over the flat land in springtime. Farmhands who worked the land as tenants would lose their income. Farm supply businesses would lose their customers.

By sheer chance however, Nargang had some powerful leverage: clean drinking water for the city of Crookston. The city was struggling to find a new source. A combination of local geology, competition from farm irrigation wells, and contamination by decades of exposure to agricultural chemicals had made clean, adequate water a scarce commodity. In scouting the area for possible groundwater sources, the city's hydrologists had found old irrigation pivots on the Glacial Ridge site, meaning there was water below. Keith Mykleseth, the city's water superintendent, called Nargang to ask permission to hunt for water, and they got to talking. Nargang had just lost his project director, and Mykleseth told Nargang he'd get a lot more done if he hired someone local. A month later Mykleseth had the job.

Mykleseth had no experience with ecology or grasslands, but he was a charming extrovert with deep connections in the community, including six kids in the school system. He gave the project instant credibility with local residents and public officials. He recognized immediately that a new source of water alone could be enough to win the endorsement of county and city officials. Not only would the aquifer below Glacial Ridge give Crookston a long-term water supply, the surrounding grasslands would forever protect the water from agricultural contamination. Nor would the city have to compete with irrigation and other demands by farmers and livestock owners. "We had them locked in," Mykleseth said.

The loss of property taxes, however, was enough to cripple the project. As a nonprofit, The Nature Conservancy doesn't have to pay them. The Fish and Wildlife Service, as a government entity, wouldn't have to pay them either when it eventually took over the property. Taking land off the tax rolls is one of the major barriers to conservation everywhere; it's one reason why, in rural areas that still have land left to protect, government officials and environmental groups can be viewed with suspicion and resentment. Affeldt acknowledged that the land "probably should never have been broke in the first place." But turning Glacial Ridge into a refuge would not only cost the county tax revenue, it would embitter a lot of residents who might see their own taxes rise to make up the difference.

In the end, Nargang cut a deal with the county that was uniquely suited to the property and the community. The Nature Conservancy agreed to set up a trust fund of $2.5 million that would provide enough income to replace the lost property tax revenue in perpetuity. The money in the trust fund would come from revenue generated by the gravel pit, owned and operated by The Nature Conservancy, as well as rental income from local farmers who leased the land before it was replanted as prairie. Overall, the strategy solved a multitude of problems for the community: a generous payment for lost taxes, jobs for local people to do the restoration work, cropland available to local farmers until it was converted, and a continuing supply of material for road construction. Letting farmers work the land for a time, in fact, was also critical to the restoration: It had to be farmed to suppress weeds and invasive plants that would quickly overtake the fledgling prairie when it was planted.

That set the stage for the new drinking water wells. The city of Crookston installed nearly twelve miles of pipe from its water treatment plant to a site that would forever be protected from agricultural contamination and overuse. Two wells would provide up to 24,000 gallons a day of clean drinking water. In 2002, after two years of planning and negotiating, the deal was done.

JASON EKSTEIN WAS hired by The Nature Conservancy in 2002 to bring back the prairie. Today, he's the roving habitat crew supervisor for the Minnesota Department of Natural Resources, but he still works out of the same office he did when he first came to Glacial Ridge. It's in a tiny white building in the shadow of a looming grain bin—another relic from the days when the land was farmed. On the south side of the bin and the office, a dirt road runs alongside a broad meadow where a pair of sandhill cranes have taken up summer residence. Their bobbing, red-capped heads are a common site as they hunt for frogs in the roadside ditches. Inside Ekstein's office, visitors sit on a sagging leather couch against a wall covered in green maps. When he first arrived at Glacial Ridge, he had a staff of two, one pickup truck, and one ATV. His job was to turn 24,000 acres of degraded farmland and pasture back into a landscape that had vanished decades earlier. He had ten years to do it, and two thousand acres of good native prairie to start with, but no road map and no idea if it would work.

Ekstein and his crew started by restoring small sections in phases. In the first year, they rented nineteen thousand acres to a local farmer. He agreed that in the last year of the contract for each section, he would plant Roundup Ready soybeans, the variety that's been genetically altered to survive the herbicide that kills every other kind of plant. When Ekstein was ready to tackle the next section, the farmer would plant soybeans, use Roundup to kill the weeds, then harvest as usual in fall, leaving the field bare. Meanwhile, Ekstein and his staff used 1970s-era farm combines to cut the nearby patches of remaining prairie and separate out the wild seeds—the same way a farmer would harvest corn or wheat from a crop. The seeds were a key piece to the ultimate success of the project. They came from native plants—grasses and flowers—that had lived on that land for thousands of years, and were uniquely adapted to the soil and the climate of that precise place in the northern tallgrass prairie. After the farmer's crop was harvested, Ekstein's crew sowed the bare fields with giant planters that drilled the prairie seeds a quarter inch into the soil— the same way a farmer would plant a crop. They soon discovered that was too deep for prairie seeds—they didn't germinate well. So instead, they scattered the seed on top of the ground in fall and spring—pretty much as it happens naturally. They pressed it into the soil with massive rock rollers, machines that drag long heavy cylinders that farmers use to push rocks into the ground. Then, like the gardeners they were, they waited to see what came up.

Often what emerged first were invasive weeds such as Canada thistle and ragweed. But with patience, they found that in two or three years the prairie plants became strong enough to squeeze out the invaders. And, to their delight, the land produced other prairie plants that hadn't been included in their seed mix. Ekstein began to appreciate the depth of the "seed bank"—wild prairie seeds that for years had been hibernating in the soil below the crops, waiting for the right conditions to return.

That left the drainage ditches. Most of them had been privately dug and were owned by farmers who had used the land, and they were removed with the help of federal funding designated for restoring wetlands. But some twenty-six miles of public ditches were controlled by the local watershed district, not the county, and were governed by some of the oldest laws in the state statutes. That means that once dug, they are extraordinarily difficult to remove. That's especially true in northwest Minnesota and along the Red River, where the land is flat, and ditches are considered vital structures for farming and to protect against flooding. Without native wetlands or deeply rooted prairie grasses to soak up early season rains and winter melt, in springtime water flows in broad sheets across the top of bare farm fields and through thousands of miles of tile lines and ditches until it reaches the notoriously flood-prone Red River. In recent decades, floods have increased exponentially along the Red, in part because of increasing rainfall from climate change, and because of the agricultural practice of getting the water off the land as fast as possible. Protecting communities along the river is a chronic problem that has cost taxpayers billions of dollars. Even now, a $2.75 billion project is underway to construct a twenty-mile dike, a series of dams, nineteen bridges, and a thirty-mile canal, all to protect 235,000 people around the flood-worn city of Fargo, North Dakota.

Removing the ditches on Glacial Ridge meant re-creating the slow, natural flow of water across the land, through wetlands, and along the lazy prairie streams. Rain and snow would again filter through sandy ridges and collect in the low spots between them. But farmers had built up a century's worth of a reverence for drainage as a critical tool in managing water on the land, and dismantling the ditches meant convincing them of something that was deeply counterintuitive—taking out the ditches would reduce flooding, not increase it.

Every time Ekstein wanted to remove a public ditch he had to make a formal request to the watershed district. It was Tim Cowdery's job to do the explaining at a public hearing before the watershed board, and any landowner who had an interest. An expert hydrologist with the U.S. Geological Survey, Cowdery understood the dynamics of groundwater flows, water storage in aquifers and soil, and the cycle by which water moves from the land to the air through evaporation and transpiration by plants. He attended meeting after meeting with the ditch boards and worried farmers who believed, he said, "that any water upstream on the landscape is just sitting there waiting to get you." It took patience, and a lot of flip charts, but over time, most of the farmers came to appreciate that more prairie above them in the watershed meant less flooding on their land. Ekstein and the local workers he hired ultimately filled in 126 miles of public and private ditches, some of them thirty feet deep.

In place of ditches, Jason Ekstein and his crew re-created wetlands. They drove every yard of the property on an ATV, measuring the dips and folds in the land and looking for low spots that would tell them where the wetlands used to be. They measured in increments of six inches to find which way the water would flow and where it would collect. Even though they were re-creating a wild place, sooner or later all the water would encounter roads and the ditches that ran alongside them. And there, at that intersection between wild and the structures of civilization, it had to be controlled.

ON A BRIGHT, windy day in late July, Ekstein was showing off the restored prairie to a handful of visitors who wanted to learn how it had been done. The group included a member of the grassland conservation group Pheasants Forever, another land manager from the state Department of Natural Resources with an intern in tow, and two representatives from nearby Soil and Water Conservation Districts. Ekstein led them to a three-hundred-acre wetland that abuts a county road. This spot had been an animal feedlot on the old farm, and had been contaminated by some powerful chemicals used on sheep. After it was turned back into a wetland, the biological processes in the soil and plants broke down the contaminants, and an earthen berm was built to hold the water back from the road. The grass between the berm and the road towered head high, almost too thick to walk through. Along the other side of the berm in the wetland, a stand of cattails had taken root and grown even taller than the grass. Ekstein is not happy with the cattails—the species is invasive and now crowds into many of the 256 wetlands he has restored for the state, and its only fan seems to be red-winged blackbirds. After searching through the grass for a while, the group found a concrete-block structure. The series of blocks acts like a culvert; it allows water to flow out the bottom of the wetland and into a ditch along the road. Twenty years after it was installed, it's still working fine, a testament to the idea that simple is often best.

The last stop of the day was an open meadow that had been a potato field when Ekstein began his work. For a time the meadow had lain fallow, and invasive Kentucky bluegrass and brome had taken over before it was reseeded with native prairie plants. This meadow had been intentionally burned earlier in the spring—a common practice that mimics the wildfires that were once a key player in the grassland ecosystem. Here, the results were spectacularly evident: the lead plants were in full, breathtaking bloom, their long purple flowers and silver-gray leaves stretching almost to the horizon like a field of French lavender.

As the group of visitors spread out among the flowers taking photos and laughing with delight, Ekstein beamed with pride. He explained that on that spot, rain and melting snow drains through the sandy soil of the ancient beach ridge and flows downhill to create a calcareous fen, a rare type of wetland fed from below with water rich in calcium and magnesium. He strode across the grass, leading the group toward a clump of dogwood a few hundred yards away and quickly disappeared in its depths. Passing through the curtain of shrubs, the group emerged into a small clearing. This is one of the three hundred or so Minnesota fens that support rare plant species: white lady's-slipper, arrow grass, and others. This spot, he said, used to be open to the sky, but steadily increasing rainfalls over the last twenty years are bringing back shrubs and trees across Glacial Ridge, especially where water naturally collects. Now the fen is like a secret garden behind its wall of dogwoods. Ekstein explained all this while bouncing gently on the soft, peaty ground. The fen always makes him feel like he's a kid on a trampoline, he said. The others joined in, bobbing ever so slightly as the ground flexed gently beneath their feet.

Glacial Ridge has become a part of the fabric of this corner of Minnesota, anchoring a mosaic of wildlife refuges and other protected areas scattered across this part of the state. Visitors' bureaus in Crookston and the nearby town of Grand Forks advertise it as a destination for tourists, birdwatchers, and hunters. But resentment over the displacement of farm life still hovers close to the surface. The tiny nearby town of Mentor is mostly a town in name only. What was once a school district is now wild prairie. John Swanson farms next to Glacial Ridge on land that his family has owned for six generations, but these days he has new neighbors. At night he sees the headlights of his car reflected in the glowing eyes of the deer who congregate in the refuge. Bringing back wildlife like those deer was always the intent of creating the refuge, but he wonders how many are eating farmers' corn. Sure, Swanson said, much of Glacial Ridge was too wet or too dry for good crops. But not all of it. Who, he asked, is going to feed the world's growing population if good farmland is returned to nature?

It's an economic argument that often dominates the debate over preserving and restoring wild lands. But it ignores the other part of the equation: the largely unrecognized benefits from restoring converted landscapes back to their original condition.

One analysis published in the journal *Nature* found that restoring just 15 percent of forests, grasslands, and wetlands around the globe would avoid 60 percent of wildlife extinctions and store nearly three hundred gigatons of $CO_2$—a third of the total increase in the atmosphere since the Industrial Revolution.

And then there is the often ignored benefit to water. Crookston's municipal wells pump 24,000 gallons of clean drinking water every day for residents, even if they have no idea where it comes from. Glacial Ridge provided the rare chance to actually measure how restoring grassland changed water for the better in other ways as well. Tim Cowdery, the hydrologist, spent more than a decade tracking the flow and the quality of water on Glacial Ridge, first while it was farmland and, later, after it was restored prairie. Initially the water that came off the farmed property was contaminated with fertilizer. One fourth of the samples he took had concentrations of nitrates that were consistently higher than the safe drinking water standard set by the EPA. Some showed the highest concentrations he'd ever seen. In groundwater and ditchwater that eventually flowed into the Red River, he also detected herbicides and pesticides, many of them toxic. Just like Des Moines, Iowa, cities that use the river for drinking water had to pay to clean it up.

But once the land was returned to grass and farmers stopped applying fertilizers that were carried away by drainage systems, the water improved dramatically. Within a year or two, nitrate concentrations in the groundwater dropped by an average of 78 percent, and in ditches and wetlands they fell by half. The impact on potential flooding was even more dramatic. Between 2002 and 2015, rainfall on Glacial Ridge increased by 14 percent, but even so, the volume of water in the ditches surrounding the property dropped by a third. Even more important, the surges of water in the ditches after big storms that cause devastating flash flooding almost disappeared. The porous ground and deep-rooted grasses acted like a sponge, holding on to the water longer and releasing it slowly over time. Cowdery and many other hydrologists across the Midwest argue that if more land were returned to prairie, especially land that is steep or sandy, the country could prevent serious flood damage. Cities like Fargo might not have to spend billions on massive flood control construction projects. The land would do much of the work for free.

Whether prairies on a scale like Glacial Ridge can be re-created in other places is an open question. Compared to other states, Minnesota had—and continues to have—some unique advantages. For one, the beach ridges left behind by Lake Agassiz, and the long hills of gravel, sand, and rock created by glaciers on the western side of the state, are not good for farming. That meant more native prairies, and their precious stores of seed, were left compared to other heavily farmed midwestern states—about 2 percent in Minnesota compared to 1 percent elsewhere. Minnesota is also unique in that it has a sizable flow of public money to preserve and protect wild lands. In 2008, after a long campaign by environmental and conservation groups, Minnesota voters overwhelmingly passed a referendum known as the Clean Water Land and Legacy Amendment. It added three-eighths of 1 percent to the state sales tax and dedicated the revenue to three pots: clean water, the arts, and land protection. The tax produces more than $120 million every year to permanently restore and protect natural landscapes. The referendum is often held up by conservationists in other states as an ideal solution, one that sidesteps political gridlock and appeals directly to voters, and, as in Minnesota's case, their love of the outdoors. All of those factors contributed to the creation of another rare asset: a detailed database of where native and non-native grasslands remain, and a twenty-five-year plan for restoring some of the state's prairies, a rarity in Corn Belt states.

The state's prairie plan, once executed, would expand on what Glacial Ridge accomplished. Minnesota once had eighteen million acres of tallgrass prairie. Today, what's left of it, marked on a state planning map as a series of tiny islands, totals 250,000 acres. The state's prairie plan would create bridges connecting those precious remnants with restored prairies like Glacial Ridge, grazing lands, and privately held land forever protected by environmental easements. If completed, Minnesota would have a continuous prairie and wetland complex stretching along the western side of the state—long corridors of grass and wetlands where animals and plants could migrate and evolve with a changing climate.

The key to that ambitious plan, and the state's highest priority for protection, are its last remnants of native prairie. Few people standing on such a piece of virgin grassland would recognize how rare it is because it looks like, well, grass. But it's ground that, for one reason or another, hasn't been touched for millennia. Those patches still retain all the evolution that occurred there across the centuries, and the hundreds of species of plants and insects that figured out a way to survive together.

Humans still don't altogether understand what's there, and they certainly can't make native prairie the way evolution did, even in places like Glacial Ridge. Once plowed, it's gone. Even if plowed ground is allowed to lie fallow for decades, prairie ecologists have found that it still produces only half the number and variety of plants that grow on prairie that's never been touched. Some species never come back. They found that a century after farming has ceased, the complex web of microbes, fungi, nutrients, and carbon in the soil are still so depleted that it would take another seven hundred years for it to return to its original state.

Despite wide recognition of how valuable they are, only about half of Minnesota's precious native prairies are permanently protected. As for the rest, the state of Minnesota and conservation groups know where they are and who owns them, but it's up to individual landowners to make the decision to permanently protect them or not. And that's not happening often enough. In the last decade the 2 percent of native prairie left in Minnesota has slowly been disappearing, at a rate of 200 to 250 acres a year.

Some was lost to gravel mining and housing developments. But most of it, 62 percent, was converted to row crops. The same is true for the other types of grassland—those lower-quality prairies that were once plowed, overgrazed, or overrun with invasive plants. In recent decades, Minnesota has lost millions of acres of grasslands. Even with a powerful land protection infrastructure in place—a constitutional amendment to ensure robust funding, wide public support, stellar examples of restoration like Glacial Ridge, an ambitious state-sponsored prairie plan, and an army of dedicated experts in government agencies and nonprofits—Minnesota's prairies are still losing ground.

The case of eighty wet acres in Lac qui Parle County in western Minnesota helps explain why. The county is deep in Minnesota's farm country. About 95 percent of its expanse is planted in corn and soybeans, and as a result, almost all of its lakes and rivers are contaminated with farm chemicals. Still, Lac qui Parle is one of Minnesota's top destinations for outdoor recreation. It has some big lakes, a state park, and large areas of protected wetlands, and the Minnesota River runs through it, creating a natural wildlife corridor that's a significant part of the state's prairie plan.

A few miles south of the river is a state-owned parcel, mostly wetlands, known as the Baxter Wildlife Management Area, popular with duck hunters. Next to it are eighty acres of cropland that, like the wildlife management area next door, were once mostly wetland. In 2018 the state agreed to buy the land from its owner, a South Dakota farmer named Phillip Sonstegard, who wanted to sell it because it was still too wet to produce good yields. The DNR planned to turn it back to nature and hunters. But the Lac qui Parle County Board of Commissioners blocked the sale. They gave a long list of reasons: The parcel could still be farmed; too much land in the county was already in public hands or under environmental protection; the county would lose property taxes; and local public opinion ran against it.

But behind the official reasoning ran the much deeper conflict over what land is for—economic gain or nature? In Minnesota, a major agricultural state that also has a well-funded constitutional amendment to protect wild land, that conflict flares up repeatedly. Todd Patzer, chair of the county board, explained it one summer day while hauling a borrowed thirty-five-seat picnic table to the Lac qui Parle County Fair. The annual farm-focused event is a high point of the summer for local people and a celebration of their farming way of life. Patzer, a county commissioner for eighteen years, is a fourth-generation farmer, growing corn and soybeans on sixteen hundred acres handed down through his family. He's not against conserving land—he grew up with a hunting rifle in one hand and a fishing pole in the other. But, he says, while environmentalists worry about the monarchs and river otters, he worries more about the people and the economy in his county.

Rural places like Lac qui Parle are emptying out, he says. Minnesota's Legacy Amendment has generated enormous sums of public money—he called it a firehose—just to protect wild land forever. But that hurts rural communities. It's farmed land that drives rural economies and keeps small towns healthy. Yes, he said, farmers are also their own worst enemy: They have failed to stand up against the economic forces that made agriculture increasingly efficient—the drive toward commodity crops that needed bigger farms, larger equipment, and far fewer people. The population of Lac qui Parle today is half what it was a hundred years ago. But in Patzer's view, that makes it all the more urgent for Sonstegard's eighty acres to remain as farmland, and maybe as a place to launch a new, young farm family. As a wildlife area, he says, it just becomes a place that people visit on a few nice Saturdays in the fall.

Sonstegard, the owner of the land, sued the county in order to force the sale of his land. He had his own set of ironclad principles—property rights—which are the foundation of the country, he argued. The county judge who presided over his lawsuit against the county agreed with him and dismissed every one of the county's arguments. Ultimately, Minnesota's top three elected leaders, who have the final authority to approve land sales to the state, were asked by the parties to resolve the dispute. After a long hearing in which Patzer, Sonstegard, attorneys, and conservationists all had their say, the sale to the Department of Natural Resources was approved. The reasoning of the panel, which included the governor and 2024 Democratic vice presidential candidate Tim Walz, the state attorney general, and the state auditor, had nothing to do with restoring wild places in the heart of the state's farm country. They sided with Sonstegard: Property rights prevail over all else.

In the end it was a hollow victory in the larger effort to revive Minnesota's prairies. In order to improve its fractious relationship with the Lac qui Parle County commissioners, the state agreed to give them early notice when land in the county is being considered for protection. And, in a concession that was far more important to the board than eighty acres of marginal cropland, the state streamlined its process for approving ditch projects. Patzer said the world's hunger for corn, soybeans, and biofuels is a powerful signal that it needs all the well-drained cropland it can get. And that, he said, is "the hill we are metaphorically willing to die on."

THAT SAME RESISTANCE to protecting wild land is playing out in rural communities across the Midwest and West, and in recent years has coalesced into more of an organized political movement. On April 22, 2022—Earth Day—former republican governor Pete Ricketts of Nebraska took the stage at a Stop the 30 x 30 Land Grab summit, described by its organizers as a gathering of national leaders and frontline fighters in the battle against land conservation. The conference was initiated by a right-wing organization called American Stewards of Liberty and sponsored by some of the leading organizations who deny that climate change is real, including The Heritage Foundation and The Heartland Institute. American Stewards, which gets much of its funding from the petroleum industry, got its start by fighting the protection of endangered species. It has funded legal battles waged by ranchers and other landowners over the EPA's efforts to preserve habitat for rare and threatened spiders, burying beetles, birds, and other species. Today American Stewards embraces a larger mission of "protecting private property rights, defending the use of our land, and restoring local control."

In 2021 President Joe Biden gave the group a new, bigger target when he launched a nationwide project to preserve one-third of the nation's land and water to slow climate change and protect biodiversity. The America the Beautiful Challenge would coordinate government and private landowner efforts to reach its 30 percent goal by the end of the decade—the "30 by 30" campaign.

The American Stewards of Liberty mounted a resistance that was immediate and fierce. The group's executive director, Margaret Byfield, labeled the Biden plan "a federal land grab," and describes it as the beginning of a socialist movement to take property rights away from landowners. The organization crafted a multi-pronged campaign with Ricketts, now a U.S. senator, as its public face. It teaches local government officials how to stop permanent land protection, whether through public acquisition or private conservation easements. American Stewards of Liberty advises county boards directly, and provides templates for local laws designed to hamper conservation on public and private lands. As of 2023, roughly 134 counties had adopted them, mostly in grassland states such as Nebraska, South Dakota, and eastern Montana.

The American Stewards of Liberty is perhaps the most potent force in a trend that frightens conservationists. Once a value shared by people of all political stripes—Republican Teddy Roosevelt made it his legacy—land protection has instead become another partisan political cudgel used by conservatives and farm groups in a profoundly divided country.

American Stewards is not alone. Across the Midwest, powerful agricultural organizations also resist grassland protections, and nowhere is that more evident than in Iowa. There, a handful of nonprofits are working to protect and restore fragments of tallgrass prairie in a state that once contained thirty million acres of it. More than a decade ago, voters in Iowa passed a constitutional amendment, similar to Minnesota's, that would devote three-eighths of a cent of the sales tax to land and water protection. But the amendment has never been funded by the state legislature. Every time it comes up, agricultural groups such as the Iowa Farm Bureau defeat it in part by stoking the fear that if it came to pass, farmers could no longer compete to buy land.

Nonetheless, there are islands of prairie across the Midwest—some tiny native patches, some larger landscapes—that provide the seeds for conservation in the future. After the Dust Bowl, the federal government acquired four million acres of land, mostly abandoned farms, and converted them to permanent grasslands. A portion of that became the Cimarron National Grassland, a federal preserve open to the public, in Kansas. The Iowa National Heritage Foundation, the state's leading land conservation group, protects or restores about five thousand new acres a year. The nearly twenty-thousand-square-mile Nebraska Sandhills, one of the largest intact native prairies in the country, is largely protected by ranchers who chose to keep it in grass forever through environmental easements.

And though Minnesota is still losing grasslands, it's losing them much more slowly than it used to, and bits of farmland are being restored back to prairie. On a foggy November day Cory Netland, a wildlife manager for the state's Department of Natural Resources, looked out over a bare farm field that curved into the distance, marked by few stands of trees. The farm in the middle of the state had been sold by a conservation-minded landowner to Pheasants Forever, which had, in turn, donated it to the state for restoration to prairie. The satellite photos in Netland's hands clearly showed the low spots that were once tiny wetlands and shallow lakes, and the higher rises that were once covered with grasses and flowers. In Netland's imagination he could see the beautiful place it would become. His title may be wildlife manager, he said, but it's really prairies that he manages. First, he would dredge the low spots and move the soil back to the hilltops where it sat before water and wind washed it away. In the bottoms, he would sow a mix of wetland plants, forty-five kinds of flowers, and fifteen types of grasses and sedges. On the hilltops he would seed fifteen grasses and fifty-seven species of flowers for the bees.

Prairie managers like Netland have learned a lot since Jason Ekstein first set foot on what would become the Glacial Ridge National Wildlife Refuge. They know how to create natural seed mixes, and what kinds of plant grows where. Using local consortiums of prairie owners, they buy and trade rare native seeds that evolved in specific regions of the state. After all, an aster from southern Iowa wouldn't fare well in the cooler climate of central Minnesota. Netland had a list of seeds and knew precisely where each came from—wild bergamot, showy tick trefoil, Maximilian sunflower, switchgrass, and dozens more. Bringing prairie back is an expensive proposition—$4,000 to $5,000 per acre to buy the land, another $1,000 per acre for native seeds—much of it paid for with Legacy Amendment money from state sales taxes. He pointed toward the horizon where a pair of giant oak trees emerged from the fog as silhouettes against the dark sky. "I think I'll leave those," he said. They were what's left of an oak savannah that stood there before Europeans showed up, a grove of gnarled trees and grass that was once a common sight on the prairie landscape. Now the acorns from those two trees would be left to grow. And when the tiny acorns become giant, spreading trees in a century or so, the restored prairie around them might resemble what it once was, too.

CHAPTER ELEVEN

## *Farmers*

ERIC AND KELLY HOIEN WERE FLYING HOME FROM A WINTER VACATION some years ago when they had an epiphany over the Gulf of Mexico. Eric recalls looking out the airplane window and seeing what looked like a giant mud puddle off the Louisiana coast. "It was brown, not blue, and really ugly. Then I realized I was looking at the Dead Zone."

The Hoiens had read about hypoxia in the Gulf of Mexico and the role of midwestern agriculture in creating it. They knew about nitrates fouling Iowa's rivers and Bill Stowe's battle to clean up Des Moines's drinking water. And they had just bought 129 acres of prime farmland near their home in northwest Iowa—land that drained into the watershed of the Mississippi River.

"So whatever ran off our land was going to wind up polluting the biggest river system in North America," Eric Hoien recalls. "It was one of those moments when you realize you're part of the problem."

On a steamy July morning, Eric and Kelly are leading a tour of conservationists and curious farmers across their property. In a field behind Eric their corn stretches off toward the eastern horizon, already six feet high and a luxuriant green. Three hundred yards in the opposite direction, across an asphalt frontage road, vacationers are frolicking in Big Spirit Lake, whose sparkling waters attract families from all over Iowa for swimming, fishing, and water-skiing. And in between, along the western edge of the Hoiens' cornfield, runs a broad strip of tall grasses and flowers that forms a barrier between the crop and the lakeshore. In conservation lingo it's known as a prairie strip or buffer strip. Bees are buzzing from blossom to blossom and swallows dart over the grass tops. Eric walks along the strip, naming the plants: "This is big bluestem, it could get seven feet high. This is rye, and this is switchgrass—notice how thick that stem is. There's some bergamot. The golden blossoms are black-eyed Susans."

The Hoiens are hosting a field day, an on-farm demonstration of conservation techniques sponsored by a remarkable little nonprofit called Practical Farmers of Iowa. From its headquarters at a small office in Ames, the group promotes farming practices that protect water and soil, spreading its message from farmer to farmer during walking tours and buffet lunches in barns and machine sheds. The group's cheery personality is reflected in the merchandise for sale at headquarters—cherry-red T-shirts and baby onesies emblazoned with the PFI logo. But they are dead serious about the mission, and after four decades of rapid growth they are showing, farm by farm, how to change the face of Iowa agriculture.

PFI represents one strand in a debate known as "sparing versus sharing" that has engaged, and sometimes divided, conservationists around the world. The strategy called "sparing" leaves wilderness alone or restores native landscapes such as Glacial Ridge in northwestern Minnesota, on the argument that wild plants and creatures do best in the absence of humans. The alternative, "sharing," allows continued human use—farming, ranching, perhaps recreation—but in a way that protects wildlife and mimics the ecological benefits of a natural landscape. For those parts of the prairie where farming is pretty much here to stay—highly productive farming states such as Iowa, Indiana, Illinois, Minnesota, Nebraska, Kansas—groups like Practical Farmers of Iowa and the sharing strategy may be the best hope for protecting the environment by creating a form of agriculture that mimics the performance of a native prairie.

PFI was born almost accidentally during the farm crisis of the 1980s, the worst recession in rural America since the Great De-

Dick Thompson, an Iowa farmer who studied conservation practices with statistical precision, helped found Practical Farmers of Iowa. *Photograph by Renee Jones Schneider, courtesy of the* Star Tribune

pression. Larry Kallem was executive director of a group that represented Iowa co-ops, and he was organizing a conference for farmers trying to survive the global collapse in commodity prices. He had heard about a local farming couple, Dick and Sharon Thompson, who were known for embracing unconventional practices while still running a successful farm. In particular, the Thompsons were experimenting with conservation techniques that could improve their soil and their crops while saving money on chemical fertilizers and herbicides. In other words, techniques that could cut the cost of producing a crop and reduce the size of their bank loans—and help farmers survive the 1980s. He invited them to give a talk and he wasn't disappointed.

"They drew a big crowd, and as I watched from the back of the room I could see that every person in the room was paying attention," Kallem recalls.

Dick had a degree in horticulture from Iowa State University and liked to apply scientific methods to their land. He also knew Wendell Berry, the great conservation writer and farmer from Kentucky, and read widely, including authors such as Henry David Thoreau and John Muir. The Thompsons were early adopters of several conservation practices on their own land, and they had received a grant from the Rodale Institute in Pennsylvania to study the results. They compared harvests and profits from one field to the next and partnered with a grad student at Iowa State to help them run in-field trials. They had the numbers to prove what worked. "Dick did the talking and Sharon ran the slide projector," Kallem recalls. "I was completely knocked out."

Kallem noticed that the Thompsons were a hit with the audience as well. When the Thompsons finished their talk, he approached them and asked if they thought a wider audience would appreciate their ideas. As it happens, the pair were scheduled to speak to another group the following week, and Dick said he would ask the audience who wanted to know more about conservation farming. A week later he called Kallem. Every hand in the room had gone up. The three began discussing how to form a nonprofit.

By 1990 PFI had a staff of seven and a growing budget. They got a grant of $50,000—a giant sum for them—from the Iowa Farm Bureau, the massive lobbying and service organization that had become the voice of big farmers and the industry that served them—and would help defeat the Des Moines Water Works lawsuit years later. The organization's leaders easily could have sneered at their conservationist philosophy but they didn't. "The

Farm Bureau could have crushed us like a bug," Kallem says. "But they saw that we were on the side of farmers, and we had a lot of respected farmers in our membership. They gave us money, and we never stopped thanking them."

Another early grant came from heirs of Henry Wallace, an Iowa legend who founded the Pioneer Hi-Bred Corn Company, served a term as Franklin Roosevelt's vice president, and ran for president in 1948. Membership stood at several hundred and was growing steadily.

At the heart of PFI's message is a trio of well-known and widely studied conservation techniques, much the same practices that Gail Fuller adopted in Kansas and Matt Leisinger employed in South Dakota. The first is no-till planting, or drilling seed directly into the ground amid stalks and stubble left from the previous fall's harvest rather than plowing open new furrows. The practice greatly reduces soil erosion and moisture loss, allows plant stubble to protect the ground and feed the soil as it decays, and avoids disturbing the underground insects and microbes that build soil health. A second is planting cover crops after the fall harvest—usually grains or legumes—and leaving them to grow over the winter. They serve as a blanket across the naked ground, preventing erosion, returning nutrients to the soil when they die, and providing the year-round living roots required to nourish the underground community of bacteria, fungi, and bugs. In the spring, before planting their cash crop, farmers can let livestock graze on the cover crop, harvest it for livestock feed, or "terminate" it with a chemical herbicide or by plowing it under. A third technique is diversified crop rotations. By supplementing the typical corn-soybean rotation with grains and legumes—oats, barley, rye, alfalfa, and peas, crops that were once common across the Midwest—the practice suppresses parasites and allows the soil to recover from one year to the next. And some cover crops, such as alfalfa and peas, are the legumes that actually restore, or "fix," nitrogen in the ground.

These techniques are not the same as organic farming. PFI members still use glyphosate and other herbicides to control weeds or terminate a cover crop. Nitrogen fertilizer is not off-limits, nor are genetically modified seeds. But by mimicking the traits of native prairie—plant diversity, healthy roots, year-round ground cover—the techniques greatly reduce a farmer's use of chemicals because they use natural forces to build soil fertility and fight pests, much like the practices promoted by Jonathan Lundgren and the Purdue insect researchers. They also greatly reduce erosion and the volume of chemicals that run off into local waterways. PFI's staff and members generally prefer the term "regenerative" to describe their farming practices.

From the very beginning PFI showed remarkable political acumen. Iowa is a state where farmers tend to vote Republican and many are quite affluent. Few states embraced the revolutions in farm machinery and chemicals more fully, so that by 2020 farms of one thousand acres were not uncommon and many had annual sales over $1 million. At $190,000 a year, the net income of the average Iowa farmer was well above the average American household income. In that environment, Kallem and the Thompsons knew they could not come off as preachy liberals or dreamy environmentalists. They emphasized that their conservation practices help farmers save money on fertilizers and pesticides, and that lower costs lead to higher profits. They kept politics off their website, and they were diplomatic in the way they critiqued conventional agriculture—the kind practiced by most of their neighbors and promoted by the big farm groups and the food industry.

The same philosophy prevails today under a new executive director, Sally Worley. "We're a big-tent group," Worley says. "If folks are interested in practical, regenerative farming, we're not going to turn them away."

Following the Thompsons' example, PFI also cultivated a close relationship with Iowa State University and the University of Minnesota, premier agricultural research schools where influential scholars such as Matt Liebman, David Tilman, Jason Hill, and Natalie Hunt have produced study after study examining the ecological and economic benefits of regenerative farming practices. PFI cites Iowa State research at their field days and often invites young scholars to participate.

Next to the registration table at the Hoiens' field day, the PFI staff had planted a tall blue banner that lists the proven benefits of prairie strips like the one on their land. The banner reflects a landmark study at Iowa State led by a landscape ecologist named Lisa Schulte Moore. In the mid-2000s, wondering how farmers could reduce their environmental footprint and still earn a living, she began to investigate whether they could use the natural strengths of native plants to offset the damage caused by intensive agriculture. The deep, strong roots of native species such as big bluestem, blazing star, and leadplant can soak up excess fertilizer, form dense underground mats that hold soil in place, and build soil structure so it becomes a subsurface sponge.

In 2017 Schulte Moore and several colleagues put their theory to the test at a set of small laboratory fields near Ames. They chose fields with a gentle slope, typical of about 40 percent of Iowa farmland, because they are most likely to shed water full of chemicals. Each "catchment" field was planted in corn and soybeans, then given varying prairie strip treatments. On one set of plots they set aside 10 percent of the land for a strip at the bottom of each field, which Schulte Moore now calls the "diaper" treatment. On a second set they planted buffer strips across the middle of the field—as "speed bumps"—also using 10 percent of the land. On the third, they devoted 20 percent of the field to the strips. The fourth set had none. Then they set up gauges to measure soil erosion and the volume of farm chemicals running off the fields as well as bird and insect populations.

The results were stunning: Buffer strips produced environmental benefits far out of proportion to their small size. They reduced soil erosion by 90 percent and cut runoff of nitrogen and phosphorus by 75 percent and 70 percent, respectively. Pollinator populations in the fields rose more than threefold and bird species more than doubled. But the knockout finding was the inflection point: Big gains came when as little as 10 percent of the field was devoted to grasses and flowers.

"Up to that point, there was something of a stand-off between farmers and environmentalists," Schulte Moore says. "For farmers, 10 percent was sort of the upper limit—beyond that and they would say we're impinging on their ability to make a living. For conservationists, the lower limit was 25 or 35 percent of the land devoted to buffer strips. Below that, they said, and it's just land for production."

By demonstrating that midwestern farmers could achieve major environmental gains without taking large tracts of land out of production, Schulte Moore's work broke the standoff and created common ground for farmers and conservationists. The Iowa State findings were celebrated across the state by leaders of both political parties. When Congress wrote a new Farm Bill in 2018, it made buffer strips eligible for certain subsidies. The amount of land planted in buffer strips nationally soared from five hundred acres in Iowa in 2019 to more than fifteen thousand acres in fourteen states in 2022.

The study put Schulte Moore on the national map of conservation agronomists and, in 2021, won her a MacArthur genius grant. It also endeared her to the leaders and members of PFI because it integrated conservation ideals with farming pragmatism. She served three years on the group's board and remains a fan. "PFI rocks," she likes to say.

The avenue that PFI opened between farmers and scholars runs in both directions—that is, the scholars often learn from the farmers. In 2014, an Iowa State plant pathologist named Alison Robertson was approached by a colleague with a puzzle about cover crops. Farmers were finding that if they planted winter cereal rye—a common cover crop in Iowa—they sometimes suffered reduced yields in their subsequent corn crop. The effect was sporadic and unpredictable, but enough to give cover crops a bad name among farmers. Robertson, who specialized in a plant fungus called pythium, immediately suspected that cereal rye might be transmitting the parasite to the corn during its spring germination.

Like Schulte Moore, Robertson had become a fan of PFI and thought its members might lend their fields to her for a set of trials. The first experiment: What would happen if a farmer waited ten days between killing the rye and planting the corn? Checking their hypothesis at university test plots and on actual farmers' fields, they found the delay reduced the incidence and the severity of the pathogen in a farmer's corn while increasing the corn yield. The result was important, because it both pinpointed a cause of yield loss and offered a solution. But some farmers warned Robertson that they couldn't always wait ten days in Iowa's brief spring before planting their corn. In a second experiment, again working with PFI farmers, Robertson set out to test whether "social distancing"—planting the rye in rows fifteen inches from the corn—would stop the pathogen. This test, too, found less pythium transfer and less "yield drag" in the corn.

Running trials in a real farm field has its drawbacks. In a lab or on university test plots, the researcher can control every variable—precipitation, temperature, pests, and so on. On a real farm, researchers might have a dry spring, or an impatient farmer might not wait ten days to plant the corn. But it also has major benefits: It can convince skeptics that a certain practice works in the real world, not just under laboratory conditions, and it often gives farmers the confidence to endorse a certain practice with their neighbors. Robertson cites research showing that farmers are more likely to adopt a practice if they hear about it from a neighbor than if it comes from an expert. "You know—'If it works for Frank, it will probably work for me.'"

Since American farmers began wider adoption of these conservation techniques in the 1990s, agronomists at universities across the country have conducted voluminous research on their effectiveness. The results are not always encouraging. No-till planting, for example, generally requires the use of herbicides because the farmer isn't using a plow or disc to kill weeds as they sprout. Cover crops can soak up soil moisture that otherwise would have been available to the cash crop in the spring, and they require additional work to plant and terminate. And because they shade the ground, they can slow the soil's spring thaw, a big disadvantage in northern climates with short growing seasons. Finally, both practices can entail substantial upfront costs—special equipment for no-till planting, seeds for cover crops—and it can take several years before they pay for themselves in higher yields or lower fertilizer bills.

But as researchers at Iowa State and the University of Minnesota launched major studies of the practices, the evidence began to swing in their favor. A series of definitive papers by Natalie Hunt, Jason Hill, and Matt Liebman found that diversifying a crop rotation by periodically adding clover, oats, and alfalfa produced huge reductions in erosion, chemical runoff, and herbicide use while delivering crop yields that were equal to or higher than conventional crop rotations. An additional study by Hunt, Liebman, and colleagues in 2020 found that the diversified crop rotations also produced major reductions in consumption of fossil fuels and greenhouse gas emissions because the farmers needed less nitrogen and diesel fuel. A long-running study by David Tilman at the University of Minnesota found that test plots with greater plant diversity had more success at surviving drought and were less likely to "leak" chemicals and fertilizers into surrounding soil and water. A series of studies by Jerry Hatfield, the former director of the USDA's Agricultural Research Service station in Iowa, found that adopting no-till cultivation and cover crops in a controlled long-term study led to dramatic improvements in soil fertility, soil moisture, and soil structure. Specifically, it increased carbon and moisture in the farm's soil, which in turn raised yields with less fertilizer, improved harvests in dry years, and reduced in-field flooding in wet years. "It weather-proofs your land," Hatfield says.

As a bonus, there is some evidence that healthier soil from these practices can store more carbon, meaning that they might also help in the battle against climate change. The distinguished agronomist Rattan Lal has calculated that soil could pull a significant amount of carbon from the atmosphere if agriculture was managed properly—enough to make a difference in global warming. Jo Handelsman estimates that a 10 percent increase worldwide in what's called soil organic matter—carbon-rich plant and animal residues that are decomposing—in the next few decades could reduce atmospheric carbon dioxide by a remarkable 25 percent. This field of work has generated a huge research literature on the subject—and some challenges to the practical promise of soil carbon sequestration. For one thing, carbon stored in the ground could be released instantly if a farmer changed practices and reverted to conventional plowing. For another, storing extra carbon in the ground could require adding nitrogen because soil needs a balance of the two elements—and nitrogen has its own well-known problems, including polluted water and release of nitrous oxide into the atmosphere. Nonetheless, healthier soil has myriad benefits—and it's no coincidence that these practices mimic the ecology of native prairie plants: They build biological diversity, suppress pests, hold soil in place, and nourish the underground microbial community. Charles Rice, the Kansas State scholar working on soils and climate change, takes pains to say he is not anti-agriculture. "I just want agriculture to perform more like native plants—more efficient use of water, more efficient use of fertilizers."

In 2013 PFI hired its own research director, an Iowa State PhD named Stefan Gailans, and today the group conducts its own in-field experiments to supplement academic research. These are less elaborate than university trials but still use rigorous techniques such as randomization and controlled variables—and they have the benefit of producing evidence from real fields farmed by real farmers. Generally they continue to show the benefits of regenerative practices. A farming couple named Kellie and A. J. Blair worked with PFI to see if they could reduce the use of nitrogen on their oats crop. Testing three fields—one with twenty-five pounds of nitrogen per acre, one with fifty pounds, and one with seventy-five pounds—they found that increased nitrogen raised their yields very little but cost them thousands of dollars. Comparing revenue to costs, they found that their peak profitability came in on the field with the least nitrogen. Another experiment, conducted on three Iowa farms, found that use of the controversial pesticides called neonicotinoids produced no increase in soybean yields. Growing soybeans without the insecticide saved money and gave the proprietors a higher return on each acre without any increase in pest damage—and reduced the use of a chemical that has been linked to pollinator deaths. A third study found that cover crops, although they soak up soil moisture in the early spring, actually help the soil conserve moisture by midsummer, when crops need it most.

Despite its affinity for science, the real genius of PFI is that it knows how to talk farmer. Farmers dominate its board of directors and develop the research agenda. They testify at the Capitol when an important issue is before the Iowa legislature. Farmers also design and host the field days. In this way, the group is very much a throwback to an earlier era of rural cooperatives and agrarian organizing. As far back as the 1890s, rural leaders were creating farmer-to-farmer educational forums such as the Farmers Institute, which organized short conferences during the winter months and sponsored lectures by agronomists and, later, university extension agents.

Although midwestern farmers have a reputation as independent and taciturn, anyone who has spent a morning in a small-town café knows that many are curious, eager for company, and happy to share their experiences. PFI events, typically held in someone's barn or machine shed, generally feature a homey buffet lunch and often take on the congenial atmosphere of a country wedding or birthday party. At the Hoiens' field day in July, when participants broke for lunch the buffet line buzzed with questions, advice, and anecdotes. A man in a Pheasants Forever cap asked his neighbor if he had tried hairy vetch in his cover crop mix. Another asked her lunch companion if aggressive grasses in his buffer strips encroached into his corn. Nearby, another table was debating the best planting depth for alfalfa to get good seed-soil contact. In other words, it was the colloquy of people who are, in effect, practicing applied science.

WILL CANNON WAS an undergraduate at Iowa State University when a school field trip took him to Dick and Sharon Thompson's farm. He was taken by their conservation philosophy and pragmatic approach, and when he finally got his own farm about a decade later, his first decision was to adopt variations of no-till cultivation. Today he and his wife, Cassie, farm about one thousand acres near Prairie City, Iowa, growing mostly corn and soybeans. And they are PFI evangelists. They say no-till cultivation has steadily improved the structure and nutrient content of their soil while also improving its ability to hold moisture. Before long they began experimenting with cover crops, generally cereal rye, mostly to reduce soil erosion from winter winds and spring rains. Soon they were trying a wider crop rotation that includes cereal grains as feed for Will's brother's cattle.

The Cannons are typical of a younger generation experimenting with new practices across the Midwest. They're willing to challenge the customs of their parents and grandparents, but practical enough to know that new practices have to pencil out. Will is also open-eyed about their experiments; the new techniques don't always improve their returns or work out as he hoped. He and

Cassie experimented with winter peas and hairy vetch as cover crops but found that they weren't very winter-hardy. Finding the right way to sow his cover-crop seed also required some trial and error before he settled on an old-fashioned harrow modified slightly for his fields. But he tracks his financials carefully and says no-till saves the farm huge amounts of money. Because he makes only one pass across his fields in the spring—instead of two or three to plow and disc the soil—he saves many hours of labor and thousands of dollars in diesel fuel and machinery repairs. And because he sows seed into a narrow strip rather than a broad furrow, he can target his fertilizer carefully rather than broadcasting it across his rows. In a region where farmers can spend $50,000 to $100,000 a year on nitrogen fertilizer, this produces significant savings. Likewise, his cover crops help regulate soil moisture so fields are less muddy in the spring and less parched in August; he and Cassie have had several seasons when they could begin planting before their neighbors, a big advantage because extending the growing season by even a few days can measurably improve yields.

The Cannons were able to gamble on these practices in part because PFI's corporate partners share the cost. The Cannons sell their corn mainly to Cargill, the giant Minnesota grain merchant, which makes corn sweeteners for PepsiCo. They sell their soybeans mostly to Archer Daniels Midland, which processes them for Unilever, the parent of Hellmann's mayonnaise. Both PepsiCo and Unilever donate to PFI, which uses the money to provide cost-share grants to farmers. Under pressure from consumers, shareholders, and advocacy groups, all these corporations are trying to reduce their chemical and carbon footprints, and PFI turns out to be the perfect partner in that effort.

At the same time, the Cannons move cautiously because they have to grow crops for a proven market. Although they are experimenting with a more diverse crop rotation, including millet, field peas, and a hybrid cereal rye, they rely on corn and soybeans for their main income. "Other crops can't compete with corn and soybeans because their price is so much better," Will says one evening while doing the books. "As a farmer looking at the economics, I have to look at which crops will support my family."

The caution of farmers should come as no surprise, given their huge outlays for seed and fertilizer, the daunting loans they must take out just to put a crop in the ground, and the volatility of grain prices in international commodity markets. To break out—to plant sunflowers or hairy vetch, or leave corn stubble on the ground all winter—isn't just a financial gamble. It also risks mockery and derision from one's neighbors. Peer pressure isn't unique to farm country—think of the suburban homeowner with too many dandelions—but stigma can carry extra weight in a small community where one planting mistake can risk losing the family farm. Rob Larew, president of the National Farmers Union, the more left-leaning of the two major advocacy associations, points out that farm lending practices and federal subsidies both tend to discourage novelty and experimentation. Most farmers can't obtain a spring operating loan unless they have secured federal crop insurance, and the government's underwriting rules, designed to minimize risk, have the effect of binding farmers to the crops and practices they have used in the past. "We have a system that is hell-bent on predictability and efficiency," Larew says. "To ask a farmer to make significant changes is a huge risk."

To reduce that risk, PFI offers a suite of incentives and financial cushions. It will pay a farmer $10 to $20 an acre to plant cover crops, perhaps 20 to 50 percent of the overall cost. With funding from PepsiCo, it also offers a fertilizer yield warranty, which protects farmers if using less fertilizer reduces their yields.

For PFI, the formula seems to work. Its membership has nearly quadrupled in the last decade, to roughly 6,300 farmers in 2023, and hundreds more attend its field days and conferences. Its annual revenues have grown to nearly $4 million, and in 2022 it distributed $1.2 million in cost-share grants to farmers. Its donor list includes Clif Bar and the Walton Family Foundation—

organizations that have placed the battle against climate change among their priorities.

Has it transformed Iowa agriculture? No, but it's making inroads. PFI members are a small percentage of Iowa's 85,000 farmers. But their land represents over three million acres, or about 10 percent of Iowa farmland, enough to make a measurable change in water pollution and soil erosion. It is also widely emulated in other states by groups such as the Kansas Soil Health Alliance, the Pennsylvania Soil Health Coalition, and the Midwest Cover Crops Council.

PFI also offers a window into the glacial pace of change in farming nationally. By almost any measure, farmers like PFI's members represent a small footprint in American farming. Organic operations, which are different from regenerative farming but a fair proxy, still represent less than 1 percent of the nation's farmland. Land in cover crops represents only about 5 percent of the nation's harvested acres. But federal and industry data show that these practices are spreading rapidly. The number of acres under no-till cultivation more than doubled in the last two decades. And the amount of land that farmers enroll in the Conservation Stewardship Program, one of the government's biggest conservation programs, has risen sixfold in the last decade.

For farmers like Will and Cassie Cannon, the biggest challenge isn't fighting the weeds or harvesting the corn. It's finding someone who will buy their novel crops and someone who still sells seeds for crops that have become scarce. Sally Worley notes that these farmers often have to look far and wide to find seed and markets for unconventional crops such as flax or barley. "They have to create their own system from beginning to end," Worley says.

IT'S A WEDNESDAY morning in January, and Luke Peterson is handing out slices of whole wheat bread from a folding table at the entrance of a Minneapolis food co-op. Peterson is young, bearded, and handsome enough to be a fashion model for Patagonia or Blundstone boots. But today he's a farmer. A few years ago he left a job with a Minnesota state agency to try his hand at organic agriculture under the guidance of a Minnesota pioneer of the practice named Carmen Fernholz, who was approaching retirement. On their A-Frame Farm in west-central Minnesota, Luke and his wife, Ali, grow a huge variety of crops, including corn, soybeans, buckwheat, flax, sunflowers, oats, alfalfa, an ancient wheat called emmer, and an experimental wheat called Kernza—all organic.

The Petersons soon found themselves in a larger community of conservation-minded farmers that includes PFI and the South Dakota Grassland Coalition. As his network grew, Luke stumbled upon Baker's Field Flour & Bread, a small Minneapolis grain miller that specializes in organic and alternative grains and now buys Luke's wheat. Baker's Field's head baker/miller Patrick Wylie, another handsome Gen Xer, is at the co-op with Luke with loaves of his bread and samples of whole-grain flour. Buttering a slab of bread with local butter, Peterson jokes that his summer job is running the farm but his winter job is building the customer base. "This is the work you have to do," Peterson says.

As Luke Peterson's experience suggests, changing the nation's food system is slow work. A handful of small food companies that emphasize their ecological bona fides have led the way. Niman Ranch, which is known for its high-quality pork and beef and counts Chipotle among its customers, sets environmental guidelines for its hog farmers and pays them a premium to cover the extra costs. The outdoor garment company Patagonia launched Patagonia Provisions, mainly by choosing niche suppliers who abide by strict environmental practices and are paid accordingly. Wisconsin's Organic Valley, the world's biggest organic farmer cooperative, sets organic standards for its milk, egg, and beef suppliers and pays them a premium. But these efforts are tiny in a $2 trillion food industry. "We joke that every time a new Chipotle opens we can add one more farmer," says Alicia LaPorte of Niman Ranch.

Changing the larger food industry in a way that would transform American farming is astonishingly difficult. By one estimate, 90 percent of the nation's corn and soybeans go into livestock feed or biofuels such as ethanol, not into grocery products such as Corn Flakes or Doritos. So even if consumers demand more organic products from Kellogg's or PepsiCo, it wouldn't change farming practices across most of the Corn Belt. Then too, most federal farm subsidies are built around conventional crops such as corn, wheat, and soybeans and have the effect of rewarding farmers for sticking to what they've grown in the past. Even when food manufacturers want to buy from regenerative farmers, expanding their source network can be maddeningly difficult. In an industry defined by ferocious price competition, companies want a supply chain that is streamlined, dependable, and cheap. "They're not going to send two trucks to western Minnesota every month, one for Luke's corn and one for his neighbor's," says Elizabeth Reaves of the Sustainable Food Lab in Vermont. And woe to a company like General Mills if it misses a delivery to Kroger because three small farmers had a poor Kernza harvest. More than one corporation has found that its chief sustainability officer, rewarded for green practices, is at war with the chief procurement officer, rewarded for finding low-cost producers.

Working with PFI gives food corporations a way to help change farming practices without overhauling their procurement systems. Rather than expanding their supply network to reach alternative farmers, they try to change the practices of farmers who already are in their supply chain. Unilever, which buys huge volumes of soybean oil for its Hellmann's mayonnaise, partners with PFI to pay midwestern farmers $10 per acre to plant cover crops. If a new farmer needs convincing, Unilever will even go "40 by 40," that is, pay $40 per acre on a forty-acre experiment. Its cover-crop network has grown from twenty-five farmers to five hundred in the last decade and now covers nearly 200,000 acres. PepsiCo, which uses huge volumes of corn sweetener, joined Unilever a few years ago and offers similar subsidies through PFI.

As a result, many of the big consumer food companies support regenerative agriculture by writing checks to groups like PFI rather than by reinventing their product lines. Paying farmers to change their practices on existing acres is simpler and cheaper than launching new products and persuading American consumers to change their eating habits. But it's not the kind of structural economic change in the food industry that would drive sweeping changes in farm country. "We have a philanthropic solution but we don't have a market solution," says Alexis Bonogofsky of the World Wildlife Fund.

And who can blame them? In a capitalist economy that prizes financial performance, CEOs who drift too far toward social responsibility may get punished by stockholders. The case of Danone Foods is an object lesson: The French conglomerate that produces Evian water, Dannon yogurt, Horizon Organic milk, and other packaged foods became a global sustainability star after Emmanuel Faber became CEO in 2014. Its board even rewrote the corporate bylaws to establish Danone as a "purpose-driven enterprise." But in early 2021 Faber was ousted by the board of directors after a period when Danone underperformed its rivals. The shareholders who forced Faber's ouster said the reason was poor financial performance, not its environmental agenda, but the episode dethroned an executive widely seen as a role model for doing both.

This leaves reformers caught in a vicious circle. Farmers won't gamble with novel crops or new practices unless they have the support of food companies and grain merchants. The big food companies won't commit to alternative suppliers or inputs unless it's clear that consumers will buy the resulting products, perhaps at a premium. And consumers, although they tell pollsters they want climate-friendly products, have so far shown only a grudging willingness to pay higher prices for organic food or try novel items that might flop at the family dinner table.

How could the country break that cycle and change the way America uses land to produce food? A simple and powerful step would be to reform federal farm subsidies, which Congress updates every five years when it writes a sprawling piece of legislation known as the Farm Bill. In recent years Washington has paid American farmers some $22 billion in annual subsidies—mainly through price-support programs that cushion farmers against a drop in market prices, and through crop insurance, which indemnifies farmers against hail, drought, and other natural disasters. In some years, federal payments have accounted for as much as 40 percent of farmers' net income, so Washington has huge influence on farmers' decisions. And while funding for conservation in the federal Farm Bill has risen sharply in the last three decades, it is dwarfed by subsidies that reward farmers for planting more acres and growing more commodities. While farm policy is eye-glazingly complex, the major environmental groups generally agree that a few simple steps could reward conservation practices rather than pure volume of production. One would be to simply increase funding for programs like the Conservation Stewardship Program, which subsidizes cover crops and other conservation practices on working farmland. There is no shortage of takers. In recent years the CSP has turned away roughly half of all applicants for lack of funding. Congress could also reform federal crop insurance, which is now by far the largest safety net in the Farm Bill. The Department of Agriculture could offer farmers a discount on crop insurance premiums for adopting proven conservation practices; it could also change the practice of tying insurance payouts to a farmer's past yields, which only rewards more plowing and planting. In the 1985 Farm Bill, Congress took a stab at protecting grasslands with a program called Sodbuster, which said farmers can lose their crop insurance if they plow up highly erodible land to plant crops. But Sodbuster doesn't apply to every state and, by most accounts, is poorly enforced on the ground. Making Sodbuster national and tightening up enforcement would dramatically cut the incentives to break grass for crops. A related step would be to pass the American Grasslands Conservation Act, which proposes an ambitious set of federal grants to reward landowners for preserving and restoring prairie habitat. The bill was written by Democratic senators—Ron Wyden of Oregon, Michael Bennet of Colorado, and Amy Klobuchar of Minnesota—but attracted bipartisan support.

The single most immediate change in federal farm policy would come from eliminating, or just reducing, the federal subsidy for corn ethanol. The so-called biofuels mandate, which requires oil companies to incorporate a certain amount of crop-based fuel into their gasoline, raises the price of corn and encourages farmers to plow more land to grow more of it. As South Dakota's farmers have learned, it has made farmers players in the nation's energy economy as well as the food industry. While voting for repeal of the mandate is considered political suicide in pivotal states such as Iowa, Wisconsin, and Illinois, it could certainly be reduced or reformed.

A second force that might break the vicious circle is technology. While history provides reason to be cautious about technological breakthroughs, several agronomists across the Midwest are trying to harness science in ways that could reward farmers for protecting soil and water. At the University of Minnesota's farm campus in St. Paul, researchers are experimenting with a handful of new crops that could change the arithmetic of regenerative farming. Agronomists here have a long track record of bringing new crops to market, including new soybean varieties in the 1960s and then a game-changing variety of turf grass. Today, in a project called Forever Green Initiative, they're chasing an ecological goal: to breed a new generation of winter crops and perennial grains that would protect the soil all winter, much like cover crops, but that would also turn a profit for farmers. "Farmers won't plant a crop if they can't make money on it," says Don Wyse, the institute's director. The Forever Green Initiative has more than a dozen new crops under development, the most promising of which are advanced oilseeds such as pennycress and winter Camelina and the perennial wheat known as Kernza. Pennycress and Camelina can soak up excess fertilizer and protect the ground all winter, then produce an oil that could be refined into jet fuel. Major airlines and aviation companies, many of which have thrown their support behind ethanol, welcome the prospect of another climate-friendly fuel and have invested in the research. If these experiments succeed, they would instantly make cover crops a profit center rather than a cost and, potentially, hugely increase their adoption.

Perhaps the most ambitious of these efforts is The Land Institute, the Kansas research nonprofit founded by a visionary agronomist named Wes Jackson. Scientists here embrace the philosophy of making agriculture function more like native prairie, but they take the mission to an almost futuristic level. Today their research focuses on an elementary fact of botany familiar to every gardener: the difference between annuals and perennials. Annuals are plants that complete their life cycle in a single year and must be planted new each spring. Because of this, annuals put most of their energy into their seeds, which explains why a kernel of wheat or an ear of corn contains so many carbohydrates useful to cows and people. The world's major food crops—wheat, corn, sorghum, rice, and soybeans—are annuals, a fact that condemns farmers to spending so much money on seed and fertilizer to put in a crop every spring—and why erosion and chemical runoff continue to plague much of the Farm Belt.

Perennials, by contrast, survive year after year, generally regrowing from their roots. Crucially, perennials, including most prairie grasses, tend to have deep roots that stay alive all year, holding the soil in place and cycling nutrients through their ecosystem. Even better, the farmer doesn't have to plow open the soil every spring to replant them. In other words, they protect and build healthy soil much as native prairie plants did.

By patient breeding and testing at its Salina research station, The Land Institute aspires to develop an agriculture based on perennial versions of these food crops. Their most famous progeny is a perennial wheat called Kernza, which is undergoing test cultivation at several sites around the world, including the University of Minnesota's farm campus. You can buy Kernza flour in the grocery store—at a hefty premium over conventional flour—and drink it in a pilsner beer brewed by the much-loved Dogfish Head brewery. The institute's perennial rice has produced several harvests over more than four years in China and has been served at ceremonial meals at food conferences.

The rap against Kernza and its siblings is that they're still in development and might not be competitive with conventional annual grain crops for decades. Tim Crews, The Land Institute's research director, acknowledges that it will be years, possibly decades, before their crops can compete as food sources. But he points out that it took centuries for humans to develop the highly refined wheat varieties and corn hybrids that feed the world today. The fabled Roman and Egyptian farmers, for example, used selective breeding to produce wheat with large kernels and uniform height. Kernza, by contrast, only came on the scene in 2008.

It's hard to be patient under the pressures of climate change, water pollution, and species loss that are ravaging what was once the tallgrass prairie. But if Don Wyse, Tim Crews, and their peers prove right, they could put the world's eating habits in line with nature's imperatives and help farmers become true stewards again.

CHAPTER TWELVE

## *Ranchers*

IN 1987 A PAIR OF GEOGRAPHERS FROM RUTGERS UNIVERSITY PUBlished an essay proposing a radical new vision for the American West. Deborah and Frank Popper described the region's hostile climate, recurring boom-bust cycles, and history of federal bailouts. They noted that its rural counties had been losing population since the 1920s and that dozens of its communities had become ghost towns. And they asked this question: What if we simply let the Great Plains continue to depopulate and turn it back over to the charismatic ungulates who once called it home? They called their essay "The Buffalo Commons."

The paper was a surprise hit with environmentalists and intellectuals across the country. It offered a chance to rescue a beloved American creature and restore a badly damaged ecosystem. And because it proposed working with Native American leaders to restore bison on tribal land, it presented a chance to heal some of the wounds of American frontier history. The Poppers—she a graduate student at the time, he a young professor—were stunned by the essay's popularity.

To residents of the Great Plains, however, the essay was about as welcome as an invasive weed. Ranchers and civic leaders saw it as one more example of outsiders telling them what to do with their land. They pointed out that the people who remained in this remote landscape loved it and had no desire to leave. They declared themselves resourceful and tenacious people who weren't about to abandon the cattle ranches that their parents and grandparents built with decades of sweat and heartache. When the Poppers went on speaking tours through Nebraska, South Dakota, and other western states, their hosts occasionally hired sheriffs' deputies for security—a step that sometimes proved wise.

The Poppers' war with the West eventually ended in a charming detente. At one forum after another, they listened patiently and admitted they had much to learn about rural America. Local audiences, for their part, discovered that the Poppers were gracious and curious people—scholars who wanted to help them think about the future, not attack their way of life. At most appearances they found that one or two locals would sidle up to them at the end and confide that their essay was a useful wakeup call or admit that local communities needed to reinvent themselves to survive. McCook, Nebraska, took the matter to heart and dubbed itself the capital of the Buffalo Commons. "At worst, they gave us credit for being good sports to come to their part of the country," Frank recalls. Eventually their essay became less of an intellectual hand grenade and more of a touchstone for discussions of prairie conservation in the West.

To this day the essay casts a long shadow over conversations about the West. It distilled questions that have gone unanswered ever since frontier farmers and ranchers began moving west in the era of Jefferson and Jackson: Are these wild spaces public or private? Who decides the fate of their magnificent wildlife—the nation as a whole or only those who live there? And now that the western grasslands in states such as Montana and South Dakota are falling to the plow, should every acre be working land that turns a profit, or should Americans carve out some piece of the prairie and leave it to nature? In the end, the essay served exactly the purpose the Poppers had intended: a framework to help Americans think about the future of a land that holds a central place in American history and, now, a key role in the survival of the planet. Almost by accident, the essay wound up influencing what happened next in Montana.

IT'S A WEDNESDAY morning in cattle country and the sound of lowing cows carries across a grassy meadow in the heart of Phillips County, Montana. Silhouetted against the rising sun, three figures on horseback ride out of the east, herding two hundred head of Black Angus toward a compound of corrals and weather-beaten barns. This is fall roundup on the Matador Ranch, an annual ritual when new calves and their mothers come down from the high summer pastures for an appointment with the veterinarian. The moos grow louder as the sun climbs in the sky—now you can tell the mothers from the calves—and soon a happy black-and-white Australian shepherd bounds into the ranch yard, too excited to wait for the herd. The cowboys are riding slowly to avoid panicking the cattle or separating the calves from their mothers. It's been a chilly night, and dewdrops sparkle on the shortgrass prairie—a low, dense carpet of blue grama, western wheatgrass, and alfalfa.

Before long one corral is full of Black Angus adults, the cows shifting restlessly against the steel bars. By 8:30 the cowboys have begun sorting calves from cows. A woman in blue jeans and a black fleece vest, her salt-and-pepper hair tucked behind a pink baseball cap, dismounts from her horse and trots toward one of the corrals. Conni French is looking for her husband, Craig, who has just arrived on his four-wheeler after rounding up a pair of stray calves. They confer on how to move the herd through the corrals and where they've left their coffee thermos, then Conni takes up duty at one of the gates.

The cowboys, dressed in Carhartt coveralls and worn barn jackets, dismount and begin leading the animals through a succession of corrals. One by one, the adults are ushered toward a squeeze chute, a rectangular cage of steel bars that holds a cow still while the veterinarian performs her tasks. Standing on one side of the chute, she gives a furry black cow a shot of vaccine and an oral dose of minerals. On the other side three women move in a well-practiced quadrille—one gives the cow a vaccine booster while another applies fly repellant and a third operates the hydraulic machinery of the chute. Their movements are smooth and rapid; they can "work" a cow in forty-five seconds, which is good because a stressed-out Black Angus will begin to buck and snort after even a few seconds pinned inside the metal bars.

If the operation seems highly cooperative, that's because Matador Ranch is a sort of cooperative. The land is owned by The Nature Conservancy, which maintains the infrastructure, employs a staff of conservation managers, and develops a grazing plan for healthy grass. But it is operated day-to-day by about a dozen local ranching families who graze some of their cattle here for several months of the year and who monitor the grass and manage the herds.

Conni and Craig French were drawn to Matador by the spirit of shared endeavor and its ranching philosophy. Much like Practical Farmers of Iowa farther east, Matador aims to keep families on the land, earning a living, while partnering with major conservation groups to protect grass, soil, water, and wildlife. But they also like the company of their neighbors and the idea that they are helping to introduce a new approach to ranching in Phillips County. Craig, who has the build of a linebacker and the beard of a lumberjack, is soft-spoken, curious, and quietly funny. Conni, whose piercing blue eyes fix a visitor in conversation, has the energy of a natural organizer and a gift for articulating deeply held values about community and nature. In their four years at Matador they have, in a way, become its informal ambassadors—spreading the conservation message to other ranchers and explaining the operation to visitors.

Today, with plows approaching from Montana's east and a fabled private wildlife preserve expanding in the south, they find themselves squarely in the middle of a debate that transcends Phillips County and even Montana: how best to save the continent's remaining grasslands on the western Great Plains.

Craig French has become a leader, along with his wife, Conni, in the grassbank movement of conservation ranching in Montana. *Photograph by Josephine Marcotty*

CATTLE CAME TO Montana in the late 1800s, first longhorns from Texas and then shorthorns from Oregon and the East. It was late in an American ranching epoch known as the Open Range, when ranchers could graze their cattle for free on unoccupied federal land—and much of the West was unoccupied by white people because it was mostly unsuitable for homesteading farmers. Railroads had finally tied the East and West Coasts together at Promontory Summit in Utah, providing fast transport from the northern plains to stockyard towns such as Abilene and Dodge City. With demand for beef rising in the nation's fast-growing cities, wealthy easterners poured money into cattle ranches. As ranchers ran out of free land in Texas and Oklahoma, Montana beckoned. The northern Great Plains were too dry, and the growing season too short, for the row crops that settlers were planting farther east. But the short native grasses—blue grama, western wheatgrass, and needle and thread—provided nutritious and abundant forage for cows. And so the band of shortgrass prairie that runs north from Texas through Wyoming and Montana and into Canada became cattle country and the canvas on which a nation sketched a new cast of heroes. Wyatt Earp and Buffalo Bill Cody lived here, as did a breed of itinerant cattle wranglers who, in just a few decades, established themselves as an enduring and romantic American icon: the cowboy. Teddy Roosevelt, wracked by the death of his young wife and his mother on the same day in 1884, came West to heal himself by raising cattle and contesting with the rugged land of the Dakota Territory. Young men of the Gilded Age elite came here to make their fortunes or test their manhood. Robert Louis Stevenson and Frederic Remington came here to capture this exotic place for audiences back home—Stevenson finding the locals rustic and uncouth, Remington turning them into romantic heroes of an untamed West.

The vast expanses of free grass and massive investment by East Coast speculators triggered a cattle boom. Montana's population more than tripled between 1880 and 1890, while the number of cows reached one million. Soon, even in Big Sky Country, ranchers were competing for land, overgrazing the prairies, and warring with farmers. But like countless other pioneers of the nineteenth century, they underestimated the violence of prairie weather. In the winter of 1886–87 tens of thousands of cattle froze to death in marathon blizzards and subzero temperatures, an event known as the "Big Die-up," and dozens of ranches went bust. The catastrophe also presaged the end of the open range. For the next several decades, Congress and state legislators sought some way to impose order on cattle ranching, mostly without success. Then in 1934, recognizing that even the West's vast grasslands required some regulation, Congress passed the Taylor Grazing Act, a landmark law that would set the template for western cattle ranching for the next century. The law divided the public grasslands into districts, and then the districts into grazing "allotments," which limited the number of cattle per acre and brought some structure to grazing rights. Typically a landowner would buy a grazing permit for an allotment adjacent to his or her ranch, then pay the federal government a grazing fee. The government required the rancher to file a management plan that specified stocking rates—the number of cattle per acre—to reduce overgrazing. Many allotments were so big that they in effect tripled or quadrupled the size of a rancher's operation. But the law had several drawbacks. It set grazing fees at artificially low levels and left oversight of grazing districts to local boards typically composed of ranchers themselves. By pairing every grazing allotment with a local parcel, the long-term leases created huge ranches, turned incumbent operations into empires, and made it hard for newcomers to stake a claim on federal grazing land. And because the grazing permits typically transferred with a ranch when it was sold or inherited, they often established long-running cattle dynasties.

By the middle of the twentieth century Montana, with the nation's biggest expanse of short-grass prairie, was among the top ten states in beef production and a place where Stetson hats were practical headgear and lassoing calves was a working skill, not a rodeo trick.

By that time the flaws of the Taylor Grazing Act had become plain. Overgrazing remained rampant, producing dead pastures, chronic erosion, and polluted streams. Although this same land had supported millions of bison for centuries, it could not endure overgrazing by cattle. Where bison move constantly, giving every patch of land a rest, cattle largely stay put and will crop grass down to the dirt. Lacking the innate cooling skills of bison, which roll in mud baths and pant to discharge heat, cows tend to linger in ponds and streams, fouling the water with urine and dung and trampling the reeds and grasses that make a healthy stream bank. The Government Accountability Office periodically reported to Congress on federal management of public grazing lands, and the reports were generally bleak. In 1988 a GAO official testified that almost two-thirds of the federal allotments were in fair or poor condition and many of these were getting worse. "About 80 percent of [the] riparian areas along Idaho's . . . streams is . . . in some stage of degraded condition," GAO associate director James Duffus testified. "This results in badly eroded streambanks, radically altered streamflows, increased siltation, decreased shrub and grass growth, and lowered water tables."

However destructive cattle culture was, it did have one redeeming feature. Because ranching was profitable and the High Plains were too dry for farming, cattle in effect kept the plows at bay for decades. The result? The northern Great Plains was protected, and to this day it is one of the largest intact grasslands in the world.

Sometime in the late 1990s this fact came to the attention of a few of the world's major conservation groups. Biologists were discovering the complexity of this remote ecosystem and its remarkable diversity of wildlife. Soil scientists were exploring the land's potential to sequester carbon and control floods. Wildlife biologists had begun to believe that it was pointless to protect individual species such as pronghorn or wolves without protecting the vast landscape where they lived.

About this time, a team of researchers led by the conservation biologist Jonathan Hoekstra published a paper urging a new approach to conservation: identify the places that have lost the most habitat and that are also least protected as preserves or national parks. To everyone's surprise, the top-ranking places were not oceans or tropical forests. They were the world's grasslands.

Craig French recalls this development with a combination of awe and irony. "It's like we became a bull's-eye for every environmental group on Earth. We suddenly saw lots of people interested in our little part of the world."

One of those taking an interest in preserving the western grasslands was a West Coast entrepreneur named Sean Gerrity. Gerrity had grown up in Montana and later moved to Silicon Valley, where he founded a successful management consulting firm. When his kids started growing up, he and his wife moved back to Montana, and on a hunting trip with buddies one week he learned about an ecologist named Curt Freese, who was preparing what would become an important study of the grassland ecosystem, "Ocean of Grass." Freese belonged to a coalition of prairie conservation organizations that with help from the World Wildlife Fund hoped to create a big grassland wildlife reserve. Freese told Gerrity about the group's vision—inspired in part by the Poppers' essay—and invited him to join the board of the nascent group, then called The Prairie Foundation. As a management consultant, Gerrity had strong opinions about structuring the new organization, and before he knew it, Freese invited him to become CEO. He spent the next several years traveling the country with Freese, holding fundraisers and courting philanthropists. They hoped to establish the reserve next to the Charles M. Russell National Wildlife Refuge, a million-acre federal preserve along the Upper Missouri River Breaks, which would instantly increase the scale of their protected land. Then they waited for ranches to come on the market, bought them, and began assembling them "like pieces of a jigsaw puzzle," much as the Glacial Ridge team pieced together patches of remaining tallgrass wilderness in western Minnesota. Gerrity and Freese aimed to assemble a big, contiguous preserve of 3.2 million acres. "It was a pure wildlife play," Gerrity says. "We wanted to restore the original populations of bison, wolves, sage grouse, prairie dogs."

To drive across American Prairie today is to appreciate both the scale of its ambition and the extent of its success. The nonprofit acquired bison from semiwild herds at Wind Cave National Park in South Dakota and Elk Island National Park in Alberta, which, along with Yellowstone National Park, had played key roles in nurturing bison back from their near-extinction at the close of the nineteenth century. By the early twenty-first century they had built a herd numbering nearly eight hundred bison, one of the biggest wild herds in the United States. They replanted overworked pastures in native grasses and prairie flowers. They dismantled the barbed-wire fences that kept deer, elk, and pronghorn from their natural migration patterns. A few years later they formed a partnership with a research arm of the Smithsonian Institution, which now uses the preserve for one of the largest bison studies ever attempted. They also partnered with the nearby Fort Belknap Reservation of the Nakoda and Aaniiih Nations, which had a bison herd of its own, to start a degree program in bison management at the tribal community college.

Anyone expecting American Prairie to resemble the lush tallgrass prairie preserves in Minnesota and Iowa will be disappointed. The reserve sits amid several cattle ranches and shares similar terrain: endless rolling downs of dun-colored grass, broken by dry gullies and dotted with sagebrush. The sere landscape is so monotonous that you wouldn't know you've driven into a protected wilderness until a huge bull buffalo rises from a patch of grass and stares you down before ambling off in the other direction. The reserve is so big that a herd of three hundred bison can easily disappear into its gentle folds, and so wild that even the reserve's former supervisor, Damien Austin, had to drive miles of gravel road in his pickup truck before he could find them.

And yet it's full of unexpected delights. Strolling down one of the reserve's many hiking trails, you suddenly find yourself in the middle of a prairie dog town as big as a football field. The air rings with the signature chirps that prairie dogs use to alert their neighbors that predators are in the vicinity. Perched atop the mounded rings of their burrows, dozens of the little rodents blend into the sandy terrain so perfectly that it's hard to spot them until, one by one, they raise a paw in the species' warning sign, then dart head-first underground. Today American Prairie covers 460,000 acres, with a wildlife roster that includes beavers, pronghorn antelopes, black-footed ferrets, swift fox, and sage grouse in addition to bison.

But like a garrulous dog bounding into a room full of cats, American Prairie soon found itself surrounded by wary neighbors. Its ecological pronouncements seemed to imply that local folks didn't value nature—or know enough to protect it. Its $200 million endowment built by far-off millionaires offended people who spent sleepless nights wondering how they would pay the next feed bill. Much like Glacial Ridge in Minnesota, the project triggered suspicions among neighboring ranchers and fear that high-minded, deep-pocket environmentalists would buy the best land or dilute the local property tax base. Although American Prairie won't pay more than fair market value when a new ranch comes on the market, it's perceived locally as a competitor to any family hoping to expand. And for some, it touched the same raw nerve as the Poppers' essay: Drive the people out and replace them with bison. The gravel roads approaching the reserve are marked with large homemade signs reading "Save the Cowboy. Stop American Prairie" or "Don't Buffalo Me: No federal land grab."

AMERICAN PRAIRIE'S CURRENT CEO, Alison Fox, understands the wariness and takes pains to avoid offending her neighbors. Fox, who grew up in New England, fell in love with the Mountain West during a summer at Glacier National Park and moved there after graduate school. She joined American Prairie in 2007 and succeeded Gerrity in 2018. Although she has a graduate degree from Georgetown University and a background in marketing, she seems equally happy behind the wheel of a pickup truck bouncing across the rough terrain and looks for opportunities to connect with the ranching community. She notes several steps American Prairie has taken to be a good neighbor: It allows hunting on large expanses of the reserve; it leases back some land for ranchers to graze cattle; and it installed a costly electric fence system to keep its bison from roaming onto neighboring ranches, where they might eat the hay and disturb the cows. Still, she regrets the skepticism of American Prairie's neighbors. "I wish we had spent a little more time in the early years getting to know them and letting them get to know us," she says.

In a state as big as Montana it's hard to believe that people feel they must compete for space. Driving north from Billings toward American Prairie, even on a third or fourth trip, a visitor's eyes need time to adjust to the giant scale of the place. From a ridgetop near Roundup, you can look down on an endless plain in the valley below. At a far-off highway intersection, a logging truck rolls to a stop, its bright red cab looking like a toy in the distance. But even that isn't far measured against the scale of Montana. Beyond the truck a rolling amber plain broken only by coulees and fences stretches away until it reaches the horizon, fully thirty-five miles away. It's as if you were standing atop the Empire State Building and there was nothing between you and Connecticut but two barns and a few strands of barbed wire. Not far from Malta, a well-known family grazes cattle on sixty thousand acres—a single ranch so big that you could fit Manhattan inside it three times and still have room for part of Queens.

But the debate that rages here isn't really a competition for space. It's a war of ideologies. Today's ranchers descend from the frontier settlers of the nineteenth century, who believed the New World was here to be acquired, improved, and harnessed to the cause of national prosperity. Now they find themselves confronting the intellectual heirs of Teddy Roosevelt and his ally, the early conservationist Gifford Pinchot. The ranchers sense that America is shifting, decade by decade, toward the camp of Roosevelt and Pinchot.

At about the same time that American Prairie was being conceived, a group of Montana ranchers and grassland specialists at The Nature Conservancy began thinking about a different strategy to save grasslands. First, they penciled out some revealing arithmetic. Even if American Prairie achieved its ambitious vision—a nonprofit wilderness of three million acres—it would be a mere postage stamp against the fifty million acres controlled by private landowners. Then, too, they got to know local ranchers, such as Dana and Bill Milton, leaders of a conservation group in nearby Musselshell County, and realized that many of them loved this land and its wildlife as much as anyone. Mindful that plows were conquering tens of thousands of acres of grass in the Dakotas and eastern Montana, they opted to form alliances with local landowners and hoped that cattle would help them save the grass. Cementing these alliances would be a novel venture: a cooperative "grass bank" that would reward local ranchers for using conservation practices. In effect, they chose the same fork as Practical Farmers of Iowa in the debate over "sparing versus sharing." Rather than set land aside as protected wilderness, they would do their best to make Montana ranch country a place working families could share with deer, elk, prairie dogs, pheasants, hawks, and other wildlife.

In 1999 The Nature Conservancy's Montana grasslands conservation director, Brian Martin, learned that a large sheep-and-cattle ranch just north of the Upper Missouri River Breaks was about to come on the market. The property controlled sixty thousand acres, half of it deeded land—that is, privately owned—and the other half a federal allotment with guaranteed grazing rights. Martin and his staff and their ranching partners waited a few weeks to see if a local buyer would make a bid, and when no one did, they bought the ranch and launched the experiment.

About that time, Craig French got a call from a neighbor who had joined the Matador project and invited him to give it a look. Craig and Conni were raising cattle on a ranch not far away and were perfect candidates for Matador. Craig comes from a respected Phillips County ranching family that has grazed cattle in the area for more than six decades. His parents, Bill and Corky, own a sprawling operation just north of Matador and were known for progressive ranching practices. Bill, now in his eighties, served for six decades on the local soil and water conservation board and jokes that his first love is grass, not cattle. "The cows are just there to eat it," he said one morning while his children and grandchildren oversaw the spring roundup. At the same time, Craig and Conni had grown wary of the bigger-is-better mentality of modern American ranching. Craig says he's happiest out in the pastures checking cattle on his four-wheeler, a small machine he prefers to a big tractor or a gas-guzzling pickup truck. They had just become grandparents too, and, hoping that one of their kids would return to Montana and take over the family operation, they were thinking about long-term stewardship. Conni, who grew up in a family that moved frequently, had found neighbors she liked and a place she could call home in Montana. She had become active in—and eventually a board member of—the Ranchers Stewardship Alliance, a band of ranchers who promote environmental practices and partner with conservation groups such as Pheasants Forever and the World Wildlife Fund. The pair share a sense of adventure and have a friendly competition to generate new ideas for their ranch, which are recorded in a notebook labeled "Bets," as in, "I bet it would work if we started raising some chickens as well as cows." With the Ranchers Stewardship Alliance, they organized a book club, hosted guest speakers, and began monitoring what the outside world said about American Prairie and Montana ranchers. In 2022, when the CBS program *60 Minutes* came to Phillips County to tape a report on American Prairie, Conni stepped forward to voice the concerns of ranching neighbors.

"Trust takes time," she told correspondent Bill Whitaker. "When a new group comes in, new neighbors, it takes some time to learn if it's legit." Craig and Conni got to like Whitaker and the *60 Minutes* crew, but the broadcast did nothing to change their opinion that journalists and other outsiders don't take local ranchers seriously.

"What all these people forget is that the folks who live here love living here," she said one afternoon while saddling up to check on the cows. "It's the sunrise. It's the birds. It's the kids playing on the haystack. We're fighting to save a culture."

On his first visit to Matador, Craig admired the vision but harbored a veteran rancher's doubts that the grass bank could succeed. He worried that it lacked sufficient water for a big herd and the infrastructure to bring it to the pastures. Then too, it would take a lot of work for him to move a part of their herd down the road to Matador every spring and bring them back in the fall. "There were too many unknowns for me to sign up in the inaugural year," he recalls.

Before long Matador raised the money to upgrade the water infrastructure. Craig noticed the ranch's pastures growing healthier and healthier and saw that the manager, Charles Messerly, was a get-things-done guy. In 2018 he approached Messerly and asked to be put on the waiting list.

Matador offers participating ranchers a simple transaction: discounted grazing fees in exchange for a commitment to follow a set of conservation practices on their own ranches and a promise to keep their land in grass. Matador's grazing fees start at 10 percent below the local average charged for private land, then Nature Conservancy staff add further discounts for each conservation milestone—a prairie dog colony, for example, or sufficient habitat for sage grouse—reached on the family's home ranch.

The Frenches value the discounted grazing fees. But the biggest practical benefit is the chance to protect and improve grass on their ranch. Skilled ranchers on the High Plains have an almost worshipful attitude toward grass; their herds literally live and die by its health. They watch as varieties evolve in case one species is outcompeting another. They take notes on when the pastures green up in the spring and whether invasives such as kochia or cheatgrass are moving in. They track which grasses and legumes the cows prefer and which seem most nutritious, using herbicides judiciously to control weeds and planting native grasses here and there. Conni and Craig found that owning more land and more cows—the goal of many conventional ranchers—became less important than protecting their grass and its nutritional value. In the dry summer of 2022, they took dozens of cows to market and reduced their herd size rather than overgrazing the grass. As Craig notes, cutting costs adds just as much to a rancher's bottom line as raising revenues, and healthy pastures save money, much as healthy soil saves money for the members of PFI. When cows are eating really nutritious forage, he says, they're less likely to need nutritional supplements or care from a veterinarian. Protecting their pastures also avoids the need to buy hay, which can run to thousands of dollars. "You make more money by not unrolling that hay bale," Craig says.

Conni and Craig also enrolled in a World Wildlife Fund initiative called the Ranch Systems and Viability Planning, which provides ranchers with technical assistance, grass management expertise, and cost-sharing for improvements such as new fences and pasture wells. The program in turn monitors participating ranches for soil carbon, water infiltration, grassland bird species, and other indicators of a healthy grass ecosystem.

Unlike most conventional ranchers, Craig and Conni generally let their cattle graze outdoors through the winter, making sure they have water and shelter from the wind and the snow. The cows feed themselves without supplemental hay and fertilize the grass as they move. Conni walked through a near pasture one winter morning and noticed that dung from the cow pies had been scattered across a large patch of grass by sparrows searching for wheat seeds in the manure. "Let the animals do the work," she says.

For its part, The Nature Conservancy has found that Matador is an efficient way to protect big swaths of shortgrass prairie. Healthy grasslands need grazing animals, and at Matador the animals come for free, provided by partner ranchers. The paid staff is minimal—roughly one-tenth the headcount of American Prairie—because the ranchers do much of the work. More important, Martin says, the grass bank spreads conservation practices "rancher to rancher" and multiplies the organization's reach. Matador covers just 60,000 acres, but it has changed ranching practices on more than 300,000 acres, counting all the land belonging to its partner ranchers. The organization also values the social bottom line. Matador supports the rural economy and sustains small communities—all the while making ranchers friends of conservation rather than enemies.

"When I signed on with Matador, some of my friends said I was joining the enemy," Messerly recalls with a laugh. "Today my phone never stops ringing with ranchers who want a slot here."

Matador has also become a laboratory where environmentalists can study grazing techniques and their effect on grassland health. The conservation plan developed by Brian Martin and his staff employs a well-known practice called rotational grazing, which simulates the migratory habits of bison by moving the cows frequently from one pasture to another to rest the grass. But The Nature Conservancy's conservation team has refined the technique to a fine art. In the early days of rotational grazing, ranchers typically used the same rotation every year, starting on the pasture closest to the home ranch, then rotating the herd to outer pastures in a giant circle until they were back near the barns by fall. But a pasture grazed repeatedly in the spring will lose its cool-season grasses, which green up early, and become vulnerable to spring weeds. A pasture grazed year after year in the summer will lose its warm-season grasses, allowing thistle and kochia to move in. Messerly's answer was simple: "Give the clock a quarter turn" every year, so that pastures take turns going first and last and have extra time to rest. The result is a mosaic of healthy pastures, some grazed short, some more lush, to create what Martin calls "structural complexity" so the landscape always has diverse habitat for various species of birds and other creatures. Even the cows provide clues to the health of the grass: Restless cattle tell the ranchers that a pasture isn't giving them the full range of nutrition they need.

This version of rotational grazing requires constant monitoring and adaptation to local conditions—one reason conventional ranchers find it burdensome. One autumn, Messerly planned to rotate Matador's herds in the third week of September. But the summer had been dry and the grass, on close inspection, was getting thin. So at the last minute he asked his ranchers to accelerate the rotation by one week.

But it's achieving The Nature Conservancy's ecological goals. Matador's bird population and diversity has grown steadily and now includes Sprague's pipits, mountain plovers, and long-billed curlews, among other desirable species. Martin's staff use drones to monitor the land's prairie dog population, mindful that the little rodents are a keystone species that supports foxes, wolves, owls, bison, pronghorn, and a variety of birds. Matador has gone from one acre of prairie dog towns to sixteen hundred on the land it owns, and probably another fourteen hundred on the federal land it leases. Even the grass itself has gotten healthier. Using a tool called the Rangeland Analysis Platform, the ranch's managers have found the pastures are producing denser growth, fewer invasives, and a greater variety of native species. "People want to bring their cattle here because we have great forage," Martin says.

To achieve that success on a larger scale, prairie conservationists will have to replicate Matador many times over, a daunting challenge. Finding the right site isn't easy, even in a place as big as the northern Great Plains. It would have to be big enough to support blended herds of five hundred to six hundred cows—a lot of acres on the thin, shortgrass prairie. It would also need sufficient water for such a herd and high-quality forage—that is, the right mix of grasses and forbs. In addition, Martin typically looks for parcels with "conservation value"—features such as clean streams, grassland birds, and a variety of healthy wildlife species. He estimates that, even if he could find such a property, it would cost on the order of $5 million. To establish an archipelago of Matadors on the Great Plains, much as conservationists are doing in western Minnesota, would require an endowment of many millions. Yet despite the hurdles, other grass banks are in the works or planning stages. A community group known as Winnett ACES, south of the Missouri River, is studying Matador and trying to assemble its own grass bank. The Nature Conservancy has started raising money to start a second ranch like Matador and is examining parcels in eastern Montana. And the Ranchers Stewardship Alliance recently sent a delegation to New Mexico to explain the concept to sympathetic ranchers there.

At the same time, by choosing the "share" philosophy of grassland conservation at Matador, The Nature Conservancy stepped into society's furious debate over the ethics of eating meat. The case against beef is powerful, and there is near unanimity among environmentalists that the planet would be healthier if people ate less red meat, especially people in the world's richer nations. Cows are a hugely inefficient way to produce protein: a ton of protein from beef requires roughly twenty times more land than a comparable amount of protein from grains and beans, and at least five times more water. The grain required to feed U.S. livestock every year would feed 840 million people on a vegetarian diet with equal calories. In today's beef industry, which sends animals to a feedlot to fatten up on grain before slaughter, cattle also require a lot of corn and the fertilizer used to grow corn. Climate change has only strengthened the argument against beef. Cows emit large quantities of methane from both ends of their digestive tracts—a byproduct of the way they digest grass—and methane is a more potent greenhouse gas than carbon dioxide.

Many ecology-minded consumers have turned to grass-fed beef—cows that spend their entire lives in the pasture rather than being finished on grain in a feedlot. Grass-fed animals are generally considered healthier, and the practice reduces the need for corn, plowed land, and fertilizer. But converting the nation entirely to grass-fed beef, even if it were possible, wouldn't solve the problem. Grass-fed cattle take more time to reach slaughter weight, and because they live longer they produce more methane than grain-finished cows. And because grain-finished cattle get bigger faster, producing the same amount of meat with grass-fed beef would require a herd that is 30 percent larger, according to Matthew Hayek of Harvard and Rachael Garrett of Boston University, and could require an impossibly large increase in the nation's supply of pasture land.

The case for cattle is a bit more nuanced. Grasslands and ungulates evolved together for millennia—bison in North America, wildebeest and antelope in the African veldt, horses and goats on the steppes of central Asia—so that grasslands need grazing animals to stay healthy. Without regular grazing and other "disturbances," such as fire, most grasslands would eventually be taken over by larger, woody plants such as sumac, dogwood, and trees. Recent studies have found that grasslands also sequester more carbon when they are grazed; the stimulus to grow accelerates their photosynthesis.

In a passionate and deeply researched book titled *Meat: A Benign Extravagance*, the British farmer and editor Simon Fairlie has made his own case for cattle. He notes that the standard calculation of carbon dioxide emissions from cattle includes carbon emitted when forests are cleared to create pastures. While that's true in some parts of the world, that's not true in places like Europe and the American West. If you take deforestation out of the equation, that drops the carbon footprint from cattle by fully 30 percent. He also argues that if beef disappeared from the human diet, the world's farmers would have to plow thousands of additional acres for food grains and legumes to replace the protein, compounding all the problems of fertilizer use and carbon loss that attend plowing more land. Finally, if the land were turned back into wilderness, it would soon produce a big population of wild ungulates—who also produce a certain amount of methane.

Like every environmental question, this one is multifaceted. Given a choice between cattle and wilderness, most environmentalists would certainly choose wilderness. But that's not the choice in places like Montana. Here land has become so valuable that owners insist that it deliver an economic return and it's unlikely to return to wilderness on any large scale. Absent a philanthropic endowment of many millions of dollars, like the one that created American Prairie, by and large the choice on the western grasslands is cows or plows.

That choice becomes simple—and ever more urgent—as the threat and science of climate change advance. Ecologists have studied many ways of using nature to slow the planet's warming, including planting more forests and increasing soil's capacity to sequester carbon. Several studies have concluded that the most effective way, by an order of magnitude, is simply to leave the world's remaining grasslands in grass.

In this sense, despite their philosophical differences, Alison Fox of American Prairie and rancher Conni French are on the same side—the side of grass. With a shared passion for grasslands and wildlife, ranchers and conservationists might just find common ground. Taking a break for coffee one morning at the Matador Ranch house, Craig French muses about having American Prairie for a neighbor. He says its arrival "made us up our game," meaning that ranchers like him and Conni questioned their own practices and intensified their efforts to protect the land and its wildlife. Mark Cool, a vice president at American Prairie, returns the compliment one morning during the bison roundup. "We have to hand it to the ranchers—if they hadn't kept this land in grass for all these decades, we wouldn't even be here."

Perhaps in a state as big as Montana, there is room for both.

CHAPTER THIRTEEN

# *Tatanka*

DAN O'BRIEN WAS A KID FROM OHIO WEDGED INTO THE BACK SEAT OF a Chevy station wagon with two brothers when he first saw the Great Plains of the American West. Towing a camper while on a family vacation, they were driving from the Black Hills in South Dakota when the land opened up before them—grass and sky, grass and sky, and more grass and sky all the way to the horizon.

"Mom," he recalls saying, "this is where I'm going to live!" His mother laughed at his excitement. It's just empty land, she told him. But it was the moment that would define his life.

Twenty years later, when O'Brien did move to western South Dakota, he started out doing what everyone does there—he raised cattle. He bought a small ranch in foreclosure, which gave him four hundred acres, a few dozen cows, and the right to a life of mortgage payments at 21 percent interest. The seller, who had run the place into the ground before giving it back to the bank, was the lonely, hard-drinking son of a rancher who had committed suicide in the barn. Its name, and the symbol on the brand that O'Brien had exclusive rights to burn into the hides of his cattle, was the Broken Heart Ranch.

Over the following years, the ranch lived up to its name. Like most ranchers out on the plains, O'Brien struggled against blazing summers, bitter winters, drought, wild swings in cattle prices, and pastures grazed down to dirt by cows transplanted to a harsh landscape. Most years, he wound up renting out the ranch to neighbors and finding work in Denver, Minneapolis, or California as an English teacher or biologist to make the payments. It struck him more than once that he was supporting the cows, rather than the other way around. There were times when he came home from a working stint off in the big city to find cows living on his porch and the pastures curtained by blowing dust. He lost his hair, his wife, and his hope, and learned to detest cattle, even the taste of their meat.

O'Brien looks and acts nothing like the tall, taciturn Stetson-wearing frontiersman of American lore. He's short and, for a South Dakota rancher, slight, with green eyes faded by the prairie light. His bare head, which he complains is too small to carry off a cowboy hat, is as brown and speckled as a duck egg. He has the Anglo skin his Irish name implies, now leathery and creased from the sun, and rimless spectacles that remind you that even after decades of wrangling fences and big animals he's at heart an intellectual. He writes books. He talks a lot.

But the biggest difference is that the mythic frontiersman—as well as many of his neighbors—have a near-religious conviction that the American West was made for them to own and tame by installing cattle to the exclusion of most everything else, including the wild animals that roamed the South Dakota plains and the Indigenous people that once depended on them. And that—the wild promise of what was once one of the richest and most diverse ecosystems on Earth, where herds of buffalo drifted like dark cloud shadows across the land—is what binds O'Brien to the plains.

Before becoming a rancher, O'Brien worked as a biologist. He spent years roaming the West helping to restore endangered peregrine falcons, which had been brought to near extinction by the pesticide DDT. He carried young chicks that had been born in breeding labs and released them all over the eastern front of the Rocky Mountains, traveling by horse, helicopter, or truck. To this day he hunts on his ranch with the magnificent black-and-brown birds, which he describes not as a hobby, but as a way of life. He trains young falcons to come to his hand and to work alongside hunting dogs. When mature, he carries them to the field on his wrist, and sends them up and soaring into the air, so high they are black specks against the sky. They hover while the dogs crisscross the grass and brush, hunting for wild game birds like sharp-tailed grouse and ducks. When the dogs flush the birds into the air, the falcon plummets for the kill in a spectacular exhibition of flight and speed.

During his years on the plains, O'Brien learned many of the native grassland species—prairie dogs, coyotes, pronghorn elk, red-tailed hawks. Cattle, on the other hand, are transplants to the West, having arrived as the plains were being emptied of buffalo by overhunting. All too often he has seen grazing cows turn the grasslands into a "desiccated and subsidized" landscape with blowing soil, and end their days in massive, stinking feedlots where they eat genetically engineered crops grown across the Midwest on what were once spectacular prairies. He understands, having experienced it himself, how ranching depends on federal subsidies like government-backed loans and guaranteed-grazing rights on public lands, and how ranchers must overgraze their own land in order to eke out a living. Like Jim and Carol Faulstich farther east, he has seen row crops encroaching on the Dakota grasslands under pressure from the relentless expansion of the industrial model of agriculture.

Those were the thoughts that churned in his mind on the days and nights when, out of angst and frustration with his life, he would get into his truck and drive for hours down the dirt roads that cut through the land around the Broken Heart. He knew he was part of a commodity meat industry that was catastrophic for the plains, and it made him furious. But it wasn't until 1997, when he was in his early fifties and struggling to come out of a deep despair, that he finally did something about it. He bought thirteen tiny motherless buffalo calves from a neighboring rancher.

"I don't know why I did it," he wrote later in his memoir, *Buffalo for the Broken Heart*. "I wanted us all to move on to a better life."

O'Brien's ranch lay north of the Black Hills, a range of fir-covered mountains, needle pinnacles, and sheer rock cliffs rising up from the Great Plains in western South Dakota. It is a sacred place for the Lakota, one in the confederation of Sioux Nations that once controlled the region, and has been occupied by people for many thousands of years. Called Pahá Sápa in the Lakota language, it was the "heart of everything that is," the place where Lakotas believe they first emerged into the world. Hundreds of years ago the Lakota lived primarily along the Missouri River Valley, but in the 1700s, their seven bands began to migrate west, leaving behind villages, agriculture, and a more sedentary life. They embraced the power of a new species that swiftly transformed their culture—the horse. They learned to breed horses, ride them with consummate skill, and use them for carrying their goods and homes. With the help of horses and in search of new hunting grounds, Lakota began moving up the tributaries that flowed from the west into the Missouri and that all pointed to Pahá Sápa. Horses made them mobile and powerful hunters, and they became almost exclusively dependent on bison for food, shelter, and trade with Europeans. They made shelter and clothing from the animals' hides, tools from their bones and horns, bowstrings from the sinews. Bison dictated where they lived, and, in conflicts over hunting land, who were allies and who were enemies.

Pahá Sápa was the center of North America's premier bison hunting range, providing the herds shade and water during the frequent droughts and protection in its ravines and valleys during winter. Pahá Sápa was a gift to the Lakota from the White Buffalo Calf Woman, the spiritual figure of their foundation story who gave them the buffalo and a sacred pipe that tied them to it. In the early 1800s, the Lakota clashed with other tribes for access to the Black Hills, forcing their rivals farther west, and aligning themselves with the northern Cheyenne and Arapaho. By the 1820s, the Lakota and their allies were wintering in the Black Hills instead of the river valleys that had protected them for hundreds of years, and Pahá Sápa was theirs.

By the second half of the century, their era was over. The buffalo that sustained the Plains Indians were in precipitous decline from overwhelming forces, both natural and human. Hide hunters like Buffalo Bill and horse-mounted Indians killed them by the thousands. Then drought struck the northern plains in the 1870s and 1880s. The buffalo had once been able to escape the dry years by moving east to wetter, more plentiful mixed-grass prairies in the river valleys and what is now western Minnesota and the eastern Dakotas. But by then they were trapped on the dry High Plain by the cattle herds, ranchers, and colonial farmers that had moved into those territories. They also competed for grass with the millions of cattle in the West.

When William Hornaday, a naturalist and director of the Smithsonian's National Zoo, conducted a landmark bison survey in 1886, he found only twenty-five of the creatures in the Texas panhandle, twenty in Colorado, ten between the Yellowstone and Missouri rivers in eastern Montana, twenty-six near the Big Horn Mountains, and two hundred in Yellowstone Park. In just forty years, between 1850 and 1890, a species that once numbered between thirty and sixty million animals across the plains had been reduced to fewer than one thousand.

The most powerful force behind the bison's abrupt demise was not hunting per se, but the nation's rapid industrialization during those same four decades. Long guns with telescopic sites made hide hunters enormously efficient in killing the animals from long distances, and bison skins were in huge demand because of a burgeoning number of factories in the East. Historian Andrew C. Isenberg described bison skins as the "sinews" of nineteenth-century industrialization because the tough leather was used to make the belts that ran the machines of manufacturing plants in the big cities. In just ten years, the value of the leather belting produced in the United States rose from $6.5 million in 1880 to $8.6 million in 1890. Western expansion of the railroads, meanwhile, created superhighways to transport the hides to massive tanneries that sprang up in the Midwest. The plains were left strewn with dried buffalo skeletons—so many that in the late 1800s they became a hindrance to the plows that moved in. Destitute Indians and early settlers struggling to get a start in the forbidding land scavenged the plains and collected the bones for sale to traders who sold them to processing plants, where they were made into pigment, material for sugar processing, and fertilizer for farms. In some years railroads shipped five thousand boxcars of bones until the plains were clear, ready for settlers, ranchers, and their cattle. The most dominant species on the Great Plains had become raw material for the nation's industrial revolution and the growing empire of agriculture.

Whether eradicating the bison was a deliberate federal strategy to crush the western tribes is still a matter of debate among historians. What is clear is that federal officials knew full well that the buffalo were at risk of extinction. In 1874 Congress took up a bill that would have outlawed bison hunting by non-Indians. The measure was supported by animal protectionists, including the American Society for the Prevention of Cruelty to Animals (ASPCA), as well as Indian humanitarians who hoped to restore the integrity of the peace treaties the federal government had recently signed. "I am not in favor of civilizing the Indian by starving him to death," said the bill's sponsor, Illinois Republican congressman Greenbury L. Fort, in remarks to his colleagues in

1874. But the anti-Indian forces in Washington, D.C., were too great, and President Ulysses S. Grant killed it with a pocket veto. A second attempt to pass the legislation in 1876 died swiftly after Custer's humiliating defeat at Little Bighorn. Overall, many white Americans saw the demise of bison as an inevitable consequence of the triumph of civilization over savagery—necessary for the elimination of the powerful nomadic tribes to make room for ranchers, farmers, miners, and railroads. As Isenberg wrote, "The engine of advancement of European Americans into the plains was the ability of an industrial society to thoroughly destroy the bison herds and thus deny their use to the nomads."

The bison, however, hadn't disappeared entirely. A few tiny herds survived, scattered across the West, mostly in private hands, and a few lived at the Bronx Zoo in New York City. The largest, last remaining wild herd, about two hundred buffalo, sheltered in Yellowstone National Park, which had been established in 1872. In 1902, despite new laws against hunting them, their number inside the park had dwindled to just twenty-two. The federal government added eighteen animals from private ranches outside the park, and began to manage the herd more carefully.

By then private nonprofit groups and conservationists like Hornaday had launched a campaign to protect the species. He founded the American Bison Society in 1905, with President Theodore Roosevelt as honorary president. Both were avid hunters with a passion for wildlife, and through two terms, Roosevelt used the power of the presidency to promote conservation in general and of the bison in particular. Starting in 1907 the Bronx Zoo, home to the American Bison Society, began shipping bison out West in a mission to repopulate their historic range. They found homes on public lands at the Wichita Mountain Reserve, Wind Cave National Park, and the National Bison Range in Montana.

Today buffalo are no longer at risk of extinction. Their numbers have recovered to roughly 400,000 in North America, half in the United States and half in Canada, scattered across the Great

Plains on ranches and in parks. But the vast majority are not wild animals. They live on ranches and are raised as livestock, a development that reflects the confused history of bison in America after 1900. During the last hundred years, an entire industry tried to breed bison with cattle, while selecting them for docile personalities, separating them from predators, and isolating them from the natural forces that would make them fit to survive in the wild. The ranchers who settled the West would tolerate bison only as fenced livestock, and in the early part of the twentieth century, it became the only way that state and federal regulators allowed their numbers to increase outside a few reserves like Yellowstone. Many states have designated bison as livestock, so that they are regulated by the same state agencies that oversee cows and pigs. Today they are the only native species classified either as livestock, a wild animal, or both. As a result, most bison today live only a little like buffalo and a lot like cattle: raised in pastures and soon sent off to slaughterhouses. Nonetheless, with most of the western grasslands in private hands, these bison are the heart of a tiny but growing industry considered by many as one of the best ways to revive the species and the grasslands that shaped them. Just as some conservation groups and ranchers have come to view cattle as a revenue-producing way to keep the grasslands intact, many see bison as a way to revive the grasslands—though, for now at least, on a much smaller scale.

What Dan O'Brien had in mind when he started replacing his cattle with buffalo was much more than reviving a species. He wanted to save an ecosystem: the grasslands and all its creatures. To do that, he intended to let his bison live, as much as possible, as the wild animals they once were, roaming with few restraints on the prairie, eating grass from the day they were born until the day they died.

At the heart of O'Brien's strategy was the ancient symbiosis of bison and the grassland ecosystem, which had evolved together through centuries of changing climates, predators, and human cultures. Buffalo are exquisitely adapted to the difficult climate and the available forage of the plains, which explains how a giant animal that walks thirty or forty miles a day can survive on a diet of grass. Bison have four stomachs and 100,000 kinds of gut bacteria, making them 20 to 30 percent more efficient than cows at extracting nutrients from grasses and sedges. The fierce winters of the plains have produced a series of remarkable adaptations in the species. In cold months, their metabolism slows down by as much as a third, allowing them to survive on fewer calories. They use their hooves and massive heads to dig through deep snow to find the grass below, which, along with their stored fat, is enough to get them to spring. Their winter coats have double layers of fur—dense and lamblike close to the skin, long and sturdy on the outside—which provide so much insulation that snow can accumulate on a buffalo's back without melting. They store fat in a layer just under their skin, not marbled through their muscles, which creates yet another layer of insulation. In a blizzard, bison will turn headfirst into the storm, passing through it quickly, whereas cattle will often let a storm blow them downwind, where, unable to escape the cold, they can freeze to death trapped against a fence. In a story that has become folklore among wildlife biologists, a team of Canadian park rangers were conducting a winter wolf count by airplane using an infrared camera to pick up the predators' body heat. Later, the researchers realized they had passed over an entire bison herd without knowing it: the camera could not detect any heat from the well-insulated creatures.

The grasslands' swift predators molded bison as well. Their massive heads and humped spines make them look ungainly, but a buffalo can pivot with breathtaking speed when threatened and jump a six-foot fence with ease. They are also remarkable endurance runners: Wildlife biologists believe running replaced fighting as a bison's chief defense about the time that giant solo predators like saber-toothed cats were replaced by gray wolves, which attack in packs from many angles. Their powerful chests and long bodies create a springlike effect in their spines, giving their forelegs huge power in each stride. During a sustained chase, a buffalo's tongue will drop forward, opening its throat to collect more oxygen. Flowing across rolling terrain at thirty miles an hour, a herd can easily outlast a wolf pack or hunters on horseback—a breathtaking sight to anyone lucky enough to see it in person. When cornered by predators—or ranchers—a bison herd will instinctively form a circle, with calves and mothers at the center surrounded by mature cows and bulls and then young males on the perimeter, horns facing outward. At calving time in the spring, bison births are typically synchronized within a herd so that all the youngsters are on their feet within a day or two and the group can move on.

The relationship between bison and the grasslands is reciprocal—they are not only exquisitely designed for their environment, they play a critical role in keeping it healthy. In their long evolution with grasslands, bison became a keystone species, meaning that they have an outsized impact that cascades through the ecosystem, influencing the land in a way that allows other life to flourish as well. For one, they wallow—thrashing around on the ground to get rid of flies and other parasites, creating clouds of dust and leaving depressions in the earth. The depressions—millions and millions of them across the prairies—become little ecosystems in their own right. Rainfall, when it comes, fills them with water, making a home for frogs, birds, insects, and other aquatic animals. The moisture seeps through the ground, giving life to the prairie plants that need more water. These flowering plants bloom, attracting pollinators and other insects, which, in turn, attracts birds. The area around buffalo wallows has been found to be the most diverse in the grasslands. When they graze, which they do constantly, they stimulate new growth in the grass. In the shortgrass prairies of the West, grazing produces thicker grass, more plants per acre. In a tallgrass prairie of big bluestem and Indian grass, grazing allows rain and sun to reach the shorter grasses below, building diversity in plants and animals. Many species of grassland birds, for example, nest in the short grasses bison leave behind, and they all use buffalo fur for their nests. Unlike cows, which stay in one place and can eat a pasture down to dirt in a matter of days, bison move constantly so that any grazed patch might get one or two years of rest before the herd returns. The shorter, cropped grass also attracts prairie dogs because it allows them to see predators coming, and because they can eat the more nutritious new growth. The endangered black-footed ferret, in turn, survives on prairie dogs alone and lives in their abandoned burrows. And when bison die, their carcasses provide rich dining for scavengers like coyotes, birds, and insects. When they roamed in the millions, bison shaped the way fire, water, and food moved across the landscape.

Bisons' importance as a keystone species in a valuable and threatened ecosystem is what inspired not only O'Brien, but others in a growing community of ranchers inclined to let bison be bison—meaning wild. Media magnate Ted Turner, who owns more buffalo than anyone else in the country, in many ways set the standard for the industry. Turner bought his first bison in 1976, and over time added more to the ranches he owns scattered across the West. He argued that the best way to protect the species was to let them earn their keep. Today he owns fourteen ranches, totaling nearly two million acres in six western states, that are home to 10 percent of the continent's bison. His operations emphasize conservation of wildlife and grasslands and careful genetic breeding to protect the species' wild characteristics. But they also turn a profit by selling bison to other ranchers and supplying meat to Turner's forty-unit restaurant chain, Ted's Montana Grill. Like all members of the National Bison Association, Turner's ranches follow an ethical code that prohibits many of the practices common in the cattle industry, such as routine use of antibiotics, growth hormones, and artificial insemination.

Turner and, increasingly, other bison ranchers aim for customers who are attracted to a regenerative meat production system with a uniquely North American story—saving the grasslands and a beloved species. And then there's the quality of the product itself. Unlike beef cattle, free-ranging bison are an almost perfect source of protein—high in the heart-healthy omega-3 fatty acids as well as other dietary acids that reduce the risk of cancer, diabetes, and obesity.

Turner's operation, which has a significant influence on the bison industry, continues to evolve in O'Brien's direction. Five years ago all of Turner's bison ended their lives in a feedlot eating corn. Today 40 percent of his 45,000 animals eat just grass. His ranch managers say they aspire to have a purely grass-fed business, but argue that they can't sell what people won't eat. Ultimately, the way that most of America's bison live and end their lives—and whether they represent a food system tied to healthy grasslands—is up to the fat-loving American consumer.

It was this paradox, treating a wild animal like livestock, that kept Dan O'Brien up night after night after he brought home the thirteen little golden balls of fluff he named "the gas house gang." In O'Brien's view, a buffalo that doesn't move a few miles every day, a buffalo that doesn't live and die with its own herd, isn't really a buffalo. It's like a peregrine falcon with a broken wing. But like every other rancher in the West, he had to make the land pay. He had a mortgage. And he had to figure out a way he could stomach to raise bison and sell their meat, and that meant keeping them out of a feedlot and a commercial slaughterhouse. It meant he had to find a way to humanely kill and slaughter them where they lived, on the grass.

The winter after O'Brien brought home the gas house gang he spent nights with spreadsheets and calculators, drawing red circles on maps to mark the ideal grazing lands for the bison ranch of his dreams. He pored through U.S. Department of Agriculture meat processing rules, which were so complicated and restrictive that they seemed designed to make commerce impossible for an independent rancher. Then one night, he found the loophole he needed. Buried in the federal meat inspection act, a single paragraph stated that a rancher can slaughter animals in the field, rather than at a commercial slaughterhouse, as long as the carcass is delivered to a meat locker and checked by a meat inspector within two hours.

With those few sentences, the Wild Idea Buffalo Company was born. When he first started harvesting his animals in the field, O'Brien drove out into the herd in a pickup truck with a sharpshooting neighbor. They would shoot one or two bulls, hoist the carcasses into the truck, and drive them to a processing plant in nearby Rapid City. Without the added costs of long-distance transportation, feedlots, and slaughterhouse bills, plus the premium price bison meat commanded in the market, his profit margins were much higher than they were for cattle—enough to pay the interest on the loans it took to buy them. When he married Jill O'Brien, a restaurateur from Rapid City, she brought marketing and cooking expertise that helped expand the customer base to restaurants, grocery stores, and home cooks. To meet that growing demand, they needed to grow the herd and process more buffalo. But they refused to increase their efficiency—at the expense of their animals—by doing it the way 90 percent of the livestock industry does it. Their buffalo would be stressed and terrified by being forced into a truck and sent to a slaughterhouse hundreds of miles away.

Instead, they decided to bring the slaughterhouse to the buffalo.

On a cold November day in 2014, a giant white semitrailer truck rumbled across one of O'Brien's pastures, just a speck against the blue-and-pink sky of a South Dakota dawn. It rolled slowly down a dirt road and turned through a gate into a pasture of gold winter grass and patches of snow. It was Thursday—harvest day at the Wild Idea Buffalo Company ranch. A pickup truck moved off into the nearby cluster of bison and stopped. A rifle barrel slowly emerged from the driver-side window, and with one shot a young bull dropped to the ground. The other animals didn't seem to notice—there was no bellowing or stampeding away. The bison was hoisted by its feet with a winch on the back of the truck and carried slowly to the waiting semitrailer. This vehicle is the ingenious heart of Wild Idea Buffalo Company, a mobile harvesting unit that allows O'Brien to kill and butcher bison in the pasture. It takes the crew of three about thirty minutes to take apart a one-thousand-pound animal. By the time the sun hovered over the western horizon, the crew had processed eight bison. The truck pulled out of the pasture, leaving behind buckets of entrails and huge stomachs like beach balls full of grass. The blood that flowed out of the truck all day left a deep purple stain where it soaked into the frozen ground, enriching the grassland where it belongs.

Today, the declared mission of the Wild Idea Buffalo Company is to "regenerate the prairie grasslands, while improving our environment and our food supply by bringing back the buffalo." O'Brien is mostly retired now, and lives in a brown wooden ranch house about thirty miles south of Rapid City, tucked up against a hill and next to the barn where he keeps his falcons and his dogs. He can stand on the deck and look out over the valley from which his new ranch got its name—Cheyenne River Ranch.

When Dan and Jill O'Brien moved to the larger ranch to accommodate their growing business, there were no buffalo on the Buffalo Gap National Grassland across the river from their property, a fact O'Brien found deliciously ironic. He caused a lot of confusion when he first applied for a grazing permit—federal officials didn't know what to make of his request to graze bison on their historic range. Now every winter his herd of eight hundred buffalo disappears into 24,000 acres of the protected federal grasslands. Sometimes they roam all the way to the Stronghold Table, a mesa that rises above the land in the hazy distance and is sacred to the Sioux. It is believed to be where the Lakota performed their Ghost Dance before the Wounded Knee Massacre in 1890, a spiritual ceremony they hoped would sweep away the white colonialists and bring back the buffalo.

In springtime, herding the animals back from the draws and creases between the hills in the national grasslands is much harder than letting them go into Buffalo Gap in the first place. Much of that work is done by the next generation of bison ranchers, Jill's daughter and son-in-law. As Wild Idea expanded, the company also built its own meat-processing facility in Rapid City. The small plant is staffed mostly by tribal members from nearby reservations and ships ground bison, roasts, and steaks directly to customers. They've also expanded their supply chain to other ranches, including Ted Turner's. Each year they drive their mobile slaughter truck around the West to harvest some nine hundred animals at other ranches and reservations that raise their bison the same way—wild. The family has recruited new investors, including the outdoor apparel company Patagonia, which sells Wild Idea buffalo jerky and $400 buffalo-hide boots alongside the fleece jackets and down coats in its catalog.

Wild Idea is still tiny, and almost unique among the nation's seventeen hundred bison producers. O'Brien compares it to a boutique winery. And like a boutique winery, its products are pricey—its ground bison sells for $17 a pound. Other bison ranchers say it's an impossibly purist model that few others could replicate. While some other ranchers do sell grass-fed bison, most say his structure—the mobile harvesting truck and the small meat plant in Rapid City—is too expensive to scale up to the larger market. It's true that O'Brien is selling a "wild idea" to a niche audience: a better food system producing a healthy product that can heal damaged grasslands and restore many of the creatures that lived there before Europeans arrived. Ever the conservationist, he says, "We are selling species diversity." But O'Brien thinks the skeptics are wrong. A decade ago he said that Wild Idea and the bison industry will have succeeded when Ted Turner runs his business the way O'Brien runs his. "It'll kill me," he said. "But what a way to go."

DURING THE SAME period when O'Brien was building Wild Idea, a parallel and larger movement was restoring bison to Native reservations across the Great Plains and as far away as Alaska and northern Minnesota. These growing tribal herds, now numbering roughly eighty, may offer the greatest opportunity to return buffalo to their land in significant numbers, and, if successful, could restore far more than the grassland ecosystem.

As far back as the 1940s, a few Plains tribes managed to keep some bison on their land. Then in the 1970s, inspired in part by the wave of Native American pride that sprang from the American Indian Movement and its epic showdown with federal agents at Wounded Knee in 1973, several more tribes tried to return small herds to their reservations. In the beginning these projects, including herds at the Rosebud Sioux reservation in South Dakota and the Fort Belknap Reservation in Montana, represented an effort to restore Native practices and culture, and they had an uneven record. Raising buffalo requires special expertise, lots of land, and expensive fencing. And because bison multiply rapidly in the absence of wild predators, bison ranchers need a plan to cull the herd—selling animals to other ranchers, managing hunts, or processing and selling the meat. Many tribes have members who are cattle ranchers—one of the few ways to earn a living in those remote places—and introducing bison can generate the same conflicts that erupt in other ranching communities. Why should the tribe turn land over to animals that cost money rather than earn it? Other efforts failed for economic reasons, like the discouraging example of a Crow tribe in South Dakota that built its own slaughterhouse with grand ambitions, only to see it fall into bankruptcy a few years later.

In 1992 these disparate efforts coalesced into the InterTribal Buffalo Cooperative, a nonprofit headquartered in Rapid City, dedicated to providing support for tribes who wanted bison. At first, the cooperative saw bison primarily as a commercial proposition, and marketed the meat and calves on behalf of about fifteen member tribes. It later changed its name to the InterTribal Buffalo Council, and became a source of expertise, and, eventually, a voice for Native Americans in state capitals and Washington, D.C.

In 1997, the tribes' alliance with bison exploded onto the national stage when a band of Native activists came to the defense of the iconic herd at Yellowstone National Park. Protected by park boundaries and with few predators to control their numbers, the once tiny Yellowstone herd had grown to several thousand. Every winter, when the snow and cold made grazing difficult, many of the animals would move across park boundaries to find forage. And although every year park officials would capture hundreds and send them to slaughter, Montana ranchers regarded the remaining animals as a threat to their cattle. Bison compete for grass and can carry brucellosis, a serious bovine disease that can cause cows to miscarry, although Montana had never documented a case of bison-to-cattle transmission. Every spring, hunters and sharpshooters would line up along a road just outside the park's boundaries and begin shooting as soon as the bison set foot off federal land. To tribal activists and conservationists, the annual slaughter seemed to be a vestige of the cultural war that drove buffalo and Indians off the land more than a hundred years ago. The number killed varied from one year to the next, but every year the bloody ritual would prevent the buffalo from following their instinct to move, making them the only wild animal in Yellowstone National Park that is constrained by artificial boundaries. The InterTribal Buffalo Council had argued for years that the National Park Service and the state of Montana should let its members take the excess bison to reservation lands, where they were wanted. But they were continually rebuffed, not least because of Montana's powerful ranching interests.

The winter of 1996–97 was especially severe, but this time Native bison activists rose up in protest against the Yellowstone slaughter. A Lakota Sioux elder and college instructor in South Dakota, Rosalie Little Thunder, led a five-hundred-mile "spiritual march" from Rapid City to Yellowstone, an outcry that succeeded in drawing national attention to the plight of the Yellowstone herd. It also ignited years of acrimony involving the National Park Service, wildlife conservation groups, the InterTribal Buffalo Council, cattle ranchers, and the state of Montana. Eventually, litigation led to an agreement among park officials, Montana regulators, and eight tribes that reduced the size of the Yellowstone herd and allows permit hunting, primarily by tribal members, of some of the animals that leave the park. The agreement didn't entirely resolve the controversy. In 2023, another bitter winter and an extraordinary year of migration, hunters killed twelve hundred of the six thousand bison in Yellowstone, eliciting national news coverage and protests from multiple animal conservation groups. Between hunting, culling for slaughter, and wounded animals, approximately 2,300 bison died—nearly a third of the herd.

The slaughter controversy also produced a remarkable project that has created a clear and enduring path for Yellowstone bison to find their way onto reservation lands. Not long after Rosalie Little Thunder's protest march, a Montana state wildlife biologist named Keith Aune proposed a novel way to save the priceless animals. He would build a quarantine facility at Yellowstone to test a certain number of bison each year to prove they were brucellosis-free. With the help of conservation groups such as Defenders of Wildlife and the World Wildlife Fund, Yellowstone would then transfer healthy bison to other sites around the state where they would be used to build satellite herds. To make the project work, Aune needed partners with two features—lots of land and a fondness for bison—and it occurred to him that his natural allies would be Montana's tribes and the InterTribal Buffalo Council. Both quickly embraced the idea, but Montana's cattle interests launched another legal battle to block any transfer to tribes. The four-year struggle finally reached the Montana Supreme Court, which ruled in 2013 that the state had no authority to block a transfer of bison to sovereign tribal Nations. The following year, Aune's quarantine facility and the Defenders of Wildlife began shipping healthy bison in small numbers to the Fort Peck Reservation in far northeastern Montana, setting the stage for what is now one of the biggest and most influential bison recovery operations on the northern Great Plains.

On a winter day in 2015 a red livestock truck blazed against the brilliant white snow of the Montana prairie, rumbling up a gravel road on the Fort Peck Reservation. As the truck backed up to a small steel corral, the thirty bison inside snorted and heaved like a storm in a box. When the rear door opened, two massive bulls peered out suspiciously, threw up their massive heads, and inhaled the clear frigid air. Another group of Yellowstone bison had arrived. One by one the wary animals lumbered down a steel ramp and circled the small corral at a restless trot. But as soon as they saw the open gate at the far end, they hurtled through and galloped into the vast open prairie beyond. In a few seconds dozens of bison had vanished into the shadowed creases of the winter landscape—as if they had lived there all their lives. "They know they're home," Robert Magnan said to the cheers of a crowd of Indian children and elders who had gathered to see a glimpse of their heritage returned to them.

Magnan, fish and game director for the Fort Peck tribes, was among the first Native leaders to step up for Aune's Yellowstone bison project, and he has made Fort Peck a leader in returning them to tribal lands. The reservation, created in 1866, lies hard against the Canadian border and includes both Sioux and Assiniboine people, both among the nomadic Plains tribes whose lives were intricately tied to bison. The reservation today is poor and concentrated in the small towns of Poplar and Wolf Point. But it is rich in land, encompassing two million acres of prime buffalo range at the top of Big Sky Country. Magnan, now in his late sixties, typically wears a camouflage sweatshirt and pants, and a billed cap with a long, gray braid hanging down his back. He's talkative and easygoing, and he's used to visitors—reporters from *The New York Times* and *National Geographic,* documentary filmmakers, dignitaries, politicians, and tourists—who come to learn about the bison at Fort Peck. Like many Native Americans of his generation, Magnan spent his boyhood in a boarding school that suppressed his Native culture. Despite the hardship, he says, the one he attended, a Catholic school in North Dakota, taught him discipline and self-reliance. After spending several years in Denver, where his parents found work, he returned to the reservation and went into tribal law enforcement. Before long he took over the tribe's Fish and Game Department and first brought bison to Fort Peck at the urging of tribal elders who wanted to restore the two Nations' cultural identity. When he heard about Aune's proposal to transfer Yellowstone bison to tribal lands—and the legal and political forces arrayed against it—he spent years driving across Montana to testify at the state legislature or in court to make the case for tribal stewardship. Building the Fort Peck herd became his life's purpose, an obsession that took all his time and his heart and cost him his marriage of twenty-three years. "The creator wants to know how much I want them," he says. "He's testing me."

Even with support of the elders, Magnan knew he would have to navigate reservation politics, which can shift every time there's an election for tribal council. He learned to build support in the community, particularly among the influential female elders' Pte Group, named for the female buffalo in the Dakota language.

Tribal members who raised cattle, just like non-Indian ranchers, weren't comfortable with the idea of "free-range" buffalo, so Magnan learned to use the term "wide-ranging." He had to persuade the reservation's elected leaders that spending money on the herd was a worthy investment. And it has been.

Today Magnan oversees two bison herds. The tribe's cultural herd, now approaching four hundred animals, is used to educate schoolchildren, support tribal rituals, and supply meat for the reservation's school lunch and elderly feeding programs. Every senior class at the reservation's high schools gets its own bison hunt at graduation. They accompany tribal hunters into the field, witness the hunt, and watch elders perform praise rituals handed down over the generations. Dyan Youpee, Fort Peck's cultural resources director, says the bison have become the center of the tribes' work to preserve history and languages for future generations. "It's not about the past, really, it's about building a future so that our young ones grow up strong," she says.

Magnan also built a smaller "business herd" that generates revenue from livestock sales and hunting. Fort Peck hunting permits, each good for taking one buffalo, go first to members of the tribe; and then to outside sportsmen from all over the world who pay thousands of dollars for the privilege of shooting a buffalo on its native land. And even though bison are extraordinarily docile creatures unless they feel threatened, Magnan doesn't make the hunt easy. The hunters have to hike miles into the grasslands to find their prey, and if they succeed in shooting a buffalo they have to field dress the carcass on the grass and pack out the meat.

With the two herds established and a steady pipeline from Yellowstone, Magnan also fulfilled his ambition of making Fort Peck a distribution hub for other tribes that want the priceless bison. Fort Peck has the only quarantine facility outside Yellowstone that can accept the park's bison and certify them as brucellosis-free. In the first year, it received five bison, then twenty, and eventually fifty bison a year. Since 2019 he has sent animals to some fourteen other reservations in several states, including the Alutiiq tribe on an island in Alaska, the Shoshone-Bannock in Idaho, the Modoc and Cherokee in Oklahoma, and the Oneida in Wisconsin. And in one gift echoing with historic justice, he sent eight bison to the Bronx Zoo in New York.

On a dry, sunny day in late September, Robert Magnan leans on his truck next to the maze of quarantine fencing that holds a

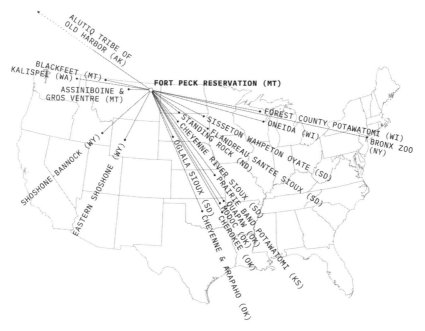

Since 2013, the Fort Peck Reservation in Montana has served as the hub for a growing network of tribal buffalo herds, rescuing threatened bison from the historic herd at Yellowstone National Park, quarantining them to assure their health, then sending them to other reservations across the country.
*Graphic by Alexander Hage*

few buffalo eating hay and waiting for their release. He's old enough to quit working, Magnan says, and sometimes he thinks about it. But challenges keep coming up. The tribal council just agreed to provide $2 million to build a livestock processing plant on the reservation, so the tribe can supply its members with meat and build a new business selling it to others. He wants to see bison returned not just to tribal lands, but also to public lands like the nearby million-acre Charles M. Russell National Wildlife Refuge, a prominent Montana wilderness preserve and a place that hasn't been grazed since bison were exterminated there a century ago. The urge to keep working is strong in him, and the West still has a lot of wide-open land where buffalo could go. They took care of his ancestors for thousands of years, he says. "Now it's our turn to take care of them."

As of 2022, some 25,000 buffalo have taken up residence in cultural or livestock herds with eighty tribes in twenty states, most of them members of the InterTribal Buffalo Council. Troy Heinert, the council's executive director, spends much of his time coordinating the transfer of bison across the country to tribes starting new herds. Heinert is a member of the Sicangu Lakota Nation and lives on the Rosebud Indian Reservation in southwestern South Dakota. During the Covid pandemic, the reservation's small grocery stores were often depleted because of supply chain problems and shoppers who came from off-reservation to find food. It made no sense, Heinert says, that the Rosebud Reservation, with a million acres of grass for cattle and bison, could not provide food for its people. Since then, tribes across the country have begun building their own bison herds to ensure culturally true food sovereignty, and the InterTribal Buffalo Council is working with several who want to build small processing facilities. One in New Mexico won a grant to buy a mobile slaughtering system like O'Brien's, perfectly suited to their small, remote community.

To help tribes overcome the obstacles in building a bison herd—political, financial, cultural—in 2014 a group of Native leaders, ranchers, and philanthropists established a nonprofit called the Tanka Fund. Its leader, Trudy Ecoffey, notes the entwined history of bison and the Plains tribes. "How do you subdue a people?" she asks in a reference to United States policy toward bison in the nineteenth century. "You take away their economy." Despite the practical challenges they face, the connection between buffalo and the tribes remains powerful, one that is reflected in the names each tribe has for their ancient kin. In the Blackfeet language the word for buffalo is "iinii," which means "taking hardships away." The Lakota use the word "tatanka," and call themselves the "pte oyate"—or buffalo people.

Troy Heinert, too, sees parallels in the histories of bison and the Plains tribes. Both were driven off their land by the forces of colonialism and then constrained—on reservations for the tribes, behind fences for bison. Indians lost their culture and their economic survival, and bison lost their keystone role in the grassland ecosystem. What happens to one, happens to the other, Heinert says. Finally, it's a more hopeful time for both.

Despite its successes—or because of them—the movement to restore bison has created pockets of resistance, especially in Montana, where the cattle industry has made it a political flashpoint. In 2020 the state elected a Republican governor, Greg Gianforte, who made an anti-buffalo crusade part of his campaign. He led the state's unsuccessful fight to deny American Prairie a set of federal grazing rights on public land. He has signed legislation that aims to block the introduction of more bison in Montana for ten years, though the state has no jurisdiction on federal or tribal lands. When a coalition of tribes and conservationists proposed reintroducing bison to the federal Charles M. Russell National Wildlife Refuge in east-central Montana, that same anti-buffalo rhetoric was deployed to stop the effort. Even though cattle are not permitted to graze there because it's dedicated to wildlife, one irate Republican legislator issued a statement saying buffalo at the refuge would jeopardize grazing land for livestock and "threaten the livelihoods of ranching families."

In many other states, from Utah to Wisconsin, bison are being welcomed—even celebrated. There are some fifteen thousand buffalo in small conservation herds scattered across the country in national and state parks, wildlife refuges—even a few in the Neal Smith Wildlife Refuge in Iowa—and nonprofit grassland conservation sites owned by organizations like American Prairie and The Nature Conservancy. The new appreciation marks the success of a century of careful conservation that has returned buffalo from the edge of extinction. A recent analysis by the International Union for Conservation of Nature concluded that, based on their number and available habitat, buffalo have achieved 17 percent of their potential recovery as a wild species. To wildlife conservationists, that's not nearly enough: Bison are still functionally extinct on the American grasslands, meaning their numbers are too small to perform their role as the keystone species, anchoring the web of wildlife and plants in the enormous expanse of their ecosystem. They are still not permitted to be wild in the same sense as wolves, deer, antelope, elk, bears, and mountain goats. Many of those species reached the edge of extinction and then recovered, and are free to move as their instincts tell them to. Buffalo are not free to roam, to migrate from Canada to Texas as they once did, fight off predators, or interact with the dozens of other grassland species that once depended on them. For now, no matter how big their pasture, sooner or later they all run into a fence.

That's why bison are at the heart of the "wildlife play" that Sean Gerrity described at American Prairie, today one of the most significant bison restoration efforts in the West. Like Dan O'Brien, American Prairie understood that owning land in the West would give it grazing rights on adjacent federal lands, and its bison could one day mingle with bison on the neighboring Charles M. Russell National Wildlife Refuge. It is one of the few places in the country where bison can be as wild as is possible in the midst of cattle country.

American Prairie is also one of the few places where scientists can begin to understand bison as a wild animal. Hila Shamon, a landscape ecologist with the Smithsonian's National Zoo and Conservation Biology Institute, now runs one of the largest bison research projects ever undertaken, using the American Prairie herd. She has spent several years studying the herd and its impact on a place that has known only cattle as the primary grazer for a century.

On a frigid, gray winter day at American Prairie, Shamon was one of several dozen people bundled up in insulated overalls and heavy boots, moving around an oval of high-fenced corridors erected to gather bison for the reserve's annual roundup. She stood next to a giant hydraulic squeeze chute—one that's bigger and stronger than the kind used for cattle—her breath creating clouds of steam in the cold, dry air. One by one, bison moved through the maze of steel fencing and stepped suspiciously into the chute, which then clamped tight around their sides while she drew blood, snagged some of their tail hair for DNA testing, and attached a small yellow solar-powered GPS transmitter to their ear. Despite the best efforts of the staff to move swiftly and calmly, the buffalo trapped in the vise were terrified, their sides heaving and their eyes rolling. Shamon and the graduate student at her side worked fast and silently, bagging samples and recording data on a spreadsheet. Each time the squeeze chute opened, the buffalo inside would leap into the air and lunge for the next and final holding pen. It was a long, cold, stressful day, sadly punctuated by the death of two of the animals. One had been gored by another buffalo while the animals milled restively in a corral; the other's death was under investigation by American Prairie's bison staff. When work resumed the next morning the bison operations manager, Pedro Calderon-Domingues, gathered the young crew into a circle and reminded them how little is known about the animals in their care. "No one under the age of ninety-five has all the lessons on bison," he said. "We are a long way from doing the job right."

In only a few years of research, Shamon has assembled one of the most comprehensive studies on bison behavior, tracking data that reveal just how different bison are from cattle—and how they change grasslands for the better. Buffalo preserve riparian areas, the green protected stream banks and riverbanks that are important corridors for bears, mountain lions, and other wildlife. Buffalo move in to drink the water, then leave, while cattle typically stay by the water, trampling banks and creating mud holes. Through their droppings buffalo spread grass and flower seeds, allowing plants to germinate over large areas. It appears that bison also respond to the smell of smoke, following it to places where prairie fires burned because they know that after the flames have passed they will find young green shoots thrusting through the blackened grass. They might even move in response to thunderstorms they can sense on the horizon because they know the clouds mean rain. Shamon's genetic studies tell stories about the matriarchal bison families and how they behave with one another. But there's a lot that can't be learned, even at American Prairie, because bison still can't behave as the nomadic animals they once were. Surrounded by fences, they can't follow the thunder as far as they might, and prairie fire as an ecological force no longer exists.

On a small scale, however, that may change. North America's prairies will never have tens of millions of buffalo wandering freely from Alaska to Mexico and the East Coast as they once did. But buffalo in more significant numbers may soon step back out into a wilderness where they can move unrestricted across at least a few million acres of prairie. Conservationists call it "the second recovery." The concept was hammered out in 2006, when a few dozen buffalo experts from Indian tribes, wildlife nonprofits, and federal and state agencies gathered at Ted Turner's Vermejo Park Ranch in New Mexico. There they drafted a document, now known as the Vermejo Statement, that describes what a second recovery would look like. "Over the next century, the ecological recovery of the North American bison will occur when multiple large herds move freely across . . . their historic range, interacting in ecologically significant ways with the fullest possible set of other native species, and inspiring, sustaining, and connecting with human cultures."

In 2023 the U.S. Department of the Interior announced a significant step toward making that a reality through an order called the Restoration of American Bison and the Prairie Grasslands. It creates a Bison Working Group to coordinate the five federal agencies that manage bison and grasslands, and includes the federally recognized tribes as full partners in developing a long-term strategy to ensure the success of the species. The order asserts that bison are wild animals and will be permitted to behave like it. The twelve federally managed wild herds on 4.6 million acres of public land in twelve states would be managed as one, meaning that the animals will be shared, moved when necessary across herds and geography to maintain the full genetic potential of the species. The federal herds, and the expertise of the federal employees who manage them, will also be a resource for other entities—state, federal, tribal, and nonprofit—who want to establish wild herds.

The plan includes $25 million in federal funds, some of which will be used to create an apprentice program to train tribal members in bison management. It will include opportunities for Native youth to work with wild herds at national parks and wildlife refuges. And for the first time, the federal government will support "shared stewardship" of the buffalo, meaning that it will develop partnerships with states, nonprofits, and, most critically, tribes to manage large wild herds across jurisdictions for multiple objectives—grassland conservation, hunting, and tribal culture, to name a few. For example, tribal members, some of whom are state legislators, have urged the Department of the Interior to allow the tribes of the Rocky Mountain Tribal Leaders Council to manage any future herd established at the Charles M. Russell National Wildlife Refuge. They suggested that another herd established by the Blackfeet Nation on the Rocky Mountain Front could range to Glacier National Park next door. These vast prairie regions were once shared by Indian tribes and buffalo, the legislators said. "Both the grass and the tribes are still present; only the buffalo are still missing."

Brendan Moynahan, a bison expert with the National Park Service who heads the federal Bison Working Group, said it will be years before wild buffalo are a common sight in the West, rather than an occasional roadside sight behind fences. But in twenty years' time there might be at least five herds larger than a thousand animals roaming across millions of acres, their number controlled by hunters and natural processes, and their survival determined by natural selection. If it doesn't happen, Moynahan says, it won't be because of the buffalo. It's the perspectives, values, and vision of people that will determine the fate of the national mammal. After all, the return of bison to the American grasslands as a wild animal—and as livestock—is still evolving, as is the place they hold in American hearts. Long a symbol of the conquest of the American West, the buffalo is becoming a cultural force for a more hopeful future—one that embodies historic justice for Indigenous people, a climate-resilient economic opportunity for the West, a healthier food system, and a wild savior for the most threatened ecosystem on Earth.

IN HIS MEMOIR, *Buffalo for the Broken Heart,* Dan O'Brien tells this story. A few days after he brought home the gas house gang of buffalo calves, the smallest and weakest one—which he'd named Peatry—died. Rather than bury the carcass, O'Brien set the tightly curled little body atop a snowy ridge. That night he sat under the stars with a glass of scotch in one hand and a viewing scope in the other, and he waited. All winter he'd been hearing the song of a new family of coyotes, and now he was wondering if they would appear. He blinked, and the mother coyote was suddenly there right beside Peatry, staring O'Brien down through the other end of the viewing scope. "With just a glance of those knowing eyes, the old bitch had me," he wrote. It dawned on him that he might be the first person in a hundred years to witness an event that no one thought would ever occur again. The coyote gave a yip and three squeals, and "in seconds Peatry was covered with hungry, squalling yearling pups. The mother did not move. There would be a few bones left for her. Not since the time of Crazy Horse had there been a banquet like this. It was a thing to savor."

EPILOGUE

ROBERT MAGNAN IS STANDING NEAR FORT PECK'S BISON CORRAL ON A September morning when a big yellow school bus comes over the rise behind him. A dozen high schoolers tumble out, their faces showing that sweet adolescent mix of wariness, cool, and curiosity. Magnan is an unpretentious man, given to practical observations rather than speeches. But when the students have assembled in a half circle he begins an expert lecture on the Fort Peck Indian Reservation, grassland ecology, and the shared history of bison and the Sioux and Assiniboine people. When he finishes, the students board a Land Rover to go find one of the tribe's herds and learn about life as their ancestors lived it. Magnan has spent more than a decade building the bison herds at Fort Peck, but these students represent his larger goal. They are the future of a people who, together with the bison, are fighting their way back from near extinction.

The return of wild bison to the West is just one of many signs that the prairie ecosystem, too, might yet survive its brush with extinction. The proliferation of nonprofits such as Practical Farmers of Iowa, groundbreaking alliances between conservationists and western landowners, the rapid adoption of conservation farming practices, the Biden administration's decision in 2022 to pour more than $19 billion into conservation agriculture and grassland restoration—all suggest that rescuing the prairie is not impossible. In fact, it's already happening.

What would it take to declare the prairie saved? It depends on how you frame the question. To revive bison as an ecologically functioning species, along with the grassland creatures that flourish around it, would require several herds of one thousand buffalo each on hundreds of thousands of acres—numbers that are no longer out of reach. To restore clean drinking water in communities such as Des Moines, or to shrink the Gulf of Mexico's hypoxic zone? That would mean retiring something like 10 percent of the Corn Belt's cropland and planting it with buffer strips and conservation wetlands. To cut soil erosion to sustainable levels, or to restore a healthy population of bugs and pollinators? That would mean tripling or quadrupling the number of farmers who embrace soil-conservation techniques. Lofty goals all, but no longer inconceivable.

The cost of such undertakings would run into the billions of dollars. But then we already spend billions of dollars supporting the farm economy every year. Quite apart from money, it would also require Americans to adopt a new attitude toward the land, and it would mean unwinding key strands of American agricultural history. Farmers would have to abandon the highly efficient corn-soybean monoculture that dominates the Corn Belt today and return to the diverse, healthy mix of crops—alfalfa, oats, barley, rye—that their grandparents sowed. It would mean using fewer chemicals to fertilize the land and fight pests, relying instead on natural diversity and regenerative farming to control bugs and build soil health, much as our ancestors did. It would mean taking cattle off some of the western plains, making space for wild creatures and diverse plant life. It might mean forfeiting some of the remarkable productivity gains agriculture has delivered in the last century. But then the United States can afford that; American farmers produce such a volume of corn and soybeans today that we have to invent new uses for them. Moving toward a food system that is kinder to the land doesn't mean that Americans have to forsake all meat or subsist on organic kale. A menu built around beans, grains, poultry, and modest amounts of red meat—including grass-raised beef and buffalo—can deliver an inexpensive diet that is far healthier for people and the land. It would also help if the government subsidized healthy foods rather than huge volumes of commodity crops. Finally, making this case doesn't mean being anti-technology. It simply means agreeing with the great nature writer Aldo Leopold that technology sometimes has diminishing returns. Defenders of industrial agriculture like to brag that Americans enjoy the cheapest food in the world. Today it's hard to know if that's a boast or an indictment.

When this nation sent frontier farmers onto the prairie in the nineteenth century, the United States had an urgent set of problems to solve. It had to feed a booming population, and soon parts of Europe. It aspired to provide a piece of land and a better life for wave after wave of immigrants. And, at least in the mind of Thomas Jefferson, it had to stake a claim to the West and fend off any remaining colonial ambitions of the European powers who still controlled big parts of the continent. In all this, nineteenth-century agriculture delivered: It produced astonishing bounty, gave land to landless immigrants, and built a prosperous rural society while establishing the West as a uniquely American place.

Today we face an entirely different set of national problems: climate change, water shortages, tainted wells, vanishing wildlife, polluted rivers, the lasting repercussions of genocide. The historian Patricia Nelson Limerick has observed that private property and free markets, which colonialists embraced in the nineteenth century, are less suited to these problems of the commons. And if we face a future of diminishing petroleum supplies, a model of agriculture built on fossil fuels will only grow more problematic.

For pragmatic reasons alone, rethinking our relationship with the land is critical.

Two or three decades ago, expecting Americans to embrace Aldo Leopold's land ethic might have seemed wildly unrealistic, given our fondness for cheap food and fascination with new technology. Today there are signs that Americans are having a change of heart. The growing market for organic food, the rapid spread of Community Supported Agriculture, the fact that farmers are signing up for conservation programs faster than Congress can fund them, the popularity of farm-to-table restaurants—even Americans' sudden love affair with bison—all seem to signal a new affection for the land that feeds us. These are tiny developments in a giant food system. But then sociologists say that a small number of committed reformers can trigger tsunamis of social change. We could already be in the middle of a movement and not know it.

In his famous 1893 lecture, the scholar Frederick Jackson Turner argued that the western frontier played a central role in forming the American character and shaping American history. Surviving on the edge of (perceived) wilderness made Americans rugged and self-reliant. Living in remote places gave them an independent streak and a suspicion of government. The frontier, endlessly receding and constantly beckoning, made us a restless people. Turner argued that when the western frontier closed in 1890—that is, when the Census Bureau declared that the West had no more unsettled counties—it marked a defining turn in American history. Turner's thesis has lost favor among subsequent generations of historians, who point out that cities, Native Americans, women, and immigrants also played crucial roles in directing the nation's history. Even so, the perception of an open frontier may have contributed to another American trait: the belief that nature is infinite and land is disposable. "This perennial rebirth, this fluidity of American life, this expansion westward with its new opportunities . . . furnish the forces dominating the American character," Turner wrote. If the tobacco crop has exhausted your soil, if the timber is gone—simply move west and start over. Today one might argue that climate change is our version of the closed frontier: It teaches us lessons the Lakota knew centuries ago—that we must give as well as take, that land is not something to be measured and consumed, and that we pay a price for exploiting the Earth—especially on the American grasslands, a region we have exploited almost to death.

When Deborah and Frank Popper published "The Buffalo Commons" in 1987, readers seized on two of its radical propositions: that depopulation of the Great Plains was a natural development and that bison should replace people. But the essay had a third radical idea, implied but no less powerful: that the entire nation has a stake in this region and deserves some voice in shaping its future. Now that climate change is upon us and the plows are moving west at a rapid pace, that proposition is more urgent than ever. To it we might add that rights entail responsibilities. If as a nation we want to rescue this priceless landscape—its people, its creatures, and its contribution to a healthy planet—it's up to us to change the way we eat, the way we support agriculture, and the way we think about who owns the land.

In a memorable phrase, the Poppers also called settling the West a "monument to our national self-delusion." Shedding those delusions would require us to reverse the polarity of American history—to understand that nature has limits, that cheaper isn't always better, that we are part of the natural world, not its master. Wendell Berry used the word "agrarian" to describe this philosophy of the land and wrote, "Agrarians understand themselves as the users and caretakers of some things they did not make, and of some things they cannot make." Having relentlessly sacrificed things we did not make—the land, its Indigenous people, and its creatures—to achieve remarkable national prosperity, perhaps we're rich enough to restore some of what we've taken and heal the scars of our history.

ACKNOWLEDGMENTS

This book grew out of a newspaper series, The Vanishing Prairie, that Josephine wrote for the *Star Tribune* of Minneapolis in 2012, when we worked there together. We continue to be grateful to our *Star Tribune* editors—Rene Sanchez, Suki Dardarian, and Eric Wieffering—who supported that work and its offshoots and always made our journalism better. We are also grateful to the Alicia Patterson Foundation for a 2014 fellowship that allowed us to deepen that reporting by traveling the northern Great Plains and visiting tribal and conservation bison herds. That is also when we first met South Dakota bison ranchers Dan and Jill O'Brien, who shared some of their delicious roast bison with us and gave us a glimpse of what a healthier food system could look like.

Turning a newspaper series into a book proposal is a big job, and we couldn't have done it without generous advice from two old friends, Sue Halpern and Bill McKibben, and then sage guidance from the wonderful Sam Freedman, who has supported uncountable fledgling writers at Columbia University and produced excellent books of his own. Jennifer Carlson, our wise and brilliant agent, saw the merit in our proposal, made it much better, and then found us the perfect publishing home. At Penguin Random House our editor, Molly Turpin, immediately recognized the promise of the book, then helped us fulfill it with insightful questions, patient observations, and endless faith that we could achieve its potential. We are also indebted to her colleagues at Penguin Random House: Monica Rae Brown, Michael Hoak, London King, Marni Folkman, Rebecca Berlant, Richard Elman, Dennis Ambrose, Liz Carbonell, Alison Rich, Erica Gonzalez, and Brianna Jackson, who made this book a reality.

Of course, you can't write about the prairie without spending a lot of time there, and you can't understand it without expert guides who can explain what's right in front of your nose. In Bozeman, Montana, Martha Kauffman and her staff at the World Wildlife Fund provided invaluable help in understanding the depth of the current environmental crisis on the northern Great Plains and introductions to working landowners who are trying to save the western prairie. Patrick Lendrum, in particular, gave us an ongoing education in prairie ecology; Curtis Freese, an ecologist who has written his own fine book on grassland restoration, offered useful critiques of one draft; and the amazing Bill Milton showed us how people can make change happen on the ground. At The Nature Conservancy, Christopher Anderson, Joe Fargione, Neal Feeken, and Brian Martin introduced us to the organization's deep research on why grasslands matter and gave endless hours to help us understand the prairie and its wildlife; their colleague Sasha Gannet offered invaluable observations on the modern American food system. Dave owes a special debt to Jonathan Proctor at Defenders of Wildlife, who introduced him years ago to some pioneering efforts in bison reintroduction projects, and to an old friend, Scott Faber, whose work at the Environmental Working Group was invaluable. At The Land Institute, Tim Crews, through his research papers and personal interviews, helped us understand what's at stake in the way modern agriculture has ravaged the grassland ecosystem. We are also grateful to Chris Helzer for sharing his stunning photos of the tiniest prairie creatures and to Amy Beattie, who helped us understand the values and aspirations of Bill Stowe and who sharpened our understanding of the fight for clean water in Iowa.

This book contains a great deal of history and science, and we could never have mastered it without help from the great land-grant universities of the Midwest. At the University of Wisconsin, Tyler Lark, whose groundbreaking work with Seth Spawn has documented and detailed the scale of grassland conversion, gave us endless hours of help. At the University of Minnesota, Forest Isbell, Jason Hill, Jessica Gutknecht, and Peter Kennedy, themselves brilliant scholars of grassland ecology, gave generously of their time and research. At Iowa State University, the great Matt Liebman, through his research papers and then interviews, helped us understand the environmental costs of modern agriculture and how simple reforms could deliver huge benefits. His fellow researchers Richard Cruse, Silvia Secchi, and Alison Robertson provided additional guidance. John Blair and Charles Rice at Kansas State University coached us patiently on their state's tallgrass prairie and the way agriculture has changed it. In Iowa, the incomparable Jerry Hatfield helped us bridge the gap between serious soil science and working agriculture, as did Brad Carlson, who teaches farmers about fertilizer and nitrogen for the University of Minnesota Extension. At Princeton University, Tim Searchinger helped us untangle the tangled science and politics of biofuels such as ethanol.

Douglas Hurt of Purdue University helped us understand the history of frontier agriculture; Michael A. Urban, formerly with the University of Missouri, connected the drainage of the past to the present; and Joe Spencer, who studies insect behavior at the University of Illinois, guided us through the complexity of pesticides and genetically engineered seeds. Keith Schilling at the University of Iowa helped us connect the region's drainage ditches with the Mississippi River and the Dead Zone in the Gulf of Mexico. Eric N. Powell, director of the science center for marine fisheries at the University of Southern Mississippi, explained the life cycle and slow death of oysters in the Gulf. Zeus Mateos-Fierro was just one of many hardworking, bug-loving entomologists who studied and trained in Ian Kaplan's lab at Purdue University and showed us the real world of pollinators and their role in producing our food. We are also grateful to Dan Egel and others at the Southwest Purdue Ag Center in Vincennes, Indiana, for helping us to understand the world of watermelon growers. Plant ecologist Dan Wovcha and others at the Minnesota Biological Survey provided the rich details of life, past and present, in the Red River Valley. Carrie Jennings, glacial geologist and an expert on the evolution of landscapes at Freshwater in Minnesota, gave us the satellite view of the Red River Valley and its history.

One of the great pleasures of this project was getting to know Deborah and Frank Popper, whose breakthrough 1987 essay on the buffalo commons was an early spark for Dave's thinking about rural places and who gave us hours of thoughtful insights and stimulating conversation.

Understanding science is one thing, understanding farming in real life is quite another. And whatever success we have had in describing it is due to the many farmers, ranchers, and other producers who gave generously of their time. At the Fort Peck and Fort Belknap reservations, Mark Azure, Robert Magnan, and Jonny BearCub Stiffarm welcomed two outsiders to the world of tribal bison herds, offered huge expertise, and trusted us to write about their work. Rob Larew of the National Farmers Union and Scott VanderWal of the South Dakota Farm Bureau helped us understand the practical impact of policies made in Washington. Jim and Carol Faulstich in South Dakota and Conni and Craig French in Montana invited us into their lives and spent hours explaining their vision of saving the grasslands. In addition, we thank the watermelon growers of Vincennes, Indiana, the many conventional farmers who explained how they thrive in today's agricultural model, the regenerative farmers who often operate without a standard playbook, the endlessly creative members of Practical Farmers of Iowa, the ranchers in the West who have grass in their blood, and the fishermen and fisherwomen on the Gulf. Together, they made it possible for us to connect the great grassland ecosystem to the people on the land—and water—today.

This book stands on the shoulders of other outstanding volumes on the history and ecology of the West. The works of Dan Flores, David Montgomery, Jo Handelsman, and Richard Manning were a revelation and an inspiration, and getting to know these writers personally was an unexpected bonus.

There are many friends and family members who helped us on our way, including photographer John Danicic, who was diligent in the "hard crop" he applied to our photos; Naomi Olson, who steered us to the concept of settler colonialism; and our brilliant friend and colleague Jennifer Bjorhus, who died before this book was published.

We are journalists, and have learned over the years that our work and the stories we tell about the American prairie are not our own; they belong, collectively, to the people we talk to. And to all of them in this book, named and unnamed, from the past and the present, we say thank you.

BIBLIOGRAPHY

*Introduction*

Cunfer, Geoff. *On the Great Plains: Agriculture and Environment.* College State: Texas A&M University Press, 2005.
Leopold, Aldo. *A Sand County Almanac.* New York: Oxford University Press, 1949; Ballantine Books, 1970.
Montgomery, David R. *Dirt: The Erosion of Civilizations.* Oakland: University of California Press, 2007.

**JOURNAL ARTICLES AND OTHER SOURCES**

Hoesktra, Jonathan M., Timothy M. Boucher, Taylor H. Ricketts, and Carter Roberts. "Confronting a Biome Crisis: Global Disparities of Habitat Loss and Protection." *Ecology Letters* 8, no. 1 (January 2005): 23–29. onlinelibrary.wiley.com/doi/abs/10.1111/j.1461-0248.2004.00686.x.
Perkins, Lora B., Marissa Ahlering, and Diane L. Larson. "Looking to the Future: Key Points for Sustainable Management of Northern Great Plains Grasslands." *Restoration Ecology* 27, no. 6 (November 2019): 1212–1219. onlinelibrary.wiley.com/doi/abs/10.1111/rec.13050.

Scholtz, Rheinhardt, and Dirac Twidwell Jr. "The Last Continuous Grasslands on Earth: Identification and Conservation Importance." *Conservation Science and Practice* 4, no. 3 (2022): e626. digitalcommons.unl.edu/agronomyfacpub/1549/.

Spawn, Seth, and Tyler Lark. Grasslands as a Natural Climate Solution in the U.S.: Research Summary of Fargione et al., in Science Advances, November 2018.

## *Part One*

### CHAPTER ONE: *Prairie*

Blackhawk, Ned. *The Rediscovery of America: Native Peoples and the Unmaking of U.S. History*. New Haven, CT: Yale University Press, 2023.

Cronon, William. *Nature's Metropolis: Chicago and the Great West*. New York: W. W. Norton & Company, 1991.

Dunbar-Oritiz, Roxanne. *An Indigenous Peoples' History of the United States*. Boston: Beacon Press, 2014.

Flores, Dan. *American Serengeti: The Last Big Animals of the Great Plains*. Lawrence: University Press of Kansas, 2016.

———. *Wild New World: The Epic Story of Animals and People in America*. New York: W. W. Norton & Company, 2022.

Fraser, Caroline. *Prairie Fires: The American Dreams of Laura Ingalls Wilder*. New York: Metropolitan Books, 2017.

Hämäläinen, Pekka. *Indigenous Continent: The Epic Contest for North America*. New York: Liveright, 2022.

———. *Lakota America: A New History of Indigenous Power*. New Haven, CT: Yale University Press, 2019.

Madson, John. *Where the Sky Began: Land of the Tallgrass Prairie*. Iowa City: University of Iowa Press, 1995.

Miller, David R., Dennis J. Smith, Joseph R. McGeshick, James Shanley, and Caleb Shields. *The History of the Fort Peck Assiniboine and Sioux Tribes*. Helena: Montana Historical Society Press, 2008.

Price, John T. *The Tallgrass Prairie Reader*. Iowa City: University of Iowa Press, 2014.

Savage, Candace. *Prairie: A Natural History*. Vancouver: Greystone Books, 2011.

Standing Bear, Luther. *My People the Sioux*. Lincoln: University of Nebraska Press, 1975.

Treuer, David. *The Heartbeat of Wounded Knee: Native America from 1890 to the Present*. New York: Riverhead Books, 2019.

Webb, Walter Prescott. *The Great Plains*. Lincoln: University of Nebraska Press, 1959.

Wilcove, David S. *The Condor's Shadow*. New York: Anchor Books, 1999.

## JOURNAL ARTICLES AND OTHER SOURCES

Bai, Yongfei, and M. Francesca Cotrufo. "Grassland Soil Carbon Sequestration: Current Understanding, Challenges, and Solutions." *Science* 377, no. 6606 (August 2022): 603–608. pubmed.ncbi.nlm.nih.gov/35926033/.

Isbell, Forest, David Tilman, Peter B. Reich, and Adam Thomas Clark. "Deficits of Biodiversity and Productivity Linger a Century After Agricultural Abandonment." *Nature Ecology & Evolution* 3 (November 2019): 1533–38. pubmed.ncbi.nlm.nih.gov/31666737/.

McSteen, Paula, and Elizabeth A. Kellogg. "Molecular, Cellular, and Developmental Foundations of Grass Diversity." *Science* 377, no. 6606 (August 2022): 599–602. pubmed.ncbi.nlm.nih.gov/35926032/.

## CHAPTER TWO: *Plow*

Bogue, Allan G. *From Prairie to Corn Belt: Farming on the Illinois and Iowa Prairies in the Nineteenth Century*. Ames: Iowa State University Press, 1994.

Cochrane, Willard W. *The Development of American Agriculture: A Historical Analysis*. Minneapolis: University of Minnesota Press, 1979.

Dahlstrom, Neil, and Jeremy Dahlstrom. *The John Deere Story*. DeKalb: Northern Illinois University Press, 2005.

Hurt, R. Douglas. *Agriculture in the Midwest, 1815–1900*. Lincoln: University of Nebraska Press, 2023.

Pritzker, Barry M. *A Native American Encyclopedia: History, Culture, and Peoples*. New York: Oxford University Press, 2000.

Stout, Jane. *Philemon Stout Diary 1900*. Published by author, 2000.

CHAPTER THREE: *Swamp*

Bogue. *From Prairie to Corn Belt*.

Flickinger, Robert E. *The Pioneer History of Pocahontas County, Iowa*. Fonda, IA: George Sanborn, 1904. archive.org/details/cu31924028914335.

French, Henry F. *Farm Drainage: The Principles, Processes, and Effects of Draining Land with Stones, Wood, Plows, and Open Ditches, and Especially with Tiles*. New York: Orange Judd & Company, 1859.

Prince, Hugh. *Wetlands of the American Midwest: A Historical Geography of Changing Attitudes*. Chicago: The University of Chicago Press, 1997.

Schwieder, Dorothy. *Iowa: The Middle Land*. Iowa City: University of Iowa Press, 1996.

Vileisis, Ann. *Discovering the Unknown Landscape: A History of America's Wetlands*. Washington, D.C.: Island Press, 1997.

Weaver, John Ernest. *North American Prairie*. Lincoln, NE: Johnsen Publishing Company, 1954.

Weaver, Marion M. *History of Tile Drainage in America Prior to 1900*. Waterloo, N.Y.: M. M. Weaver, 1964.

Whitney, Gordon G. *From Coastal Wilderness to Fruited Plain: A History of Environmental Change in Temperate North America from 1500 to the Present*. New York: Cambridge University Press, 1994.

JOURNAL ARTICLES AND OTHER SOURCES

Aldrich, Charles. "The Old Prairie Slough." *The Annals of Iowa* 5, no. 1 (1901): 27–32. doi.org/10.17077/0003-4827.2659.

Blann, Kristen L., James L. Anderson, Gary R. Sands, and Bruce Vondracek. "Effects of Agricultural Drainage on Aquatic Ecosystems:

A Review." *Critical Reviews in Environmental Science and Technology* 39, no. 11 (2009): 909–1001. doi.org/10.1080/10643380801977966.

Carlson, Anthony E. "The Other Kind of Reclamation: Wetlands Drainage and National Water Policy, 1902–1912." *Agricultural History* 84, no. 4 (2010): 451–78. jstor.org/stable/27869012.

Hersom, Louis, and Bruce J. Haddock. Oral histories. Earthwatch-SHSI Oral History Project. State Historical Society of Iowa. https://iowa.minisisinc.com/scripts/mwimain.dll/MK2MsEFO3tIJK8/BIBLIO/CA_NAME/Earthwatch-SHSI%20Oral%20History%20Project./$/WEB_BIBLIO_DETAIL_REPORT?JUMP.

Hofstrand, Don, Kelvin Leibold, and Ann M. Johanns. "Understanding the Economics of Tile Drainage." *Ag Decision Maker*. Iowa State University. 2023. extension.iastate.edu/agdm/wholefarm/html/c2-90.html.

Lien, Dennis, and Dave Orrick. "Minnesota Farm Drain Tiling: Better Crops, but at What Cost?" *TwinCities Pioneer Press*. August 31, 2012. twincities.com/2012/08/31/minnesota-farm-drain-tiling-better-crops-but-at-what-cost/.

McCorvie, Mary R., and Christopher L. Lant. "Drainage District Formation and the Loss of Midwestern Wetlands, 1850–1930." *Agricultural History* 67, no. 4 (1993): 13–39. jstor.org/stable/3744552.

McGowan, Kevin P., and Eric Lautenschlager. "Nineteenth Century Agricultural Drainage Technology in the Midwest." *Material Culture* 20, nos. 2/3 (1988): 57–67. jstor.org/stable/41784808.

"Paying $307,000,000 for Iowa Drainage." *The New York Times*. September 23, 1910. nytimes.com/1910/09/23/archives/paying-307000000-for-iowa-drainage-private-owners-of-farms-to-spend.html.

*Sackett v. Environmental Protection Agency*, 598 U.S. (2023). supremecourt.gov/opinions/22pdf/21-454_4g15.pdf.

Urban, Michael A. "An Uninhabited Waste: Transforming the Grand Prairie in Nineteenth Century Illinois, USA." *Journal of Historical Geography* 31, no. 4 (2005): 647–65. doi.org/10.1016/j.jhg.2004.10.001.

## CHAPTER FOUR: $NH_3$ (Ammonia)

Brender, Jean D. "Human Health Effects of Exposure to Nitrate, Nitrite, and Nitrogen Dioxide." In *Just Enough Nitrogen,* edited by Mark A. Sutton et al., 283–94. Cham, Switzerland: Springer, 2020. doi.org/10.1007/978-3-030-58065-0_18.

Follett, R. F., and J. L. Hatfield, eds. *Nitrogen in the Environment: Sources, Problems, and Management.* 2nd ed. Amsterdam: Elsevier, 2008.

Hager, Thomas. *The Alchemy of Air: A Jewish Genius, a Doomed Tycoon, and the Scientific Discovery That Fed the World but Fueled the Rise of Hitler.* New York: Harmony Books, 2008.

Smil, Vaclav. *Enriching the Earth: Fritz Haber, Carl Bosch, and the Transformation of World Food Production.* Cambridge, MA: The MIT Press, 2004.

### JOURNAL ARTICLES AND OTHER SOURCES

Cunfer, Geoff. "Soil Fertility on an Agricultural Frontier: The US Great Plains, 1880–2000." *Social Science History* 45, no. 4 (August 2021): 733–62. doi:10.1017/ssh.2021.25.

Galloway, James N., Allison M. Leach, Albert Bleeker, and Jan Willem Erisman. "A Chronology of Human Understanding of the Nitrogen Cycle." *Philosophical Transactions of the Royal Society B* 368 (2013). dx.doi.org/10.1098/rstb.2013.0120.

Gruber, Nicolas, and James N. Galloway. "An Earth-system Perspective of the Global Nitrogen Cycle." *Nature* 451 (2008): 293–96. doi.org/10.1038/nature06592.

Hill, Jason, Andrew Goodkind, Christopher Tessum, et al. "Air-quality-related Health Damages of Maize." *Nature Sustainability* 2 (2019): 397–403. doi.org/10.1038/s41893-019-0261-y.

Hu, Chen, Timothy J. Griffis, Alexander Frie, et al. "A Multiyear Constraint on Ammonia Emissions and Deposition Within the US Corn Belt." *Geophysical Research Letters* 48, no. 6 (2021). doi.org/10.1029/2020GL090865.

Jensen, Anja S., Vanessa R. Coffman, Jörg Schullehner, et al. "Prenatal Exposure to Tap Water Containing Nitrate and the Risk of

Small-for-gestational-age: A Nationwide Register-based Study of Danish Births, 1991–2015." *Environment International* 174 (2023). doi.org/10.1016/j.envint.2023.107883.

Kub, Elaine. "High Fertilizer Prices: The History and Future." *Progressive Farmer*. December 15, 2021. dtnpf.com/agriculture/web/ag/news/article/2021/12/15/high-fertilizer-prices-history.

Ribaudo, Marc, Michael Livingston, and James Williamson. "Nitrogen Management on U.S. Corn Acres, 2001–10." United States Department of Agriculture Economic Research Service. Economic Brief Number 20, November 2012. ers.usda.gov/publications/pub-details/?pubid=42867.

## Part Two

### CHAPTER FIVE: *River*

Davis, Jack E. *The Gulf: The Making of an American Sea*. New York: Liveright, 2017.

Egan, Dan. *The Death and Life of the Great Lakes*. New York: W. W. Norton & Company, 2017.

Kirchman, David L. *Dead Zones: The Loss of Oxygen from Rivers, Lakes, Seas, and the Ocean*. New York: Oxford University Press, 2021.

McPhee, John. *The Control of Nature*. New York: Farrar, Straus and Giroux, 1989.

### JOURNAL ARTICLES AND OTHER SOURCES

Hatfield, J. L., L. D. McMullen, and C. S. Jones. "Nitrate-nitrogen Patterns in the Raccoon River Basin Related to Agricultural Practices." *Journal of Soil and Water Conservation* 64, no. 3 (May 2009): 190–99. doi.org/10.2489/jswc.64.3.190.

Rabalais, Nancy N., R. Eugene Turner, and Donald Scavia. "Beyond Science into Policy: Gulf of Mexico Hypoxia and the Mississippi River." *BioScience* 52, no. 2 (February 2002): 129–42. doi.org/10.1641/0006-3568(2002)052[0129:BSIPGO]2.0.CO;2.

Raymond, Peter A., Neung-Hwan Oh, R. Eugene Turner, and Whitney Broussard. "Anthropogenically Enhanced Fluxes of Water and Carbon from the Mississippi River." *Nature* 451 (January 2008): 449–52. doi.org/10.1038/nature06505.

Schilling, Keith E., Kung-Sik Chan, Hai Liu, and You-Kuan Zhang. "Quantifying the Effect of Land Use Land Cover Change on Increasing Discharge in the Upper Mississippi River." *Journal of Hydrology* 387, no. 3–4 (June 2010): 343–45. api.semanticscholar.org/CorpusID:128529591.

Schilling, Keith E., Philip W. Gassman, Antonio Arenas-Amado, Christopher S. Jones, and Jeff Arnold. "Quantifying the Contribution of Tile Drainage to Basin-scale Water Yield Using Analytical and Numerical Models." *Science of the Total Environment* 657 (March 2019): 297–309. doi.org/10.1016/j.scitotenv.2018.11.340.

Smith, Martin D., Atle Oglend, A. Justin Kirkpatrick, et al. "Seafood Prices Reveal Impacts of a Major Ecological Disturbance." *PNAS* 114, no. 7 (January 2017): 1512–17. doi.org/10.1073/pnas.1617948114.

Sprague, Lori A., Robert M. Hirsch, and Brent T. Aulenbach. "Nitrate in the Mississippi River and Its Tributaries, 1980 to 2008: Are We Making Progress?" *Environmental Science & Technology* 45, no. 17 (August 2011). doi.org/10.1021/es201221s.

Stackpoole, Sarah, Robert Sabo, James Falcone, and Lori Sprague. "Long-term Mississippi River Trends Expose Shifts in the River Load Response to Watershed Nutrient Balances Between 1975 and 2017." *Water Resources Research* 57, no. 11 (November 2021). doi.org/10.1029/2021WR030318.

Turner, R. Eugene, and Nancy N. Rabalais. "Changes in Mississippi River Water Quality in this Century: Implications for Coastal Food Webs." *BioScience* 41, no. 3 (March 1991): 140–47. doi.org/10.2307/1311453.

———. "Linking Landscape and Water Quality in the Mississippi River Basin for 200 Years." *BioScience* 53, no. 6 (June 2003): 563–72. doi.org/10.1641/0006-3568(2003)053[0563:LLAWQI]2.0.CO;2.

United States Environmental Protection Agency. "Memorandum: Accelerating Nutrient Pollution Reductions in the Nation's Waters." April 5, 2022. epa.gov/system/files/documents/2022-04/accelerating-nutrient-reductions-4-2022.pdf.

USDA Natural Resources Conservation Service. "Conservation Practices on Cultivated Cropland: A Comparison of CEAP I and CEAP II Survey Data and Modeling." March 2022. nrcs.usda.gov/Internet/FSE_DOCUMENTS/nrcseprd1893221.pdf.

Water Resources Center. "South Fork Crow River Watershed: Water Plans." Minnesota State University. January 2014. mrbdc.mnsu.edu/mnnutrients/sites/mrbdc.mnsu.edu.mnnutrients/files/public/watershed/pm_waterplans/untitled%20folder/19_sfrkc_wp.pdf.

Yin, Shihua, Guangyao Gao, Yanjiao Li, et al. "Long-term Trends of Streamflow, Sediment Load and Nutrient Fluxes from the Mississippi River Basin: Impacts of Climate Change and Human Activities." *Journal of Hydrology* 616 (January 2023). doi.org/10.1016/j.jhydrol.2022.128822.

## CHAPTER SIX: *Dirt*

Brown, Gabe. *Dirt to Soil: One Family's Journey into Regenerative Agriculture*. White River Junction, VT: Chelsea Green Publishing, 2018.

Cunfer. *On the Great Plains*.

Egan, Timothy. *The Worst Hard Time: The Untold Story of Those Who Survived the Great American Dust Bowl*. New York: Houghton Mifflin, 2006.

Handelsman, Jo. *A World Without Soil: The Past, Present, and Precarious Future of the Earth Beneath Our Feet*. New Haven, CT: Yale University Press, 2021.

Lal, R., J. M. Kimble, R. F. Follett, and C. V. Cole. *The Potential of U.S. Cropland to Sequester Carbon and Mitigate the Greenhouse Effect*. Boca Raton, FL: Lewis Publishers, 1999.

Montgomery. *Dirt*.

Montgomery, David R., and Anne Biklé. *The Hidden Half of Nature: The Microbial Roots of Life and Health*. New York: W. W. Norton & Company, 2016.

Worster, Donald. *Dust Bowl: The Southern Great Plains in the 1930s*. New York: Oxford University Press, 1982.

———. *The Wealth of Nature: Environmental History and the Ecological Imagination*. New York: Oxford University Press, 1993.

## Journal Articles and Other Sources

Basche, Andrea. "Turning Soils into Sponges: How Farmers Can Fight Floods and Droughts." Union of Concerned Scientists. August 2017. ucsusa.org/sites/default/files/attach/2017/08/turning-soils-into-sponges-summary-august-2017.pdf.

Fargione, Joseph E., Steven Bassett, Timothy Boucher, et al. "Natural Climate Solutions for the United States." *Science Advances* 4, no. 11 (November 2018). science.org/doi/10.1126/sciadv.aat1869.

Montgomery, David R. "Soil Erosion and Agricultural Sustainability." *PNAS* 104, no. 33 (August 2007): 13268–72. pnas.org/doi/10.1073/pnas.0611508104.

Mulvaney, R. L., S. A. Khan, and T. R. Ellsworth. "Synthetic Nitrogen Fertilizers Deplete Soil Nitrogen: A Global Dilemma for Sustainable Cereal Production." *Journal of Environmental Quality* 38, no. 6 (November 2009): 2295–314. pubmed.ncbi.nlm.nih.gov/19875786/.

Pimentel, David. "Soil Erosion: A Food and Environmental Threat." *Environment, Development and Sustainability* 8 (2006): 119–37. link.springer.com/article/10.1007/s10668-005-1262-8.

Pimentel, David, C. Harvey, P. Resosudarmo, et al. "Environmental and Economic Costs of Soil Erosion and Conservation Benefits." *Science* 267, no. 5201 (February 1995): 1117–23. science.org/doi/10.1126/science.267.5201.1117.

Rice, Charles W., Paul M. White, Karina P. Fabrizzi, and Gail W. T. Wilson. "Managing the Microbial Community for Soil Carbon Management." University of Sydney, SuperSoil Conference proceedings, 2004. researchgate.net/publication/252043745_Managing_the_microbial_community_for_soil_carbon_management.

Thaler, Evan, Isaac Larson, and Qian Yu. "The Extent of Soil Loss Across the U.S. Corn Belt." *PNAS* 118, no. 8 (February 2021). pnas.org/doi/abs/10.1073/pnas.1922375118.

Thaler, Evan, Jeffrey Kwang, Brendon Quirk, Caroline Quarrier, and Isaac J. Larsen. "Rates of Historical Anthropogenic Soil Erosion in the Midwestern United States." *Earth's Future* 10, no. 3 (October

2021). agupubs.onlinelibrary.wiley.com/doi/full/10.1029/2021 EF002396.

Wright, Christopher K., and Michael C. Wimberly. "Recent Land Use Change in the Western Corn Belt Threatens Grasslands and Wetlands." *PNAS* 110, no. 10 (February 2013): 4134–39. pnas.org/doi/full/10.1073/pnas.1215404110.

Zhang, Xuesong, Tyler J. Lark, Christopher M. Clark, Yongping Yuan, and Stephen D. LeDuc. "Grassland-to-Cropland Conversion Increased Soil, Nutrient, and Carbon Losses in the U.S. Midwest Between 2008 and 2016." *Environmental Research Letters* 16, no. 5 (2021). iopscience.iop.org/article/10.1088/1748-9326/abecbe/meta.

## CHAPTER SEVEN: *Bugs*

Goulson, Dave. *Silent Earth: Averting the Insect Apocalypse*. New York: Harper, 2021.

McWilliams, James E. *American Pests: The Losing War on Insects from Colonial Times to DDT*. New York: Columbia University Press, 2008.

Milman, Oliver. *The Insect Crisis: The Fall of the Tiny Empires That Run the World*. New York: W. W. Norton & Company, 2022.

### JOURNAL ARTICLES AND OTHER SOURCES

Douglas, Margaret R., Douglas B. Sponsler, Eric V. Lonsdorf, and Christina M. Grozinger. "County-level Analysis Reveals a Rapidly Shifting Landscape of Insecticide Hazard to Honey Bees *(Apis mellifera)* on US Farmland." *Scientific Reports* 10 (2020): 797. doi.org/10.1038/s41598-019-57225-w.

Hitaj, Claudia, David J. Smith, Aimee Code, Seth Wechsler, Paul D. Esker, and Margaret R. Douglas. "Sowing Uncertainty: What We Do and Don't Know About the Planting of Pesticide-Treated Seed." *BioScience* 70, no. 5 (May 2020): 390–403. doi.org/10.1093/biosci/biaa019.

Hladik, Michelle L., Anson R. Main, and Dave Goulson. "Environmental Risks and Challenges Associated with Neonicotinoid Insecticides." *Environmental Science & Technology* 52, no. 6 (February 2018): 3329–35. doi.org/10.1021/acs.est.7b06388.

Janousek, William M., Margaret R. Douglas, Syd Cannings, et al. "Recent and Future Declines of a Historically Widespread Pollinator Linked to Climate, Land Cover, and Pesticides." *PNAS* 120, no. 5 (January 2023). doi.org/10.1073/pnas.2211223120.

Krupke, C. H., J. D. Holland, E. Y. Long, and B. D. Eitzer. "Planting of Neonicotinoid-treated Maize Poses Risks for Honey Bees and Other Non-target Organisms over a Wide Area Without Consistent Crop Yield Benefit." *Journal of Applied Ecology* 54, no. 5 (May 2017): 1449–58. doi.org/10.1111/1365-2664.12924.

LaCanne, Claire E., and Jonathan G. Lundgren. "Regenerative Agriculture: Merging Farming and Natural Resource Conservation Profitably." *PeerJ* (February 2018). doi.org/10.7717/peerj.4428.

Leach, Ashley, and Ian Kaplan. "Prioritizing Pollinators over Pests: Wild Bees Are More Important Than Beetle Damage for Watermelon Yield." *Proceedings of the Royal Society B* (November 2022). doi.org/10.1098/rspb.2022.1279.

Losey, John E., and Mace Vaughan. "The Economic Value of Ecological Services Provided by Insects." *BioScience* 56, no. 4 (April 2006): 311–23. doi.org/10.1641/0006-3568(2006)56[311:TEVOES]2.0.CO;2.

Meinke, Lance J., Dariane Souza, and Blair D. Siegfried. "The Use of Insecticides to Manage the Western Corn Rootworm, *Diabrotica virgifera virgifera*, LeConte: History, Field-Evolved Resistance, and Associated Mechanisms." *Insects* 12, no. 2 (2021): 112. doi.org/10.3390/insects12020112.

Pecenka, Jacob R., Laura L. Ingwell, Rick E. Foster, Christian H. Krupke, and Ian Kaplan. "IPM Reduces Insecticide Applications by 95% While Maintaining or Enhancing Crop Yields Through Wild Pollinator Conservation." *PNAS* 118, no. 44 (October 2021). doi.org/10.1073/pnas.2108429118.

Raven, Peter H., and David L. Wagner. "Agricultural Intensification and Climate Change Are Rapidly Decreasing Insect Biodiversity." *PNAS* 118, no. 2 (January 2021). doi.org/10.1073/pnas.2002548117.

Shaffer, Leah. "RNA-based Pesticides Aim to Get Around Resistance Problems." *PNAS* 117, no. 52 (December 2020): 32823–26. pnas.org/cgi/doi/10.1073/pnas.2024033117.

Tooker, John F., Margaret R. Douglas, and Christian H. Krupke. "Neonicotinoid Seed Treatments: Limitations and Compatibility with Integrated Pest Management." *Agricultural & Environmental Letters* 2, no. 1 (October 2017): 1–5. doi.org/10.2134/ael2017.08.0026.

Unglesbee, Emily. "New Rootworm Trait Debuts: A Look into Bayer's New Rootworm Mode of Action Coming in 2022." *Progressive Farmer.* September 30, 2021. dtnpf.com/agriculture/web/ag/crops/article/2021/09/30/look-bayers-new-rootworm-mode-action.

———. "Production Blog: What the Endangered Species Act Means for Ag Pesticide Use." *Progressive Farmer.* January 21, 2022. dtnpf.com/agriculture/web/ag/news/article/2022/01/21/endangered-species-act-means-ag-use-2.

Wagner, David L. "Insect Declines in the Anthropocene." *Annual Review of Entomology* 65 (January 2020): 457–80. doi.org/10.1146/annurev-ento-011019-025151.

Wilson, Edward O. "The Little Things That Run the World (The Importance and Conservation of Invertebrates)." *Conservation Biology* 1, no. 4 (December 1987): 344–46. jstor.org/stable/2386020.

CHAPTER EIGHT: *Water*

Hatfield, McMullen, and Jones. "Nitrate-nitrogen Patterns in the Raccoon River Basin."

CHAPTER NINE: *Plow II*

JOURNAL ARTICLES AND OTHER SOURCES

Claassen, Roger, Fernando Carriazo, Joseph C. Cooper, Daniel Hellerstein, and Kohei Ueda. "Grassland to Cropland Conversion in the Northern Great Plains: The Role of Crop Insurance, Commodity, and Disaster Programs." United States Department of Agriculture Economic Research Report, no. 120 (June 2011). ers.usda.gov

/webdocs/publications/44876/7477_err120.pdf?v=0#:~:text=En
vironmentalists%2C%20wildlife%20groups%2C%20and%20some%20
livestock%20interests,Region%20of%20the%20Northern%20Plains
%20(USGAO%2C%20Morgan).

Congressional Research Service. *The Renewable Fuel Standard (RFS): An Overview.* Updated July 31, 2018. crsreports.congress.gov/product/pdf/R/R43325/37#:~:text=The%20RFS%E2%80%94established%20by%20the,36.0%20billion%20gallons%20in%202022.

Fargione, Joseph, Jason Hill, David Tilman, Stephen Polasky, and Peter Hawthorne. "Land Clearing and the Biofuel Carbon Debt." *Science* 319 (February 2008): 1235–38. pdf.usaid.gov/pdf_docs/pnadp308.pdf.

Lark, Tyler J., Nathan P. Hendricks, Aaron Smith, et al. "Environmental Outcomes of the U.S. Renewable Fuel Standard." *PNAS* 119, no. 9 (Feburary 2022). pnas.org/doi/full/10.1073/pnas.2101084119.

O'Brien, Peter L., Jerry L. Hatfield, Christian Dold, Erica J. Kistner-Thomas, and Kenneth M. Wacha. "Cropping Pattern Changes Diminish Agroecosystem Services in North and South Dakota, U.S.A." *Agronomy Journal* 112, no. 1 (December 2019): 1–24. acsess.onlinelibrary.wiley.com/doi/10.1002/agj2.20001.

Scully, Melissa J., Gregory A. Norris, Tania M. Alarcon Falconi, and David L. Macintosh. "Carbon Intensity of Corn Ethanol in the United States: State of the Science." *Environmental Research Letters* 16, no. 4 (March 2021). iopscience.iop.org/article/10.1088/1748-9326/abde08.

Searchinger, Timothy, Ralph Heimlich, R. A. Houghton, et al. "Use of U.S. Croplands for Biofuels Increases Greenhouse Gases Through Emissions from Land-Use Change." *Science* 319, no. 5867 (February 2008) 1238–40. pubmed.ncbi.nlm.nih.gov/18258860.

Searchinger, Timothy, Tim Beringer, and Asa Strong. "Does the World Have Low-carbon Bioenergy Potential from the Dedicated Use of Land?" *Energy Policy* 110 (November 2017): 434–46. sciencedirect.com/science/article/abs/pii/S0301421517305104.

Spawn-Lee, Seth A., Tyler J. Lark, Holly K. Gibbs, et al. "Comment on 'Carbon Intensity of Corn Ethanol in the United States: State of the Science.'" *Environmental Research Letters* 16, no. 11 (November 2021). iopscience.iop.org/article/10.1088/1748-9326/ac2e35.

United States Environmental Protection Agency. "Biofuels and the Environment: Second Triennial Report to Congress." June 29, 2018. cfpub.epa.gov/si/si_public_record_Report.cfm?Lab=IO&dirEntryId=341491.

Wright, Christopher K., Ben Larson, Tyler J. Lark, and Holly K. Gibbs. "Recent Grassland Losses Are Concentrated Around U.S. Ethanol Refineries." *Environmental Research Letters* 12, no. 4 (March 2017). iopscience.iop.org/article/10.1088/1748-9326/aa6446.

## *Part Three*
### CHAPTER TEN: *Prairie II*
#### JOURNAL ARTICLES AND OTHER SOURCES

Cowdery, Timothy K., Catherine A. Christenson, and Jeffrey R. Ziegeweid. "The Hydrologic Benefits of Wetland and Prairie Restoration in Western Minnesota—Lessons Learned at the Glacial Ridge National Wildlife Refuge, 2002–15." U.S. Geological Survey, *Scientific Investigations Report 2019–5041*. pubs.usgs.gov/publication/sir20195041.

Gerla, Philip J., Meredith W. Cornett, Jason D. Ekstein, and Marissa A. Ahlering. "Talking Big: Lessons Learned from a 9000 Hectare Restoration in the Northern Tallgrass Prairie." *Sustainability* 4, no. 11 (November 2012): 3066–87. doi.org/10.3390/su4113066.

Minnesota Department of Natural Resources. "Minnesota Prairie Conservation Plan." dnr.state.mn.us/prairieplan/index.html.

Minnesota Department of Natural Resources. *Minnesota's Red River Valley and Tallgrass Aspen Parkland: A Guide to Native Habitats*. Minneapolis: University of Minnesota Press.

### CHAPTER ELEVEN: *Farmers*

Cochrane, Willard W., and C. Ford Runge. *Reforming Farm Policy: Toward a National Agenda*. Ames: Iowa State University Press, 1992.

Coppess, Jonathan. *The Fault Lines of Farm Policy*. Lincoln: University of Nebraska Press, 2018.

Frerick, Austin. *Barons: Money, Power, and the Corruption of America's Food Industry*. Washington, D.C.: Island Press, 2024.

**JOURNAL ARTICLES AND OTHER SOURCES**

Schulte, Lisa A., Jarad Niemi, Matthew J. Helmers, et al. "Prairie Strips Improve Biodiversity and the Delivery of Multiple Ecosystem Services from Corn–Soybean Croplands." *PNAS* 114, no. 42 (October 2017): 11247–52. pnas.org/doi/full/10.1073/pnas.1620229114.

CHAPTER TWELVE: *Ranchers*

Brown, Lauren. *Grasses: An Identification Guide*. New York: Houghton Mifflin, 1979.

Hurt, R. Douglas. *The Big Empty: The Great Plains in the Twentieth Century*. Tucson: The University of Arizona Press, 2011.

Licht, Daniel S. *Ecology and Economics of the Great Plains*. Lincoln: University of Nebraska Press, 1997.

Manning, Richard. *Grassland: The History, Biology, Politics, and Promise of the American Prairie*. New York: Penguin Books, 1995.

———. *Rewilding the West: Restoration in a Prairie Landscape*. Berkeley: University of California Press, 2009.

Voigt, William. *Public Grazing Lands: Use and Misuse by Industry and Government*. New Brunswick, NJ: Rutgers University Press, 1976.

CHAPTER THIRTEEN: *Tatanka*

Aune, Keith, and Glenn Plumb. *Theodore Roosevelt and Bison Restoration on the Great Plains*. Charleston, SC: The History Press, 2019.

Bailey, James A. *American Plains Bison: Rewilding an Icon*. Helena, MT: Farcountry Press, 2013.

Dary, David A. *The Buffalo Book: The Full Saga of the American Animal*. Athens, OH: Swallow Press, 1989.

Hämäläinen. *Lakota America*.

Isenberg, Andrew C. *The Destruction of the Bison*. New York: Cambridge University Press, 2000.

Lott, Dale F. *American Bison: A Natural History*. Berkeley: University of California Press, 2002.

O'Brien, Dan. *Buffalo for the Broken Heart: Restoring Life to a Black Hills Ranch*. New York: Random House, 2001.

——. *Equinox: Life, Love, and Birds of Prey*. Lincoln: University of Nebraska Press, 1997.

——. *Great Plains Bison*. Lincoln: University of Nebraska Press, 2017.

——. *Wild Idea: Buffalo and Family in a Difficult Land*. Lincoln: University of Nebraska Press, 2014.

Olson, Wes, and Johane Janelle. *The Ecological Buffalo: On the Trail of a Keystone Species*. Regina, Saskatchewan: University of Regina Press, 2022.

Wilkinson, Todd. *Last Stand: Ted Turner's Quest to Save a Troubled Planet*. Guilford, CT: Lyons Press, 2013.

## JOURNAL ARTICLES AND OTHER SOURCES

Gates, C. Cormack, Curtis H. Freese, Peter J. P. Gogan, and Mandy Kotzman. "American Bison: Status Survey and Conservation Guidelines 2010." *IUCN* (2010). iucn.org/resources/publication/american-bison-status-survey-and-conservation-guidelines-2010.

Robbins, Jim. "Mass Yellowstone Hunt Kills 1,150 Bison." *The New York Times*. April 4, 2023. nytimes.com/2023/04/04/science/bison-hunt-yellowstone-native-americans.html.

Sanderson, Eric W., Kent H. Redford, Bill Weber, et al. "The Ecological Future of the North American Bison: Conceiving Long-term, Large-scale Conservation of Wildlife." *Conservation Biology* 22, no. 2 (April 2008): 252–66. doi.org/10.1111/j.1523-1739.2008.00899.x.

Shamon, Hila, Olivia G. Cosby, Chamois L. Andersen, et al. "The Potential of Bison Restoration as an Ecological Approach to Future Tribal Food Sovereignty on the Northern Great Plains." *Frontiers in Ecology and Evolution* 10 (January 2022). doi.org/10.3389/fevo.2022.826282.

# INDEX

**A**
actinomycetes (bacteria), 125
Affeldt, Warren, 229–30
A-Frame Farm (Minnesota), 262
Agassiz National Wildlife Refuge (Marshall County, Minnesota), 227
Ag PhD Field Day, annual farm festival (Baltic, South Dakota), 214–16
agrarian(s): caretaking, 325; ideals, 44; organizing, 257; term, usage, 325
agribusiness, 215; and Iowa economic output/revenues, 180, 192; in Iowa politics, 180; and waterworks (Iowa), 170–89
Agribusiness Association of Iowa, 179, 183
agricultural and farm chemicals, xvi, 59, 113, 132–33, 146, 168, 174–76, 184–86, 216, 229, 235, 240, 251–52, 267, 322. *See also* agrochemical industry; fertilizers; herbicides; insecticides; pesticides

Agricultural Research Service (USDA), 255
agriculture: and biology, 85; and business, 53; changing, 91, 247; and civic virtue, 45; environmental consequences and costs of, xiv–xx, 23–25, 28, 85, 145, 329; environmental regulation of, 181; evolution of, 192; expansion of, 23–24; frontier and settler, 19–20, 39, 49, 329; and grasslands, xvii, 23–24, 70, 83–85, 96, 293; history of, 61, 97–98, 189; intensive, 101, 145, 252, 363; and Iowa economy, 180; and land, 112; and mechanization, 25, 27, 34, 37–38, 44, 50, 69, 216; and nature, 131; paradoxes in, 81–82, 141–42, 192; prairie, 214–15, 248, 267; productivity gains, 322–23; regenerative, 137, 141, 201, 251–52, 257, 261–66, 322–23, 330–31; and self-reliance, 45; specialized, 79; supported, 325; sustainable,

agriculture: and biology (*cont'd*): 135–37, 168; transformation of, 27, 36, 91, 214–15; and urban industry, 49. *See also* agribusiness; crops and cropland; farmland; farms and farming; industrial agriculture; plowing and plows
agrochemical industry, xviii, 93–94, 141, 147, 151–52, 154, 165–66, 215. *See also* agricultural chemicals; fertilizers
agronomy and agronomists, 15, 80, 82, 85, 124–25, 128, 131, 134, 215, 253–57, 266–67
air pollution, 80–81, 83
Aldrich, Charles, 55, 336
algae: and bacteria, 81, 96–97, 143; and cyanobacteria, 81, 172; and fertilizers, 145; and water pollution, 80–81, 96–97, 104–5, 143, 145, 172, 176
American Bison Society, 297
American Farm Bureau Federation, 181–82
*American Farmer, The* (journal), 65
American Grasslands Conservation Act, 266
American Indian Movement, 306
American Indians. *See* indigenous people
American prairie. *See* prairie
American Prairie, 277–83, 285, 289–90, 314–17
American Society for the Prevention of Cruelty to Animals (ASPCA), 296
American Stewards of Liberty (ASL), 242–44
America the Beautiful Challenge, 243
ammonia, 60–85; creation and production of, 72, 74–78; for crops, 70; experiments with, 75; for explosives and weapons during war, 76, 78; and fertilizers, 70–72, 76–78, 81; formula for, 73–74; gaseous components of (nitrogen and hydrogen), 70, 73–74, 76–77; and nitrates, 64, 71, *71,* 82; and nitrogen fixation, *71;* and nitrous oxide greenhouse gas, 81; and prokaryotes, *71;* synthesis, 75–77, 82–83; wartime use, 76, 78
Andrus, Leonard, 32
animals. *See* grazing animals; mammals; wild animals; wildlife; specific animal(s)
antelope, 9, 315; populations, 13–14; as prey, 12; pronghorn, xiv, 12, 276, 278–79, 286; and ranches, 288. *See also* pronghorns
antibiotics, 127, 148, 163, 301
archaeologists, 41
Archer Daniels Midland, 259
Armour, Philip Danforth, 37–38
Army Corps of Engineers (U.S.), 101–3, 108, 128, 132
arthropods, 125–27
ASL. *See* American Stewards of Liberty (ASL)
ASPCA. *See* American Society for the Prevention of Cruelty to Animals (ASPCA)
Atchafalaya River (Louisiana), 101
*Atlantic, The* magazine, 204
atmospheric nitrogen, 9–10, 60, 70–71, *71,* 73–74, 76, 126
Audubon, John James, 19
Aune, Keith, 308–10, 348
Austin, Damien, 278

B

*Bacillus thuringiensis. See* Bt (bacteria)
bacteria: and algae, 81, 96–97, 143; and denitrification, 71; and fertilizers, 72; and fungi, 10, 137, 250; in soil, 10, 61–62, 70–71, 80–81, 125–27, 137, 199, 250. *See also* actinomycetes (bacteria); Bt (bacteria); cyanobacteria; diazotrophs (bacteria); symbiotic rhizobium
badgers, 10–11, 227

Baker's Field Flour & Bread (Minneapolis grain miller), 262
Bar, Clif, 260–61
Basche, Andrea, 133–34, 342
BASF chemical company, 74–76
Battle of Fallen Timbers, 29
Battle of Tippecanoe, 29
Baudelaire, Charles, 19
Baum, L. Frank, xvi–xvii
Baxter Wildlife Management Area (Baxter Township, Minnesota), 240
Bayer Crop Science, 93–94, 147, 154, 179, 187, 215–16
bears, 315, 316–17; grizzly, xiv; hunted, 18; polar, 144
Beattie, Amy, 178–79, 329
beavers, 41, 224–25, 279; populations of (1830s), 19
bees, wild: and colony collapse disorder, 155; conservation of, 167; damage to, 155, 162–63; decline in, 145; insecticides killing, 141; migration of, 163; as pollinators, 144–46, 155–56, 161–67; and population declines, 145. *See also* bumblebees; honeybees
Belchim agrochemical/biological company, 215
Bell, Alexander Graham, 33
Bennet, Michael, Senator, 266
Bennett, Hugh Hammond, 121–23, 129
Bennett, Mark, U.S. District Judge, 187–88
Benz, Karl, 34
Berry, Wendell, 119, 249, 325
Biden, Joe, 243, 321–22
Biloxi, Mississippi, as Seafood Capital of the World, 106
biochemists, 170
biodiversity. *See* diversity, biological
biofuels: and carbon, 212; and corn, 213, 242, 263; critiques of, 213; and cropland, 242; defenders of, 213; and ethanol, 263, 266, 329; federal mandate, 213, 266; legislation, 212; and octane of gasoline, 211–12; politics and science of, 329; and soybeans, 213, 263; as venerable technology, 211–12. *See also* ethanol; fossil fuels
biological diversity. *See* diversity, biological
biotic community, 87
birds: evolution and survival of, 227; grasslands, xiv, 223, 227, 286–87, 301; groundnesting, xiv, 30; habitats for, xiv, 12, 41, 227, 243, 286; migrating, xiv; population increases, 252–53; song-, 39, 194–95; species, 41, 194, 252–53, 284, 301; study of, 19; variety of, 12, 41, 286; wild game, 293
bison, xiv, 294–320; behavior studies, 316–17; and bovine disease, 307–8, 311; business herds, 311; and cattle, breeding with, 298; and cattle, differences, 316–17; and cattle, replaced by for grazing, 25, 217, 275–76, 285–86, 288; and cattle, similarities, 298; as charismatic, 269; and climate change, 298–99; conservation, xix, 297, 305–9, 314–17, 321–22, 327; current love affair with, 324; demise of, 295–97; drowned, 224; ecological recovery and revival, 317–18, 322; and ecosystem, 14, 269, 298, 300–301, 315; eradicated, 296–97, 307, 325; evolution within ecosystem, 14; expansion and evolution on prairie, history of, 14–15; as extinct, 315; fate, 319; fenced, 280, 314; free-range, 302; and geography, 318; grass eating and grazing, 6, 9, 12, 14, 275–76, 285–86, 295, 298–99; and grasslands, 12, 14, 288, 298–306, 316–19, 321; habitat, 315; herds, xix, 14–15, 278, 292–322, 327, 330; hides,

bison (cont'd):
225, 295–96; history, 6, 9, 12–16, 294–99, 313–14, 321; hunted, 6, 14–16, 21, 224, 294–96, 307–8, 311, 319; and industrialization, demise due to, 295–97; as industry, 298, 301–2, 305–6; innate cooling skills, 276; as keystone species, 300–301, 314–15; and landscapes, 301; as livestock, 297–98, 302, 311, 313, 319; management, 297; management, tribal, 278, 318; meat, 14, 301–5, 307, 323–24, 327; migration, 275–76, 278, 285, 315; monitored, 286; as national mammal, 319; in national parks, 318–19; as no longer extinct, 297–98, 321; as nomadic, 14; physical attributes, 299–300; predators of, 12–14, 298–300, 306–7, 315; as prey, 12, 14; protected, 202, 289–90, 297–98; and protein, 14, 302; on public lands, restoration of, 312–13; ranches, 269–71, 286, 291–320, 327; research projects, 315–16; reserves, 278–80, 297–98, 316; restoration, 269, 277–79, 291, 294–98, 306–9, 312–22, 325–28; riparian areas preservation, 316–17; roaming, 12, 275–76, 298, 301, 305, 315; rotational grazing by, 285–86; roundups, 290, 316; satellite herds, 308–9; shared stewardship of, 318; and shortgrass grazing, 14; slaughtered and culled, 303, 307–9; survey (1886), 295; survival, 297–98; symbiosis of, 298–99; threatened, 202, 312; watering holes, 40; as well-insulated, 299; wild, xix, 297–98, 301–8, 315–22; and wildlife play, 315. *See also* tatanka (pte oyate or buffalo people)
Bison Working Group, 318–19

Black Hawk, Sauk Chief, 16, 29, 188
Black Hills (South Dakota), 21–22, 291, 294–95
Black Swamp (Ohio), 42, 54, 59
Blair, John, 124–25, 329
Blair, Kellie and A.J., 257
Blickensderfer Tile Ditching Machine, 49
Blue Dasher Farm (South Dakota), 140–41, 168
Bogue, Allan, 30, 335
Bonnet Carré Spillway (Mississippi Valley), 102–4, 107–8
Bonogofsky, Alexis, 264
Boorstin, Daniel, 30
boreal forest, 9, 89
Bosch, Carl, 76–77, 338
botany, 267
Bourke-White, Margaret, 115–16
Bradley, Ryan, 108–9
*Braiding Sweetgrass* (Kimmerer), 219
Branstad, Terry, Governor, 178–79, 188
British Association for the Advancement of Science, 72
Broken Heart Ranch (South Dakota), 291–94, 319–20, 349
Bronx Zoo (New York), bison at, 297, 312
Bryan, William Jennings, xvi–xvii, 38
Bt (bacteria): corn, 147–54, 205; -engineered seeds, 149, 154; genetic technologies, 154, 205; protein, 148; toxins, 151–53; varieties, 147–48, 151
Buckeye Trencher, 49
buffalo. *See* bison
"Buffalo Commons, The" (Poppers' essay), 269–70, 325, 330
*Buffalo for the Broken Heart* (O'Brien), 294, 319–20, 349
Buffalo Gap National Grassland (South Dakota), 304–5
Buffalo Nations Grasslands Alliance, 201–2
buffer strips, 99, 184–85, 191–92, 247, 252–53, 258, 322

bugs. *See* insects
bumblebees: commercial, 159–60; as pollinators, 156, 159; western, rarity of, 145; wild, 160. *See also* honeybees
Bunsen Society for Physical Chemistry, 75
Bunyan, John, 42
Burns, Ken, 114
burrowing: arthropods, 125; mammals, xiv; rodents, 278–79
Burton, Samuel, 43
business: and agriculture, 53; and governments, 68; and landowners, 53. *See also* agribusiness
butterflies, xiii–xiv, 144, 227; insecticides killing, 141; Karner blue, 146; monarch, xvi, 10, 146, 241; as pollinators, 156; varieties of, xiv
BWMA. *See* Baxter Wildlife Management Area (Minnesota)
Byfield, Margaret, 243

C

Cady, Mark, Chief Justice, 188
Calderon-Domingues, Pedro, 316
Cannon, Will and Cassie, 258–61
carbon, 60; and biofuels, 212; cycle, 23, 80, 212; and ethanol, 212–13; and fertilizers, loss due to, 289; and flood control, 276; footprints, 259, 289; and grasses, 198–200; loss of, 199–200, 289; in soil, 7, 10, 121, 124–28, 135, 199–201, 212–13, 218, 239, 255–56, 276, 284, 289; terrestrial, 10, 199
carbon dioxide, xv, 8–10, 81, 124, 199–200, 202, 237, 256, 288
Cargill, 90, 92, 94, 259
Carlson, Brad, 82–83, 329
carnivores, xiv–xv, 12
Carson, Rachel, 147
Cather, Willa, xvi–xvii, 1
Catlin, George, 19
Central Iowa Water Works, 190

Charles M. Russell National Wildlife Refuge (Montana), 277, 312–15, 318
cheatgrass, 284. *See also* kochia
Cheerios, xvi
chemicals, agricultural. *See* agricultural and farm chemicals; agrochemical industry
Chesapeake Bay estuary, 106, 146, 176, 191–92
Cheyenne River Ranch (South Dakota), 304
Chincha Islands (Peru), 64–65, 67
chinook winds, 7
Cimarron National Grassland preserve (Kansas), 244
Civil War, 32, 146
Clean Water, Land and Legacy Amendment (Minnesota), 238, 241, 245
Clean Water Act (CWA), 58, 111, 144, 176–77, 182–83, 185–87. *See also* water pollution
climate change, xv–xvi, xvii, xix, 6, 13–14, 81, 83, 221, 224, 233, 239, 260–61, 268, 323, 325; and beef, 287–88; and bison, 298–99; denial, 242–43; and grasslands, 199–202, 212–13; and insects, 145; and ranching, 287–89; and renewable fuels (biofuels), 212–13; and rivers, 98, 102–3, 106, 109–12; and soil, 125, 127, 202, 256; and swamplands, 40, 58. *See also* global warming; Ice Age
Clinton, Bill, 110
Cochrane, Willard W., 39, 335, 347
Cody, Buffalo Bill, xvi–xvii, 274, 295
Coen, Floyd and Dale, 114, 128
colonization and colonialism, xix–xx, 28, 31, 61, 73, 84–85, 90, 184–85, 323, 331; and grasslands, 17; and indigenous people, 295, 305, 314; and insects, 142; and prairie, 17, 23, 225; and soil, 120; and swamplands, 42, 57, 59
Colorado prairie, 10

356 | *Index*

Columbian Exposition (Chicago, 1893), 51–52
Community Supported Agriculture, 324
conservation: agriculture and farming, xviii, 94–95, 123, 137, 176, 180, 185, 209, 216, 231, 246–68, 321–22; of bees, wild, 167; bison, xix, 297, 305–9, 314–17, 321–22, 327; and ecology, xviii–xix; and environmentalists, 238; grassbank movement, 273; grasslands, xviii–xix, 196–97, 209, 213, 234, 243–44, 266, 276, 281, 287, 289, 301, 314–15, 318, 329; land, 230, 242–44; and nonprofit groups, 297; prairie, 191, 226, 239, 244, 270, 277, 286–87; and public health, 176; ranching, 196, 272–73, 281–85, 289; share philosophy of, 287; soil, xix, 122–23, 129–30, 136–38, 234–35, 282, 322; statistical precision of, 248; and tribal activists, 307; tribal lands, 201–2; water, xix, 176, 234–35, 282; wetlands, 184–85, 191–92, 228, 322; wildlife, 301, 308, 315. *See also* environmentalists
Conservation Reserve Program, 130, 214
Conservation Stewardship Program (CSP), 261, 265
Cool, Mark, 290
coral reefs, 41
corn, xv–xvi, 5, 91–92, 94, 99, 174, 176, 180, 186, 197–98, 200, 207–16, 263–67; 90-day hybrid, 207; amount risen, 79–80, 84, 110, 145, 208, 213–14, 323; and biofuels, 213, 242, 263; Bt, 147–49, 151–53, 205; and carbohydrates, 267; catchment, 252; and cattle, 287; as commodity, 84; as conventional crop, 263; and corn on corn practice, 135, 150; DT drought-tolerant hybrid, 207–8; ethanol, 210–14, 222, 263, 266; expansion of, 197–98; flour, 211; genetically engineered, 208; as global major food crop, 267; and grass-fed beef, 288; herbicide-resistant, 215; insecticides for, 147, 154; and insects, 140, 145, 147, 149–52, 161–62; as interchangeable, xviii; as livestock feed, 150, 222, 263; nitrogen-dependent, 79; and nitrogen fertilizer, 79; pathogens in, 254; pesticide-treated, 154, 164–65; planting ease, 120; pollinated, 159; and prairie, 226, 240–42; price of, 180, 213, 226, 266; as profitable, 57, 92, 186, 197, 209, 211, 218, 259–60, 266; seed technology, 207–8; short hybrid, 215–16; and soil, 135; strenuous job of producing, 209; sweeteners, 259, 264; volume of, 323; and wheat, 5, 135, 226
Corn Belt: and diverse, healthy mix of crops, 322; farming, 263, 322; insects and pests, 140, 151–52; plowed up, 200, 213; prairie, 239; rivers, 95, 102; soil, 132, 134, 138; states, 57–58, 81, 98, 151, 238; swamplands, 54; tallgrass prairie, 81; water, 192
corn borers, 140, 148–49, 151, 153, 160–61
Coronado, Francisco Vázquez de, 16–17
COVID-19 pandemic, 313
Cowdery, Tim (Timothy K.), 234, 237–38, 347
coyotes, xiii, xiv, 12, 195, 293, 301, 319–20
Crandall, Lewis, 33
Crazy Horse, Lakota Chief, 22, 320
Crews, Tim, 135–36, 139, 169, 213, 268, 328
Crookes, William, Sir, 72–73, 78
crops and cropland: alternating, 61–62; ammonia for, 70; binded

Index | 357

to, 260; and biofuels, 242; cash, 124–25, 136, 205, 250, 255; catchment, 252; climate for, 62, 110–11; cover, 94–95, 110, 112, 134, 136–38, 185, 216, 250–51, 253–61, 263, 265–67; diseases, 225; diverse, 135, 142, 186, 209, 250, 255, 259–60, 322; and dry western plains, xvii; and efficiency, 82–83; expansion of, 197–98; fall/autumn, 94–95, 138, 209–10, 217, 250; food, 63, 144–45, 155, 163, 165–66, 267–68; grain, 99, 134, 268; and grasses, xv, xviii, 23, 25, 85, 134, 193, 197–202, 207, 209, 227, 266, 293; harvesting, 94–95, 101, 134, 138, 209–10, 217, 249–50; healthy, 30–31, 186, 322; insurance for, 154, 200, 210, 226, 260, 265–66; and nitrogen, 61, 63, 210; no-till, 136–38, 206, 209, 250, 255–61; novel, xviii, 261, 264; nutrients for, 70; and pests, 149, 153, 157, 169; and prairie, 227, 242; profitable, 119, 138, 160, 197, 205, 208–9, 218, 228, 249, 259–60, 266–67; roots of, 47; rotations of, 78, 135, 138, 146–47, 150, 164, 209, 250, 255, 258–59; row, xvi, 11, 25, 79, 84, 95–96, 110–11, 192, 198, 200, 207–8, 210, 214, 226, 240, 274, 293; scientists, 205, 207; seeds for, 149, 208, 255, 261; shallow-rooted, 99–100; small grain, 99; spring, 99, 138, 209, 267; and swamplands as sterile and barren, 45; tillage practices and techniques, 123, 184–85; two-system, 150; variety of, 79, 161, 186, 209, 262, 322; and wetlands, 48, 185, 240, 322; winter, 99, 253–54, 266, 267; world's major food, 267; year-round, 62. *See also* agriculture; farmland; farms and farming; plowing and plows

Crow River (Minnesota), 90–91, 95; South Fork, 90–91, 94–95, 111–13, 341
Cruse, Richard, 85, 329
CSP. *See* Conservation Stewardship Program (CSP)
Cullen, Art, 174, 182–83
Cullen, John, 182
Cullen, Tom, 182–83
Cunfer, Geoff, xvii–xviii, 14, 205
Custer, George Armstrong, Lieutenant Colonel, 21–22, 297
CWA. *See* Clean Water Act (CWA)
cyanobacteria, 81, 172
cyanotoxins, 172

**D**

Dahlstrom, Neil and Jeremy, 33, 335
Dakotas: farmland, 38–39; grasslands, 293; prairie, xiv, xvii, 11. *See also* North Dakota; South Dakota
Darwin, Charles, 63–64
Davis, Jack E., 208, 210, 339
Daybreak Ranch (South Dakota), 194–95, 217–18
DDT (pesticide), 146–47, 150, 292–93
Dead Zone. *See* Gulf of Mexico, Dead Zone
Debs, Eugene V., xvi–xvii
Decatur, Illinois, as Soybean Capital of the World, 84–85
deer: forage for, 195; grass for, 217; grazing, 12, 194; habitat, 195; natural migration patterns of, 278; populations of (1830s), 19; as prey, 12; watering holes for, 40; white-tailed, xiv, 195; wilderness, 39
Deere, John, xvi, 26–28, 31–36, 43, 70, 83, 131, 197, 216, 335
Defenders of Wildlife, 308–9, 328
deforestation, 288–89
DeLonge, Marcia, 138
denitrification, 71
Des Moines, Iowa, as urban heart, 176
Des Moines City Council, 180

Des Moines Lobe ecoregion (Iowa), 54, 57, 90, 173, 175–76, 185
*Des Moines Register* daily newspaper, 177, 179, 181
Des Moines River tributary (Iowa), 172
Des Moines Water Works, 170, 174, 176–77, 190, 249
diazotrophs (bacteria), 70
Dickens, Charles, 43
dirt. *See* dust; soil
discipline, and self-reliance, 310
ditch-and-tile machine (Jacobson), *50*
diversity, biological, xiv, xvi, 11–12, 81, 127, 141, 196, 218, 243, 256, 286, 300–301, 322
Dow Chemical, 215
drainage: agricultural, 46–47, 50–53, 56–58, 112, 145, 175, 181, 187, 189, 225–26, 242; artificial, 57–58; authorities, 174; districts, 53–54, 175, 181, 185, 187–89; ditches, 46, 49–51, *50, 54,* 57–58, 93, 173–76, 187, 189, 225, 233–34, 329–30; and engineering, 51–52; projects, 46, 53, 57, 123, 185; revolution, 49–50; systems, 47, 81, 90, 95, 99–100, 150, 187–89, 237; and technology, 51, 54; for water management, 23, 50–51. *See also* tile drainage
dredging, 45–46, 49, 128–29, 175, 245
drinking water, xvi–xvii, 15, 52, 57–58, 80–81, 128–29, 133, 170–93, 198, 229–31, 237–38, 246, 322, 323
drones, 215, 286
drought, 58, 190, 195, 200, 203, 207–10, 216, 255, 265, 292–95; and insects, 158; and soil, 117, 119, 123, 129
duck hunters, 240
Duffus, James, 276
DuPont Pioneer, 207
dust: bowl, term, usage, 117; bowl farmers, *118,* 121; storms, 114–17, 122, 129, 198; and winds, 115
Dust Bowl, 99, 118–23, 129–30, 134–35, 208, 244; farm families, *118*

E
eagles, xiv
Earp, Wyatt, xvi–xvii, 274
Ecdysis Foundation, 140–41
Ecoffey, Trudy, 313
ecology: and conservation, xviii–xix; and cows, 288; disappearing, xviii–xix; and grass-fed beef, 288; grasslands, xviii–xix, 15–16, 59, 197, 199, 202, 230, 277, 321, 329; of native prairie plants, 256; prairie, xviii–xix, 11–13, 23–24, 239, 328; soil, 128, 256; of the West, 331; wetlands, 53. *See also* ecosystems
ecosystems: and bison, 14, 269, 298, 300–301, 315; damaged, 269; destruction, 25, 77, 192; disappearing, xviii–xix; and farming, 218; grasslands, xviii–xix, 7, 9, 25, 85, 91, 191, 196–97, 218, 227–28, 230, 235, 276–77, 284–86, 298–99, 306, 314, 328, 331; insects, 156; prairie, xiv–xv, xviii–xix, 7, 25, 123–24, 192, 321; and ranching, 218; remade, 25; services, 127, 218; soil, 127–28; threatened, 85, 319; wetlands, 41. *See also* ecology
Edison, Thomas, 33–34, 51–52
Ekstein, Jason, 231–36, 245
elk: grasses for, 9; natural migration patterns, 278; populations, 13, 19; pronghorns, 293; watering holes for, 40
Elk Island National Park (Alberta), 278
Energy Independence and Security Act of 2007, 212
Energy Policy Act of 2005, 212

*Enriching the Earth* (Smil), 83, 338
entomology and entomologists, 140–42, 144, 150–51, 158, 161, 166, 330. *See also* insects
environmentalists, 94, 137, 156, 184–86, 191, 202, 206–7, 212, 216, 230, 238, 241, 251, 253, 265, 269, 277, 279, 285–89. *See also* conservation
environmental law, xviii, 57, 176, 189, 213
Environmental Protection Agency (EPA), xvi, 58, 110, 133, 150–51, 156–57, 186, 191–92, 198, 207, 212, 237, 243
*Environmental Research* journal, 172
environmental scientists, xv–xvi, 111, 202
Environmental Working Group, 328
EPA. *See* Environmental Protection Agency (EPA)
Erie Canal, 29–30, 46
Ernst, Ferdinand, 28
estuaries, 96, 102–3, 108. *See also* tributaries
ethanol: and biofuels, 263, 266, 329; and carbon, 212–13; as climate friendly, 212; corn, 210–14, 222, 263, 266; and farming, 212, 214; and greenhouse gases, 213; as not renewable, 213; as potent grain alcohol, 211; production, 211–12; soybean, 263; support for, 267. *See also* biofuels
evolution: and diversity, 11–12; and farming, 169, 185–86; and insects, 142–43, 150, 169, 239; of landscapes, 11, 330; prairie, 9–12, 239; and survival of the fittest, 169; theory of, 63–64
extinction, mass, 143–44

**F**
Faber, Emmanuel, 264
Faber, Scott, 328
Fairchild, Ephraim, 28
Fairlie, Simon, 288–89

Fargione, Joe (Joseph E.), 212–13, 328, 342, 346
Farm Aid benefit concerts, 94
Farm Belt, 130–31, 267
Farm Bill, 58, 109, 210, 253, 265
farm bureaus, 181. *See also* specific Bureau(s)
Farmers Institute, 257
*Farmer's Magazine, The,* 66
farmland, 28–29, 38–45, 72, 200, 202–4, 246, 252, 261, 265, 343; degraded and depleted, 134, 231–32, 239; expansion of, 38–39, 69–70; expensive in eastern states, 28; and grasslands, 244; limited supply of, 61; and prairie, 223, 226, 241; restored back to prairie, 231–32, 236–37, 244; rich, 20; and rivers, 90–91, 97–99; and soil, 130–34; soil management on, 202; and swamplands, 40–45, 52–54; and water, 175–76, 183–84; and wetlands, 41, 61, 240. *See also* agriculture; crops and cropland; farms and farming
farms and farming, 246–68; abandoned, 244; autonomous, 215; bailouts, 269; bonanza, 225; changes in, 261, 263, 266; character of, 36–37; colonial, 31, 295; cost-share grants, 259–60; crises, 94, 112, 195, 211, 248; diligent, 138–39; diverse, 149; ecologically diverse, 169; economic and cultural roots, 228; economy, 322; and ecosystem services, 218; and efficiency, 82–83; efficient and destructive era in, 25; and ethanol, 212, 214; and evolution, 169, 185–86; expansion of, 38–39, 146; as family operations, 93; fertile, 119, 121; foreclosures, 117–18; frontier, pioneer, and settler, xvi, 5, 25, 29–30, 34–35, 39, 323; grassroots coalitions of, xix; hog,

360 | Index

farms and farming (cont'd): 176, 178, 262; and hydrology changes, 99; and industrial agriculture, 93–94; innovations in, 34, 216; and manufacturing, 37; and mechanization, 25, 27, 34, 37–39, 44, 50, 69, 216; new age of, 34; organic, 251, 261–65; output and growth, 36–37; profitable, xvii–xviii, 82–83; regenerative, 137, 141, 201, 251–52, 257, 261–66, 322–23, 330–31; scale of, 36–37; and science, 330; and self-sufficiency, 35, 37; and stewardship, 175–76, 188, 261, 265, 268; subsidies for, xvii, 193, 197, 210, 214, 218, 263, 265; and suicide rates, 94; survival of, xviii; transformed, 78; westward expansion of, 38–39; and yeoman ideal, 19–20, 37–38. See also agriculture; crops and cropland; farmland; plowing and plows; ranches and ranching
Farm Security Administration, 118
Faulstich, Carol, 194–98, 208, 217–18, 293, 330
Faulstich, Jim, 194–98, *196*, 204–5, 208, 217–18, 293, 330
Fernholz, Carmen, 262
ferrets, 12, 202, 279, 301
fertilizers: agricultural, *x–xi*; and algae, 145; and ammonia, 70–72, 76–78, 81; and bacteria, 72; and carbon, 289; chemical, xvi, 99, 150, 187, 195, 248–49; commercial, 134–35; consumption, 95–96; and crops, 95–96; efficient use of, 256; fewer and less, 94–95, 168, 216, 237, 255–56, 260, 322; natural, 68, 70; organic, xviii; phosphorus, 90–91, 110; as pollutants, 94–95, 98, 103, 110–12, 132–33, 145, 172, 184, 187, 192, 289; synthetic, 39, 70, 77–79, 81–83, 135, 146, 186, 200. See also agricultural chemicals; agrochemical industry; guano (huano); nitrogen fertilizer
field days, 214–16, 247, 252, 257–58, 260
Fillmore, Millard, 66–67
fireflies, 146
First Nations Development Institute, 201–2
Fish and Wildlife Service (U.S.), 228, 230
fishing industry, 109, 113, 156
Flint Hills grasslands (Kansas), 5, 227
Flood Control Act of 1928, 101
flood control and management, 45, 101–3, 107–8, 229, 238, 276
floods, xvii, xix, 41, 43–45, 58, 101, 103, 127, 133–34, 177–78, 198–99, 225, 233–34, 237–38, 256
Flores, Dan, 13, 331, 334
flowers: growth, 253; native, 195, 201; prairie, 6, 42–43, 125, 235–36, 278; seeded, 245; species, xiv; tall, 247. See also wildflowers
food: healthy, 323; industry, xviii, 251, 262–66; organic, 262–64, 324; production, 64, 69, 83, 109, 134, 142, 216, 265; supply, xvi, 23, 39, 63–64, 69–73, 78, 84–85, 159, 162–63, 186, 200, 211, 216, 222, 236, 268, 304; system, xviii, 84–85, 142, 157, 167, 262, 302, 304–5, 319, 323–24, 327–28
food chains, xiv–xv, 59, 71, 97, 143, 156
Ford, Henry, 38, 211
forests, 8, 61, 159, 199, 224, 237, 288–89. See also boreal forest; hardwood forests; rainforests
Forever Green Initiative, 266–67
Fort, Greenbury L., 296–97
Fort Belknap Indian Reservation (Montana), bison at, 278, 306, 330
Fort Laramie Treaty, 21

Fort Peck Indian Reservation
    (Montana), *312;* bison at,
    309–12, *312,* 321, 330; history
    of, 24–25
fossil fuels, 80, 83, 135, 200, 212,
    255, 323–24. *See also* biofuels
Fox, Alison, 279, 289
foxes, xiv, 12, 55, 195, 202, 279, 286
Fraley, Rob, 205–6
Freese, Curt, 277, 328
French, Bill and Corky, 282
French, Conni, 271–73, 282–85,
    289–90, 330
French, Craig, 271–73, *273,* 277,
    282–85, 289–90, 330
Fuller, Gail, 136–37, 250
fungi: arbuscular mycorrhizal,
    126–27; and bacteria, 10, 137,
    250; prevention of, 153; in soil,
    8, 10, 125–27, 199, 239;
    underground, 8, 199, 250
fur trade, 224–25

# G

Gailans, Stefan, 256
Garrett, Rachael, 288
Geiger, Robert, 117
General Mills, 90, 263
geography and geographers, xvii, 22,
    41, 125, 197, 200, 269, 318
Geological Survey (U.S.), 102–3,
    186–87, 203–4, 234
geology and geologists, 5–6, 89, 120,
    131, 135, 229; glacial, 330
Gerrity, Sean, 277, 279, 315
Gillette, Graham, 178
Glacial Lakes Energy, 210–11
Glacial Ridge National Wildlife
    Refuge (Polk County,
    Minnesota), 223, 227–33,
    236–40, 245, 247, 277, 279, 347
global warming, 256. *See also*
    climate change
Grand Kankakee Marsh (Illinois and
    Indiana), 59
Grant, Ulysses S., 297
*Grapes of Wrath, The* (Steinbeck),
    xvi–xvii

Grassel, Shaun, 201–2
grasses: adaptable, 8; and carbon,
    198–99, 198–200; and crops, xv,
    xviii, 23, 25, 85, 134, 193,
    197–202, 207, 209, 227, 266,
    293; and dry climate, 8–9;
    expansive, 9; flourished, 9; and
    food grains, 8; global, 8; healthy,
    15, 196, 272, 284–86, 302;
    hunting in, 217; infinite, xiv;
    native, 31, 146, 195, 201, 227,
    274, 278, 284; nutrients for,
    9–11; plowed, 212–14; prairie,
    9, 26, 125, 184, 193, 200, 233,
    267; seeded, 245; species of, xiv,
    8–10, 14; surviving, 5–6, 195.
    *See also* grasslands; shortgrass
    prairie; tallgrass prairie
grasslands, x–xi; and agriculture,
    xvii, 23–24, 70, 83–85, 96, 293;
    as American desert, xiii; as
    American Serengeti, 13; and
    bison, 12, 14, 288, 298–306,
    316–19, 321; conservation,
    xviii–xix, 196–97, 209, 213, 234,
    243–44, 266, 276, 281, 287,
    289, 301, 314–15, 318, 329;
    conversions, 197, 209, 213, 329;
    destruction, 25, 91, 325;
    diversity, 11–12; and earliest
    human occupants, 13; eulogy to,
    50; evolution, 11–12; expansive,
    9; exploited, 325; and farmland,
    244; healthy, 15, 196, 272,
    284–86, 302; history, xix–xx,
    5–24, 271; hunting in, 217; as
    infinite, xiv; native, 14, 227–28,
    238, 293; plowed up, 25, 39,
    119, 141, 199–202, 216,
    270–71; prairie, 7, 196, 304,
    318; and ranching, 194–205,
    218, 244, 269–86; regulation,
    274; remaining, xix, 85, 202,
    227, 273, 289; reserves, 277;
    restoration, 124, 191, 194–205,
    223, 237, 244, 265–66, 273,
    277, 281, 298, 304, 321–22,
    328; rich, 14; share philosophy
    of, 287; size, 9; species, 293;

grasslands (cont'd):
structural complexity, 286; temperate, xv, 7–9; threatened and vanishing, xv, xviii–xix, 85, 197, 293; transformed, xix–xx, 39, 217; on tribal lands, 201; and ungulates, 288–89; and water pollution, due to loss of, 192; and wetlands, 141–42, 156, 199–200, 227; and wildlife, 201, 276–77, 281, 289, 301. *See also* grasses; landscapes; prairie

grazing animals, xiv–xv, 6, 9, 12–15, 209, 284–89. *See also* wildlife; specific animal(s)

grazing lands, 226, 239, 275–76, 281, 286, 302, 314

Great Depression, 117, 248

Great Lakes, 17, 41, 221. *See also* specific lake(s)

Great Plains, 8, 13; agriculture, 203–4, 207; bison repopulated and restored on, 269, 291, 294, 296–98, 306, 309, 325, 327; environmental crisis, 328; geography, 22; grasslands, 273, 276; homesteaders on, 203–4; and indigenous people, as sacred place (Pahá Sápa), 294–95; insects, 145; isolation and emotional hardship of life on, 204; landscape, 203–4; prairie, 224; protected, 276; ranches, 269–70, 273–76, 287, 294; soil, 115–18, 122–23, 128, 135; wild bees decline on, 145. *See also* prairie

*Great Plains, The* (Webb), 203

Great Sioux Reservation, 21, 201

Greeley, Horace, 18

Green Giant, 92

greenhouse gases, xv, xviii, 81, 127–28, 200–202, 212–13, 255, 288

green revolution, 39

Gross, Doug, 179–80, 182–83

guano (huano), 65–70, 73; Chinese laborers in Chincha Islands, Peru, 67

Guano Islands Act, 67

Gulf of Mexico, 89–113; bison in, 14; coast, 96, 101, 106, 109; Dead Zone, x–xi, 85, 96–98, 105, 109–11, 113, 132–33, 192–93, 246, 329–30; and drainage ditches, 329–30; future of, 111, 113; herbicides in, 132–33; hypoxia in, 96–98, 110–11, 191, 246, 322; insecticides in, 132–33; nitrate in, 81, 172–73; nitrogen in, 97–98, 110, 172–73; nutrients in, 111, 193; oil spill in (BP, 2010), 98, 104–5, 107; oxygen depletion in, 96, 98, 105; phosphorus in, 110; pollution of, x–xi, 85, 91, 94–97, 104, 109–11, 113, 132–33, 145, 172–73, 191, 193; and shipping, 132; watershed to, 44–45, 91, 95, 113. *See also* Mississippi River

Guthrie, Woody, 117–18

**H**

Haber, Fritz, 74–77, 338

Haber-Bosch system of ammonia synthesis, 77, 80, 82–83, 134–35, 146, 200

habitats, xvi, xix, 10, 12, 156–57, 167, 195, 218, 231, 243, 266, 276, 283, 286, 315

Haddock, J. Bruce, 55–56

Hämäläinen, Pekka, 21–22, 334, 348

Hamilton, Alexander, 120–21

Handelsman, Jo, 10, 199, 256, 331, 341

hardwood forests, 6, 12–13

Harrison, William Henry, 29

Hatfield, Jerry L., 190, 255–56, 329

Hayek, Matthew, 288

Heartland Institute, The, 242–43

Hefty Brothers (Darren and Brian), 215

Heinert, Troy, 313–14

Hellmann's mayonnaise, 259, 263

Helzer, Chris, 11, 328–29
Henderson, Caroline, 117–18, 123
herbicides, 197–98, 248–51;
    chemical, 250; effective, 206;
    fewer and less, 141, 255;
    glyphosates in, 206–7, 215, 251;
    and hybrid seeds, 197; as
    ineffective, 148; judicious use,
    284; and no-till planting, 255; as
    pollutants, 132–33, 145, 237;
    -resistant seeds, 207, 215; in
    runoff, 132, 237; safer, xviii. *See
    also* insecticides; pesticides
Heritage Foundation, The, 242–44
Hersom, Louis, 55–57, 337
Hettiarachchi, Ganga, 127
Highmore, South Dakota, 196, 198, 204
Hill, Jason, 128, 212–13, 251–52, 255, 329
*History of Tile Drainage in America* (Weaver), 47, 336
Hoekstra, Jonathan, xviii–xix, 276
Hoien, Eric and Kelly, 246–47, 252, 258
Homestead Act of 1862, 21, 203–4, 225
honeybees: commercial, 160;
    damage and threats to, 155–56,
    162–63; domesticated, 146, 156,
    159, 163, 165; on farms, 140,
    161; as pollinators, 144–46,
    155–56, 159–66; and symbiotic
    relationships, 163; wild, 160.
    *See also* bumblebees
Hormel, 92
Hornaday, William, 295, 297
Horrall, Mike, 161–63, 167–68
horses, xv, 13–14, 16, 21–22, 24, 35,
    42–43, 49, 55–56, 62, 114, 116,
    225, 288, 294
horticulture, 51, 249
Humboldt, Alexander von, 65
hummingbirds, 10
Hunt, Natalie, 251–52, 255
Hurricane Katrina (2005), 104, 107
Hurt, Douglas, 36, 329
Hyde County, South Dakota, xviii, 204–5, 209

hydrogen: and nitrogen, in ammonia, 70, 73–74, 76–77; and oxygen, 60–61; from water, 76
hydrological engineering, 52–54
hydrology and hydrologists, 23, 52–54, 59, 93, 99, 175, 191, 234, 238

**I**

Ice Age, 6–7, 40, 90, 199, 221–22.
    *See also* climate change
Illinois: in Corn Belt, 98, 151; farms,
    28, 34–35, 43, 84, 152, 248;
    population growth, 30; prairie,
    xv, 10, 26, 40–41, 54
Illinois State Horticultural Society, 51
immigrants, xvi, 39, 46, 49, 54–56,
    203–4, 225, 323–24. *See also* migration
Indiana: in Corn Belt, 57–58; farms,
    53–54, 57–58, 248; population
    growth, 30; prairie, 40–41
Indian Removal Act of 1830, 20
Indian reservations. *See* tribal lands;
    specific reservation(s)
indigenous people: baskets made by,
    42; and bison, xix, 14–15, 269,
    278, 291–321; and bison,
    history of, 6, 14–16, 224,
    294–97, 307, 313, 318–19, 321,
    325; canoes and waterways, 42;
    conflicts with, 20–22; and
    conservation, 201–2; cultural
    identity, 310; foraging by, 15,
    21; forced dislocations and
    removals, 16, 20–24, 29, 70,
    225, 307; and fur trade, 224–25;
    genocide, xvii, 23, 323; historic
    justice for, 319, 325; history,
    xix–xx, 6, 14–17, 20–25, 41–42,
    294–97, 304–6, 310–11, 324;
    and horses, 294; hunting by,
    14–16, 21, 41–42, 224, 294–95;
    knowledge of, 201; and land,
    xix, 17; and natural world,
    balance with, 15–16, 30; political
    system of, decentralized, 17; on

indigenous people (cont'd):
prairie, history of, 6, 14–18, 20–25; pride, 306; proud, free character, and noble expression of, in paintings, 19; sacrificed for national prosperity, 325; self-sufficiency and self-determination, 201; societies established by, 15; wars against, 38, 42; and wild animals, 292. See also tatanka (pte oyate or buffalo people); tribal lands
industrial agriculture: and bison, 296; consequences of, xvii–xviii; defenders of, 323; as destructive, 25; and environmental problems, xvii–xviii; and farms, 93–94; on grasslands, xviii, 293; and interchangeable crops, xviii; model of, 85; pre- farmers, 61–62; and production of cheap food, 186; and row crops encroaching on grasslands, 293; and soil degradation/erosion, 134–36; urban, 49
industrialization, xviii, 25, 37–38, 68, 84–85, 145, 295–97
Industrial Revolution, 10, 69, 74, 84, 199, 202, 237, 296
Ingenhousz, Jan, 124
ingenuity, 33, 45, 52–53, 73, 76–77, 216
insecticides, 146–49; amount used, 154–55; broad-spectrum, 153–54; Bt bacteria in, 147–48; enduring use of, 168; evidence against widespread use of, 163–64; fewer and less, 141, 154, 164–68, 257; futility of, 167; increased use of, 169; indiscriminate use, 148; industry, 142; as ineffective, 148; judicious use of, 147; most widely-used, 154; and natural selection, 148–49; as pollutants, 132–33, 141–42, 150–51, 155, 161–64; resistance to, 148, 150; risk of, 161; on seeds, 153–56, 162; ubiquitous use of, 142; and watermelon fields, 157. See also herbicides; neonicotinoids; pesticides
insects, 140–69; aerobic, 121; agricultural impacts on, 145–46; aquatic, 144–45, 156; balance of good and bad, 141–42, 169; beneficial, 141; and Bt toxins, 147–48; and corn, 150; and crops, 141, 158, 161; decline, endangered, extinction, threatened, 143–46, 150, 156, 322; and diversity, 11, 143; ecosystems, 156; and evolution, 142–43, 150, 169, 239; and farming, 141, 157–59, 164, 168–69, 252, 322; and fruit, 161; habitats, 156–57; iconic, 146; management of, 142; and monocultures, 141–42, 145–47; and naturally diverse biological systems, 141; and natural selection, 148–49; and neonicotinoids, 156; paradigm shift in view of, 169; and pests, 140–42, 168–69; and plant evolution, 141, 239; as pollinators, 140–69; population declines, 145–46; populations of, 252; prairie, 156, 239; in soil, 137, 143, 156, 250, 322; as structural and functional base of all living systems, 143; as successful and diverse, 143; survival of, 148, 169; terrestrial, 144; toxic landscape for, 159; underground, 9, 250. See also entomology and entomologists; insecticides; pests; pollinators and pollination; specific insect(s)
Integrated Pest Management (IPM), 147, 164–65
International Harvester, 34
International Union for Conservation of Nature, 315
InterTribal Buffalo Cooperative, 307
InterTribal Buffalo Council, 307–9, 313

Iowa: in Corn Belt, 57–58, 98, 151;
    and environmental law, 57;
    farms, 38–39, 55–58, 84,
    175–76, 180–81, 193, 200,
    246–61; as Food Capital of the
    World, 84–85; grasslands, 227,
    244; prairie, xv, 10, 39, 41, 57,
    193; urban and rural
    communities, cultural rift, 176
Iowa Corn Growers Association,
    183
Iowa Environmental Council, 174,
    185
Iowa Farm Bureau, 181–83, 244,
    249–50
Iowa Historical Society, 55
Iowa National Heritage Foundation,
    244
IPM. *See* Integrated Pest
    Management (IPM)
Isenberg, Andrew C., 296–97, 349

J
jackrabbits, xiii–xiv, 195
Jackson, Andrew, 20, 270
Jackson, Wes, 267
Jacobson, P. G., 50
Jefferson, Thomas, 6, 19–20, 37,
    270, 323
John Deere. *See* Deere, John
John Redmond Reservoir (Kansas),
    128–32, 136
Johnston, John, 46–51, 47
Jonny Lee Bearcub Stiffarm, 24–25,
    330

K
Kaiser-Wilhelm Institute, 77
Kallem, Larry, 248–51
Kankakee River tributary, 43
Kansas: farms, 38–39, 70, 114, 248;
    grasslands, 5, 227, 244; prairie,
    xv, xvii, 5, 10, 39
Kansas State University, 124–25,
    256, 329
Karges, Blair, 216
Kellogg, 263

keystone species, 286, 300–301,
    314–15
Kimball, Darlene, 107–8
Kimmerer, Robin Wall, 219
Klobuchar, Amy, Senator, 266
Koch Agronomic Services, 179, 183
kochia, 284, 285–86. *See also*
    cheatgrass
Konza Prairie Biological Station
    (Kansas), 124–25
Kramer, Randy, 91–95, 112–13, 135,
    138, 185–86

L
Lac qui Parle County, Minnesota,
    240–42
Lake Agassiz (Canada and U.S.),
    221–26, 238
Lake Erie, 29, 42
Lake Itasca (Minnesota), 89–90
Lake Michigan, 52
Lake Pepin, 132
Lake Pontchartrain estuary
    (Louisiana), 102–4
lakes: pollution of, xvi, 132–33,
    192–93. *See also* prairie
    potholes; rivers; specific lake(s)
Lake Winnipeg (Manitoba, Canada),
    222
Lal, Rattan, 256
land: and agriculture, 112; as
    commodity, 30; conservation,
    230, 242–44; conversions,
    xv–xvi, 200; for economic gain,
    240–41; enduring attitude
    toward, 30; frontier, 203–4;
    future of, 271; grazing, 226, 239,
    275–76, 302, 314, 348; history
    of, 112; improved, 29, 51; and
    individual property rights, 30;
    investors in, 210; as limitless,
    xx; management, 99; and
    nature, 240–41, 270–71, 324;
    ownership, 30, 84, 204;
    preserving and restoring, 236;
    and private ownership/property,
    20, 184–85, 192, 241–43;
    protection, 202, 214, 236, 238,

land (cont'd):
241–43, 289–90; purpose of,
240–41, 270–71; restoration of,
236, 325; rethinking
relationship with, 323–25;
sacrificed for national
prosperity, 325; settlers
squatting on, 29; speculation
and speculators, 225;
stewardship of, 19–20; subsidies,
38; and territorial expansion,
29. *See also* crops and cropland;
farmland; grazing lands;
landowners; landscapes; public
lands; tribal lands; wetlands
Land Acts, 20
land ethic, xix, 324
Land Institute, The (Kansas),
135–36, 139, 169, 267–68, 328
landowners, 20, 202, 234, 239,
243–44, 266, 328; and
conservation, 321–22; and
grasslands, 281; and ranching,
275, 281; and soil, 123; and
swamplands, 45–46, 51–53; and
water, 175, 184, 189
landscapes, xv–xvii, xix, 10–12, 16,
84, 325; agricultural, 30, 142,
155, 168, 185, 203–4, 252;
altered, 113; ancient, 5–6;
changing, 12, 209, 223–24;
destruction, 25; diversity, xvi,
11–12; evolution of, 11–12, 330;
forest, 159; and insects, 145–46,
150, 159; native, 247; natural,
39, 238, 247–48, 363; pastoral
English, 170; prairie, xiii, 34–35,
159, 226–27, 231–32, 234,
244–45; and ranches, 270, 276,
278; remote, 270; rescuing, 325;
restored, 236, 247; river, 90,
110, 112–13; and soil, 128, 131;
structural complexity, 286; and
swamplands, 44–45;
transformed, 25, 39; and tribal
lands, 292–93, 309, 315;
underexploited, 216. See also
grasslands; prairie
Lane, John, 33

Lang, Craig, 181
Lange, Dorothea, 118
LaPorte, Alicia, 262–63
Larew, Rob, 260, 330
Lark, Tyler, xv–xvi, 197–200, 213,
329
La Trobe, Charles, 28
Leach, Ashley, 165–67, 344
Legacy Amendment. *See* Clean
Water, Land and Legacy
Amendment (Minnesota)
legacy nutrients, 111
legumes: as animal feed, 79, 284;
and clover, 71; as cover crops,
62, 250; for cows, 284; and
microbes, 126; and nitrogen
restoration, 62, 70–71, *71*,
79–80, 126, 250; and nutrients
for soil fertility, 62, 70; as
protein, 289; and soybeans, 71;
symbiosis of, 62, 70
Leisinger, Matt, 208–10, 250
Leopold, Aldo, xix, 87, 323–24
Le Sueur, Meridel, 116
Lewis and Clark (Meriwether and
William), 6
Lewis and Clark Lake, 132
Liebman, Matt, 251–52, 255, 329
Limerick, Patricia Nelson, 323
Little Bighorn, 22, 297
Little Crow, xvi–xvii
"Little Things That Run the World,
The" (Wilson essay), 143, 345
Little Thunder, Rosalie, 308
living soil, organic matter in,
xiv–xv, 7
Loess Hills grasslands (Iowa), 227
Louisiana Purchase (1803), 19–20
Low, Ann Marie, 116
Lower Brule Sioux reservation
(South Dakota), 201
Lundgren, Jonathan, 140–41,
168–69, 195, 251

# M

Madison, Minnesota, 50
Madson, John, 8, 12, 334
Magnan, Robert, 309–13, 321, 330

Malm, Richard, 185–87, 189
Malthus, Thomas, 63
mammals: burrowing, xiv; evolution and survival of, 227; extinct, 14; giant, 13–14; grasslands, 227; grazing, 12–13; organic material from, in soil, 7. *See also* ungulates; wild animals; wildlife; specific mammal(s)
Manning, Richard, 14, 204, 331, 348
manure, 62–63, 78–79, 127, 209, 284–85. *See also* guano (huano)
Martin, Brian, 281, 285–87, 328
Matador Ranch (Montana), 271–73, 282–89
Mateos-Fierro, Zeus, 157–59, 330
Maumee River (Ohio), 172
McCormick, Cyrus, 34, 37, 43, 197
Means, Tatewin, 201
*Meat: A Benign Extravagance* (Fairlie), 288–89
Medicine Lodge Treaty (1867), 21
megafauna, 13, 144
Melon Acres, 161
Messerly, Charles, 283, 285–86
metal thallium, 72
methane (greenhouse gas), 288–89
Michigan population growth, 30
microbes, 9, 68, 82, 172, 209; soil, xiv, 7, 125–27, 206, 239, 250; underground, 136, 199, 250, 256
migration: from Eastern Seaboard, 28; settler, 21, 55; Westward, 19, 21, 28–30, 54–55, 323. *See also* immigrants
Miller, Lynette, 137
Milton, Bill, 328
Milton, Dana and Bill, 281
Minnesota: in Corn Belt, 57–58, 98, 151; farms, 38–39, 57–58, 79, 84, 91, 112, 200, 228, 240–41, 248, 261–62; grasslands, 244–45; lake country, 90; as Land of 10,000 Lakes, 133; prairie, xiv–xv, 10–11, 39, 238–40; swamplands, 41
Minnesota Corn Growers Association, 83
Minnesota Corn Research & Promotion Council, 80, 81–82
Minnesota Department of Natural Resources, 226, 231, 234–35, 242, 244, 347
Minnesota River tributary (Minnesota), 221–22, 240
Mississippi Commercial Fisheries United, 108
Mississippi River, 89–113; basin, 45, 59, 84–85, 96, 99, 113, 138; coast, 91, 96, 98, 104–6, 246; environmental catastrophes along coast, 104; health of, 113; history of, 18–19; hypoxia in, 110–11, 191; nitrogen in, 98; pollution of, *x–xi,* xvi, 90–91, 94–95, 102, 113, 132–33, 172–73, 191, 193, 246; reengineering of, 101–3; silt, 132; source in Minnesota, 89–90; watersheds, 91, 99, 112–13, 246; water volume of, 98–100, 102. *See also* Gulf of Mexico
Mississippi Sound, 104, 106
Mississippi Valley, 15, 145
Missouri prairie, 39
Missouri River, xvi, 20, 84, 103, 132, 134, 137, 176, 277, 281, 287, 294–95
mites, xiv, 8, 126, 160
moles, 10
monocultures: corn-soybean, 208, 322; and insects, 141–42, 145–47, 157; and pests, 142, 146–47; and watermelon fields, 157; for world markets, 142
Monsanto, agrochemical company, 154, 181, 183, 205–8, 215
Montana: grassbank movement of conservation ranching in, 273; grasslands, 243, 270, 281; prairie, xiii–xv, 10, 309; ranches, 269–90, 307
Montgomery, David R., 120, 126, 131, 135, 213, 331
Montgomery Ward, 34
Mouzin, Dennis, 160
Moynahan, Brendan, 319

Muir, John, 249
muskrats, 40–41, 55, 224–25
Mykleseth, Keith, 229–30

**N**
N2. *See* atmospheric nitrogen; nitrogen cycle
N2O. *See* nitrous oxide
Nargang, Ron, 226, 228–31
National Bison Association, 301
National Bison Range (Montana), 297
National Farmers Union, 260, 330
*National Geographic,* 310
National Oceanic and Atmospheric Administration, 96
national parks: bison herds in, 278, 318–19; protected, xviii–xix, 276. *See also* specific park(s)
National Park Service, 307–8, 319
national prosperity, 280, 325
National Resources Inventory, 130–31
National Watermelon Association, 167
National Zoo. *See* Smithsonian's National Zoo & Conservation Biology Institute (Washington, DC)
Native people. *See* indigenous people
natural resources, 30, 106, 128, 195
Natural Resources Conservation Service (USDA), 123, 136–37, 228, 341
nature: and agriculture, 131; balance in, 169; and community, values about, 272; cycles of, and animals, 15; and expectations, xx; healing of, xx; and invincibility of challenges, 216; and land, 240–41, 270–71, 324; latent energies of, 73; as limitless, xx; and people, relationships between, 78; and ranching, 196; and science, 216
Nature Conservancy, The, 226–32, 272, 281, 283, 285–87, 314–15, 328
*Nature* journal, 202, 237

Neal Smith National Wildlife Refuge (Iowa), 10, 314–15
Nebraska: corn and soybean commodities, 84; in Corn Belt, 151; farms, 38–39, 84, 248; grasslands, 227, 243–44; prairie, xiv, xvii, xviii, 11, 39
nematodes, xiv, 9–10, 126, 160
neonicotinoids (neonics), 153–64, 257
Neosho River (Kansas and Oklahoma), 128–29, 136
Nernst, Walther, 74–75
Netland, Cory, 244–45
New Deal, 121–22, 129
*New York Times, The,* 22, 57, 310
NGOs (non-governmental organizations), xviii–xix
NH3. *See* ammonia
Niebling, Brett, 137–38
Niman Ranch, 262–63
nitrates: and ammonia, 64, 71, *71,* 82; and atmospheric nitrogen, *71;* and nitrites, 9–10, 71, *71,* 126; in plants, 82; pollution, 111, 170–76, 183–90, 237; potassium, 64; soil, 135; surface water content, 111; tolerance for, 111; and treatment plants, 189–90
Nitrate War (Peru, Bolivia, Chile), 69
nitrites, and nitrates, 9–10, 71, *71,* 126
nitrogen, 60; from air, 9–10, 60, 70–71, *71,* 73–74, 76, 126; and ammonia, *71;* commercial, 97–98, 135; and crops, 61, 63, 210; depletion of, 78; as elemental force in natural world, 64; fixing, and ammonia, 61, 70–71, *71;* and hydrogen, in ammonia, 70–71, 73–74, 76–77; in manure, 62–63; mining, 135; pollution, 80–81, 83, 99–100, 110–12, 173, 176; in soil, 61, 64, 77, 99, 121, 135, 250, 256–57; sources of, 62; synthetic, 77–78, 146, 172–73. *See also* atmospheric nitrogen

nitrogen cycle, 23, 70–71, 74, 78, 80, 82–83, 170; prokaryotes in, *71*
nitrogen fertilizer, 78–83, 90–91, 97–99, 133, 146, 186, 251, 259, 342
nitrous oxide, *71,* 81, 256
NOAA. *See* National Oceanic and Atmospheric Administration
non-governmental organizations (NGOs), xviii–xix
nonprofit organizations: and bison, 277–78, 307, 313, 318; and conservation, 247, 297, 314–15, 321–22; and farming, 249; and grasslands, 201–2, 244, 314–15; and insects, 140–41; and lobbying, 109; and prairies, 230, 240, 244, 267, 281; and sustainable agriculture, 135–36; tribal, 201–2, 307, 313, 318; and wilderness, 281; and wildlife, 317
North American prairie. *See* prairie
North Dakota, 116, 140, 205, 216, 221, 233, 310; in Corn Belt, 151. *See also* South Dakota
Northwest territory, 30
nutrients: artificial, 84; for crops, 70; for grasses, 9–11; legacy, 111; plants, 64; for plants, 9–11, 134; in rivers, pollution by, 110–11, 180, 193; in soil, 41, 96, 121, 123, 239, 267

O
Obama, Barack, 10, 191
O'Brien, Dan, 291–94, 298–99, 301–6, 313, 315, 319–20, 327, 349
O'Brien, Jill, 303–5, 327
Ohio: farms, 84; population growth, 30; prairie, 10
Ohio River, xvi, 43, 84, 98, 103
Oklahoma: panhandle, 116–17; prairie, 10
*O Pioneers!* (Cather), 1
Organic Valley (Wisconsin), 262

oxygen: and carbon dioxide, 124; depletion of in water, 96–98, 104, 145; and hydrogen, 60–61; and water, 126
oysters, 97, 103–9, 113, 192, 330

P
PACs. *See* political action committees (PACs)
parasites, 11–12, 126–27, 143, 155, 162, 250, 254, 300. *See also* pathogens
Paris climate accords, 202
Paris Exposition (1867), 34
Patagonia Provisions, 262
pathogens, 11–12, 126–27, 254. *See also* parasites
Patzer, Todd, 241–42
PCBs, as potential carcinogens, 206–7
Pecenka, Jacob R., 164–66, 344
PepsiCo, 259–60, 263–64
peregrine falcons, 292–93, 302
pesticides, 164–66, 179, 198, 215, 251; advanced, xvii; chemical, 99; complexity of, 329; controversial, 257; effectiveness of, 158; exposure to, 227; fewer and less, 94–95, 165–66; immunity to, 207; as pollutants, 84, 145, 148, 184, 186, 192, 237, 292–93; on seeds, 154, 164–65, 329; synthetic, 186; treadmill, 148, 157. *See also* DDT (pesticide); herbicides; insecticides; neonicotinoids
pests: control of, 127, 141–42, 146–47, 153, 161, 195, 218; and crops, 141–42, 149, 153, 157, 169; evolution of, 146; in grasses, 12; and insects, 140–42, 168–69; and monocultures, 142, 146–47; outbreak of, 142; prevention of, 153; primary, 153; resistance to toxins, 148; and watermelon fields, 157. *See also* insects; pesticides
Peterson, Ali, 262

Peterson, Luke, 261–63
PFI. *See* Practical Farmers of Iowa (PFI)
Pheasants Forever, 244, 258, 282
phosphorus, 64, 90–91, 96, 99, 110–13, 134, 176, 184, 186, 191
photosynthesis, xv, 9, 124, 127–28, 212, 288
Pierson, William P., 51
Pike, Zebulon, 18–19
*Pilgrim's Progress, The* (Bunyan), 42
Pillsbury, 90
Pimentel, David, 131, 134, 342
Pinchot, Gifford, 280–81
Pine Ridge Reservation (South Dakota), 201
Pioneer Hi-Bred Corn Company, 250
plants: aquatic, 62; carbon-rich, 256; diversity of, 11–12, 142–43, 200–201, 218, 227, 251, 255, 300–301, 322; evolution of, 11, 143, 239; food from, 291; invasive, 231, 240; medicine from, 291; native, 124, 146, 227, 232, 252, 256, 267–68; nitrates in, 82; nutrients, 64, 123; nutrients for, 9–11, 134; prairie, 8, 12, 124, 126, 226, 232–33, 235, 239, 256, 267–68, 300; roots and root systems of, 2–3, 8, 12, 47–48, 123–27, 139, 149, 199–200. *See also* flowers
Platte River (Nebraska), xvi
plowing and plows, xvii–xviii, 14, 26–39, 194–218; breaker, 31; cast-iron, 26, 32, 36; contour, 123, 130; for crops, 197, 200–201, 214, 226, 228; lister, 123, 125; mass production of, 27; mole, 49; singing, 26–27; and soil, 119; steel, 5, 25–28, 33–36, 131; and technology, 33–34. *See also* agriculture; farms and farming
*Poaceae. See* grasses
political action committees (PACs), 181–82
politics: of biofuels, 329; conservative, 181; and land protection, 242–43; partisan, 243; and science, 329
pollinators and pollination, xvi, 140–69; agricultural, 159–60; bees, wild as, 144–46, 156, 161–67; bumblebees as, 156, 159; butterflies as, 156; and ecological intensification, 168; and food, 159; habitats for, 167; honeybees as, 144–46, 155–56, 159–66; insects and pests as, 140–69; and neonicotinoids, 156; population increases, 252–53; self-, 159. *See also* insects
pollution: agricultural, xvi–xx, 23–25, 28, 85, 98, 132–33, 145, 189–92, 229, 329; industrial, xviii; law, 176–78, 187–88, 190; of soil, 132–36; standards, 211–12. *See also* air pollution; rivers, pollution of; water pollution
Popper, Deborah and Frank, xvii, 269–71, 277, 279, 325, 330
potassium nitrate, 64
potato beetles, Colorado, 146, 157
potholes. *See* prairie potholes
Powell, Eric N., 106–7, 330
Practical Farmers of Iowa (PFI), 247–64, 272, 281, 284, 321–22, 330–31
prairie, *iv*, x–xx, 5–25, 221–45, 321–25; as American desert, xiii; ancient, 135; cities, xvi; conquest of, 23; conservation, 191, 226, 239, 244, 270, 277, 286–87; destruction, 25, 59, 85, 192–93; diversity, 11–12, 196; and earliest human occupants, 13; eulogy to, 59; evolution, 9–12, 239; feared by pioneers, xiv; as fertile, 13; grasslands, 7, 196, 304, 318; and heroes, xvi–xvii; history, xvi–xvii, 5–24, 28–30, 324; as infinite, xiv; large-scale settlement, history, 19–21; mixed-grass, *x–xi*, xiv,

11, 119, 197, 295; and mythology, xvi–xvii; and national character, xvi–xvii; native, xv, 124, 184, 227–28, 232, 235, 238–40, 244, 248, 251, 256, 267–68; North American, xiv–xv, 6; paradox, xiv, 192; of pioneers and settlers, 5, 23, 26, 70, 119, 203–5; plowed up, 70, 119, 197–98, 225–26, 232, 270–71; preserves, 278; re-creation, 223; reengineering of, 101–2; remaining, 226, 232; restoration, 201, 223, 234, 237–40, 244–45, 266, 322, 325; as sea of grass (metaphor), 18; shunned by tourists, xiv; structural complexity, 286; symbiosis, xiv; temperature extremes of, xv, 7–8; as threatened, xv, 202; as timeless and unchanging, 12–13; transformed, 25, 39, 85, 193; as wasteland, xiv; westward expansion of (history), 18–21, 28–30, 38–39, 42–44, 54–56, 120–21, 270, 323, 325; zones, 10–11. *See also* grasslands; Great Plains; landscapes; shortgrass prairie; tallgrass prairie; wet prairies
prairie dogs, xiv, 12–14, 18–19, 195, 277–79, 281, 283, 286, 293, 301; populations of (1830s), 19
Prairie Foundation, The, 277
prairie potholes (tiny lakes), xiv, 40, 54, 58, 174
predators, xiv, 12, 14–15, 158, 165, 278–79, 298–301, 306–7, 315
Preemption Act of 1841, and surveyed public land sales, 29
Prescher, Gary, 79–82
prokaryotes, in nitrogen cycle, *71*
pronghorns: antelopes, 279; elk, 293; flower grazing, 12; grass grazing, xiv, 12; migration patterns of, 278; protection of, 276; on ranches, 286. *See also* antelope

protein: and animals, 71; and bison, 14, 302; from cows, 287; creation of, 60; from foods, 287, 302; and soil, 125–26; and sunshine, v
public health, 53, 81, 85, 148, 170–72, 176, 188, 207
public lands: and agricultural productivity, 45; bison returned to, 297, 312–14, 318; claims and sales of, 29; conservation of, 243; grazing rights on, 293, 314; in private hands, 29–30, 45; and ranching, 293; and territorial expansion, 29

**Q**
Quaker Oats, xvi
Quick, Herbert, 39

**R**
Rabalais, Nancy N., 96–97, 109–10, 132, 339, 340
Raccoon River (Des Moines, Iowa), 170–75, 178, 183–85, 189–90; watershed, *174*
railroads, 22, 50–51, 55, 225, 273, 296–97; transcontinental, 34, 38
rainforests, xiv–xv, xix, 41
Ranchers Stewardship Alliance, 282, 287
ranches and ranching, 269–90; bison, 269–71, 286, 291–320, 327; cattle, xix, 11, 208, 210, 218, 226, 270, 274–75, 278, 281, 283–89, 291–93, 306, 308; conservation, 196, 272–73, 281–85, 289; conventional, 284; descended from frontier settlers, 280; and ecosystem services, 218; federal bailouts for, 269; and grassbank movement of conservation, 273, 281; and grasses preserved, 289–90; and grasslands preserved, 194–205, 218, 244, 269–86; and hope, 218; and ideologies, 280;

ranches and ranching (*cont'd*): modern American, 282; and nature, 196; Open Range, 273; and philosophical differences, 289; and profitable crops, 209; and stewardship, 195, 218, 282, 287; and wildlife, 195–96, 272; windbreaks on, 195. *See also* farms and farming
Ranch Systems and Viability Planning, 284
Rangeland Analysis Platform, 286
Reaves, Elizabeth, 263
Redmond Reservoir. *See* John Redmond Reservoir (Kansas)
redressability, judicial standard, 187–88
Red River, 222–26, 233, 237
Red River Valley, 221–26, 330, 347
Remington, Frederic, 274
Restoration of American Bison and the Prairie Grasslands (U.S. Dept. of the Interior), 318
Reynolds, Kim, 182
rhizosphere, roots and root systems of, in soil, 126
Rice, Charles, 124–25, 127, 256, 329
Ricketts, Pete, 242–43
rivers, 89–113; landscapes of, 90, 110, 112–13; pollution of, xvi, 113, 132–33, 170–93, 198, 237, 323; water volume of, 98–100, 102. *See also* lakes; watersheds; specific river(s)
RNA interference (RNAi), 152–53
Robertson, Alison, 253–54, 329
Rodale Institute (Pennsylvania), 249
rodents, 9, 278–79, 286
Roosevelt, Franklin, 121–23, 250
Roosevelt, Theodore (Teddy), 243, 274, 280–81, 297
rootworm, 140, 149–53
Rosebud Sioux Tribe of the Rosebud Indian Reservation (South Dakota), bison at, 306, 313
Rothko, Mark, 84
Roundup herbicide, 148, 183, 205–8, 232
Russell National Wildlife Refuge. *See* Charles M. Russell National Wildlife Refuge (Montana)

## S

Saltonstall, Nathaniel, Judge, 42
saltpeter, 64, 68, 73, 76
sandhill cranes, 223, 231
Sandhills grasslands (Nebraska), 227, 244
Schoolcraft, Henry Rowe, 89
Schulte Moore, Lisa, 252–54
science: of biofuels, 329; and farming, 330; and history, 83, 329; and ingenuity, 216; and nature, 216; and politics, 329; and technology, 51–52
Science Center for Marine Fisheries, at University of Southern Mississippi, 106
Scragg's Patent Tile Machine, 48
sea of grass, 6; term, usage, 18
Searchinger, Tim, 213, 329, 346
Sears Roebuck, xvi
sedges, xiv, 125, 245, 299
seeds: for crops, 149, 208, 255, 261; genetically engineered and modified, 140, 152, 208, 251, 329; hybrid, 151, 197; in manure, 284–85; native, 245; novel hybrids, xvii, 206; prairie, 232–33, 245
self-reliance: and agriculture, 45; and civic virtue, 45; and discipline, 310; and wilderness, 324
Seneca Foods/Green Giant, 92
Shamon, Hila, 315–17
shelterbelts, and windbreaks, 122–23, 129–30
shortgrass prairie, *x–xi,* xiv–xv, 10–11, 119, 197–98, 271, 274–75, 285, 287, 300–301. *See also* tallgrass prairie
*Silent Spring* (Carson), 147
Simpson, Curtis, 161–64
Sitting Bull, Lakota Chief, xvi–xvii, 22

# Index | 373

*60 Minutes* (CBS television program), 282–83
Skrmetta, Louis, 104–5
sloughs, 41, 55–56, 58, 174–75
Smalley, E. V., 204
Smil, Vaclav, 83, 338
Smithsonian Institution, 278
Smithsonian's National Zoo & Conservation Biology Institute (Washington, DC), 295, 315
Snider, Samuel E., 24
social change, 324
Sodbuster program, 265–66
soil, 114–39; aboveground, 125–26; analysis, xiv; bacteria in, 10, 61–62, 70–71, 80–81, 125–27, 137, 199, 250; biology of, xix; black, 27–28, 77, 126, 203; clay, 90, 225; conservation, xix, 122–23, 129–30, 136–38, 234–35, 282, 322; and corn, 135; degraded and depleted, 124, 134, 213, 239; denitrification, 71; ecosystems, 127–28; erosion, xvii, 119–39, 145, 174, 209, 222, 245, 250–53, 255, 258, 261, 267, 322; and farming, 239; fertile, 13, 57, 132, 180, 188; fertility, 62, 69–70, 78, 119, 132, 135, 213, 251, 255, 338; as finite resource, 213; as fragile, 120–21; fungi in, 8, 10, 125–27, 199, 239; and grasses, 9–10; grasslands, 77; gumbo, 26, 225–26; healthy, 110, 112, 124–27, 133, 137, 146, 206, 213, 218, 250, 256, 261, 267–68, 284, 322; heavy, 26; insects in, 137, 143, 156, 250, 322; management, 202; microbes, xiv, 7, 125–27, 206, 239, 250; microorganisms in, 41; minerals in, 198; moisture, 11, 133, 255–59; nitrates, 70–71, 135; nitrogen in, 61, 64, 77, 99, 121, 135, 250, 256–57; nourished, 139; nutrients in, 41, 96, 121, 123, 239, 267; organic material in, 7; pollution, 132–36; prairie, 9, 26, 28, 30–31, 135; protected by nature, 139; protection, 247; and protein, 125–26; replenished and regenerated, 130–31; rich, 11, 30, 77, 90, 121, 143, 193, 198, 200; salts in, 100–101; structure, 252, 255; survey, 121; terrestrial carbon in, 10; and topography, 130; underground, 126–28, 199; virgin black, 27–28. *See also* living soil; topsoil
Soil and Water Conservation Districts, 123, 234–35
*Soil Erosion a National Menace* (USDA paper), 121
soil science and scientists, 82, 121, 124, 133, 276, 329
Sonstegard, Phillip, 240–42
South Dakota: farmers, 208, 240, 266; grasses, 197, 217; grasslands, 243, 270; prairie, xiv–xv, xvii, 194; prairie plow-up, 197–200; ranches, 291–94, 298, 304; tribal land grasslands, 201. *See also* North Dakota
South Dakota Farm Bureau, 214, 330
South Dakota Grassland Coalition, 196, 262
South Fork Crow River (Minnesota), 90–91, 94–95, 111–13, 341
soybeans, xv–xvi, 84–85, 94, 197–98, 207–10; and biofuels, 213, 263; and carbon dioxide, 200; catchment, 252; and cattle, 210; as commodity, 84; as conventional crop, 263; crop expansions, 99, 145, 197–98, 208, 323; and ethanol, 263; and gene transfer for immunity/resistance, 205–8; as global major food crop, 267; insecticides for, 147, 154; insecticide-treated, 157; and insects, 140, 150, 159, 161–62; as interchangeable, xviii; and

soybeans (*cont'd*):
legumes, 71; as livestock feed, 263; oil, 263; pesticide-treated, 154, 164–65; planting ease, 120; pollinated, 159; and prairie, 226, 240–41; as profitable, 57, 186, 197, 209, 218, 259–60; seed technology, 205–8; and soil, 135; volume of, 323; and water, 174, 180

sparing strategy, *vs.* sharing, 247, 281
Spawn, Seth, 200, 329
Spencer, Joe, 150–52, 329
spillways, 101–9
*Spokesman* weekly newspaper, 181
steamboats, 33, 43, 225
Steinbeck, John, xvi–xvii, 118
Stevenson, Robert Louis, 274
stewardship, 20, 175–78, 188, 195, 218, 261, 265, 268, 282, 287, 310, 318
Stop the 30 x 30 Land Grab summit (Nebraska), 242–43
*Storm Lake Times Pilot* biweekly newspaper, 174, 182
Stout, Jane, 336
Stout, Philemon and Penelope, 34–35
Stout, Philemon Jr. and Louisa, 35–36, 43
Stowe, Bill, 173–81, *177,* 185, 188–89, 191, 193, 246, 329
Stowe, Liam, 177
Superfund cleanup sites, 207
Sustainable Food Lab (Vermont), 263
Swampland Act, 44–45
swamplands and swamps, 40–59; and crops, 45; and diseases, 44, 46; fearful imagery of, 42; and landscapes, 44–45; reclaimed, 189; as refuges for indigenous people, historically, 42; as sterile and barren, 45. *See also* drainage; sloughs; wetlands
Swanson, John, 236
symbiosis: of bison and grasslands, 298–99; of honeybees, 163; of legumes, 62, 70; of prairie, xiv

symbiotic rhizobium, 9, 62, 70, 126
Syngenta, agrichemical company, 93–94, 147, 187, 207, 215

T

tallgrass prairie, *x–xi,* xiv–xvi, 8, 10–12, 21, 81, 85, 232, 329, 347; and agriculture, 35, 195–97, 247, 268; biological diversity, xiv; and farming, 59; and insects, 145–46, 159; landscape, 223–24; last settled, 54, 57; preserves, 277–78; and ranches, 277–78; restoration and protection of, xix, 244; and rivers, 90; small as islands, xix, 227, 238–39; and soil, xiv, 119; species loss in, 268; surviving, 5–6; and swamplands, 57–59; transformed, 39; and tribal lands, 300–301; and water, 184; and wetlands, 41. *See also* shortgrass prairie
Tallgrass Prairie National Preserve (Kansas), 5–6
Tanka Fund, 313
tatanka (pte oyate or buffalo people), 291–320; term, usage, 314. *See also* indigenous people; tribal lands
Taylor Grazing Act (1934), 275–76
technology: agriculture and farming, xx, 27, 33–34, 49, 100–101, 216, 266–67; of biofuels, 211–12; changes and inventions, 33–34; and drainage, 51, 54; and ingenuity, 52–53; and markets, 210; and progress, xx; and science, 51–52; and urban industry, 49
Tecumseh, Chief, 29
Tennessee Valley Authority (Muscle Shoals, Alabama), 78
Terry, Jennifer, 189
Tesla, Nikola, 51–52
Texas: panhandle, 295; prairie, 10

Thaler, Evan, 131–32, 134, 342–43
thallium, metal, 72
Thompson, Dick, *248,* 248–49, 251–52, 258
Thompson, Sharon, 248–49, 251–52, 258
Thoreau, Henry David, 249
Thornton, Alonzo, 56–57, 175
Thunder Valley tribal nonprofit, 201–2
tile drainage, 46–54, *50, 54,* 56–59, 81, 84, 90–93, 98–101, *100,* 156, 172–73, 184–85, 187
Tilman, David, 251–52, 255
Timber Culture Act, 18
topography, and soil, 130
topsoil, 118, 120–21, 129–38; blown off by windstorms, *120*
transcontinental railroads, 34, 38
Treaty of Fort Laramie, 21
tribal lands: bison eradicated from, 296–97, 307, 325; bison on, xix, 202, 269, 278, 291–320, *312,* 327; conservation, 201–2; farms on, 21; and grasslands protected/restored, 201; and landscapes, 292–93, 309, 315; protected, 201–2; water on, 202. *See also* indigenous people; tatanka (pte oyate or buffalo people)
tributaries, 90–91, 95, 129, 224, 294. *See also* estuaries
Turner, Frederick Jackson, 324–25
Turner, Gene (R. Eugene), 96–97, 109–10, 132, 340
Turner, Ted, 201–2, 301–2, 305–6, 317
Tyson Foods, 181

**U**

ungulates, 12, 269, 288–89. *See also* mammals
Unilever, 259, 263–64
Union of Concerned Scientists, 130, 138, 342
United Nations Food and Agriculture Organization, 134

urban: development, 110, 227; industry, and agriculture, 49; sprawl, 145
urbanization, 38, 69

**V**

*Vandemark's Folly* (Quick), 39
VanderWal, Scott, 214, 330
van Helmont, Jan Baptist, 123–24
Vaught, David, 36–37
Vermejo Park Ranch (New Mexico), 317
voles, xiii
von Liebig, Justus, 64

**W**

Wallace, Henry, 250
Walton Family Foundation, 260–61
Walz, Tim, Governor, 242
Ward, Robert DeCourcy, 7
Washington, George, 120–21
water, 170–93; abundance of, 188; conservation, xix, 176, 234–35, 282; efficient use of, 256; management, 95, 112; protection, 244, 247; shortages, 323; wild rice, protected, 202; and winds, 130, 222, 245. *See also* drinking water; floods; lakes; rivers; water pollution
Waterman, Thomas, Justice, 189
watermelon, 157–68, 195, 330
water pollution, 83, 170–93; agricultural, xvi, 59, 94, 113, 132–33, 170–93, 198, 202, 240, 261, 268; and algae, 80–81, 96–97, 104–5, 172, 176; and grasslands loss, 192; laws, 176–77, 187, 189–90; point sources, 176–77, 187. *See also* Clean Water Act (CWA)
watersheds, 52–53, 90–91, 95, 99, 103, 109–13, *174,* 191, 221–22, 229, 233–34, 246
Wayne, Anthony, Major General, 29
Weaver, John Ernest, 59, 336
Weaver, Marion M., 47, 336

Webb, Walter Prescott, 203, 335
wetlands: and aquifers, 41; coastal, 42; conservation, 184–85, 191–92, 228, 322; and crops, 48, 185, 240, 322; destruction, 59; and diseases, 44; drainage, 145; ecology, 53; and economic progress, 53; ecosystems, 41; eulogy to, 59; and farmland, 41, 61, 240; and grasslands, 141–42, 156, 199–200, 227; plowed up, 141, 199; pollution, 59; and prairie, 59; protection, 58–59, 228, 240; re-creating, 184–85, 234; restoration, 237; and tallgrass prairie, 41; transformed, 49–51, 61; as wicked, 42. *See also* drainage; swamplands and swamps
wet prairies, 40–46, 49, 54, 90
wheat, 35–36, 72–73, 84, 94, 205, 213–14; ancient, 262; and carbohydrates, 267; as conventional crop, 263; and corn, 5, 135, 226; emmer, 262; expansion of, 197–98; experimental, 262; exported, 73; global demand for, 225, 267; insecticides for, 154; and insects, 146, 156; Kernza, 262–63, 267–68; organic, 262; perennial, 267–68; prices, 119; and soil, 115, 119, 128; as staple, 205; winter, 99
wheatgrass, xiii, 10–11, 195, 271, 274
Whitaker, Bill, 283
White Earth band of Ojibwe (Minnesota), 202
WHO. *See* World Health Organization
Wichita Mountains National Wildlife Refuge (Oklahoma), 297
wild animals, 292, 297–98, 302, 307–8, 315, 318–19. *See also* mammals; wildlife; specific animal(s)
Wilder, Laura Ingalls, xvi–xvii
wilderness: abundance of, 188; preserves, 276, 313; reserves, xviii–xix; and self-reliance, 324; transformed, 39; and ungulates, 289
wildfires, 9, 12, 15, 235, 317
wildflowers, 12, 89, 155, 162, 222–23
Wild Idea Buffalo Company ranch (South Dakota), 303–6, 349
wildlife: abundance of, 13–14; and agriculture, xvii, 85; and climate change, 201; conservation, 301, 308, 315; DDT damage to, 147; diversity of, 276; endangered and threatened, xvii, 218, 323; and grasslands, 201, 276–77, 281, 289, 301; habitats for, 218; inventory (1830s), 19; management, 240, 244–45; migration patterns, 23; nonprofit organizations, 317; and plants, in ecosystem, 315; play, and bison, 315; prairie, 13–14, 19, 202; preserves, 273, 276; protection, 202, 247, 289–90; and ranching, 195–96, 272; refuges, 228–30, 236, 241, 314–15, 318; reserves, 277; transportation routes, 11. *See also* grazing animals; mammals; wild animals; specific animal(s)
wild rice, 41–42, 202
Wilson, E. O. (Edward O.), 142–43, 345
Wilson, Gail, 127
Wimberly, Michael C., 197, 343
windbreaks: and contour plowing, 130; and grasses, 195; on ranches, 195; and shelterbelts, 122–23, 129–30; and wind erosion reduction, 129–30
Wind Cave National Park (South Dakota), 278, 297
winds, xiii, 203; chinook, 7; and dust, 115; and erosion reduction, 129–30; and rain, 119, 129–30, 245, 258; and snow, 284; and water, 130, 222, 245; winter, 258, 284. *See also* windstorms

windstorms, *120*, 198. *See also* dust, storms
Winnett ACES community group, 287
Wisconsin: population growth, 30; swamplands, 41
*Wizard of Oz, The* (Baum), xvi–xvii
Wolf Creek nuclear power plant (Kansas), 128
Wolfe, Patrick, 23
Wolf Point, Montana, 24–25, 310
wolves, xiv, 12–14, 55, 276–77, 286, 299–300, 315
World Health Organization, 207
World's Columbian Exposition (Chicago, 1893), 51–52
World War I, xvi, 39, 76–77, 119, 142, 146
World War II, xvi, 39, 77, 146
World Wildlife Fund, xix, 197, 201, 264, 277, 282, 284, 308–9, 328
Worley, Sally, 251, 261
worms, xiv, 8–9, 126, 140, 149–53, 156. *See also* rootworm

Wounded Knee Massacre (1890), 22, 304–6
Wright, Christopher K., 197, 214, 343, 347
Wyden, Ron, Senator, 266
Wylie, Patrick, 262
Wyoming prairie, 10
Wyse, Don, 266–67, 268

X

Xerces Society for Invertebrate Conservation, 145

Y

Yellowstone National Park, xix, 278, 295, 297–98, 307–12
Yellowstone River tributary, 295
Yellowstone Valley, 14–15
yeoman ideal, 19–20, 37–38
Youpee, Dyan, 311

Z

Zabel, Mark, 112

ABOUT THE AUTHORS

DAVE HAGE oversaw environmental and health reporting at the Minneapolis *Star Tribune* for a dozen years, editing projects that won a Pulitzer Prize, an Edward R. Murrow Award, and other national journalism honors. His previous books include *No Retreat, No Surrender: Labor's War at Hormel* and *Reforming Welfare by Rewarding Work*. He is a Minneapolis native whose parents grew up in the small prairie towns of western Minnesota. He lives in St. Paul with his wife, a florist and master naturalist.

JOSEPHINE MARCOTTY is an award-winning environmental journalist who has spent her life in the Midwest. She was a reporter for the *Star Tribune* in Minneapolis where she chose to cover complex, science-based topics critical to the community, including economics, healthcare, and the environment. *Sea of Grass* is a natural expansion of the in-depth reporting she did for the newspaper on the vanishing prairie, the importance of natural landscapes, and the consequences of intensive agriculture. She lives in Minneapolis with her husband.